Biotechnology of Plants and Microorganisms

EDITED BY O. J. CROCOMO, W. R. SHARP,
D. A. EVANS, J. E. BRAVO,
F. C. A. TAVARES, AND E. F. PADDOCK

Biotechnology of Plants and Microorganisms

OHIO STATE UNIVERSITY PRESS : COLUMBUS

Library of Congress Cataloging-in-Publication Data

Biotechnology of plants and microorganisms.

  Includes index.
  1. Genetic engineering. 2. Plant genetic engineering.
3. Microbial genetics. 4. Biotechnology. I. Crocomo,
Otto J.
TP248.6.B557   1986   660'.6   86–23890
ISBN 0–8142–0375–2

Contents

Preface

In the midst of the rising tide of expectations that accompanied the "Alliance for Progress" initiative of the early 1960s, a unique partnership was established between the Escola Superior de Agricultura "Luiz de Quieroz" (ESALQ), Universidade de Sao Paulo at Piracicaba, in the state of Sao Paulo, Brazil, and the College of Agriculture and Home Economics at the Ohio State University, Columbus, Ohio.

Preliminary studies and correspondence preceded a September 1963 visit to the OSU campus by Professor Hugo de Almeida Leme, then dean of the College of Agriculture at ESALQ. Dean Leme's visit and subsequent discussions with Brazilian officials lead to an OSU Survey Team visit to ESALQ during the period of 17 November to 2 December 1963.

A contract, duly signed by the participating parties, was formally concluded effective 16 March 1964. Terms of the contract, which was funded by the United States Agency for International Development (USAID), stipulated that Ohio State University was to give assistance to the Sao Paulo University Agricultural College at Piracicaba in the following ways:

1. Integration of teaching, research, and extension in the pattern of the U.S. land-grant university system.

2. Strengthening of research extension at ESALQ to better serve the needs of the state of Sao Paulo and the Brazilian nation.

3. Advisement and assistance in planning and establishing a graduate level program, including initiation of administrative procedures for certification of graduate level faculty; development of graduate level courses and graduate research programs; establishment of a grading system; and determination of appropriate standards of achievement as well as essential records.

4. Assistance in strengthening undergraduate teaching programs in specific subject matter areas.

5. Advisement and assistance in the development and conduct of short courses, workshops, and in-service training programs.

6. Planning and development of a School of Home Economics.

7. Planning and conducting economic research designed to guide state and federal governmental agencies as well as cooperatives, farmers, and other private enterprises in the development of Brazil's agricultural industries.

8. Assistance in the coordination of state and federal agencies with ESALQ efforts to increase agricultural production, improve marketing procedures, and enhance processing, storage, and distribution of agricultural products and supplies.

9. Assistance in strengthening the professional competence of faculty members in agriculture, veterinary medicine, and home economics as well as teachers of agriculture and home economics in the secondary schools of the state of Sao Paulo.

Soon after formal initiation of the contract in March 1964, Dean Leme was appointed the Brazilian minister of agriculture, and Professor Euripedes Malavolta became dean of agriculture at ESALQ. By late summer 1964, nine OSU faculty members had been stationed at ESALQ and were busily engaged in fulfilling the contractual agreements that earlier had been agreed upon.

Much credit for both the initial and ultimate success of the project must go to members of the OSU Survey Team composed of John T. Mount, vice-president and secretary of the OSU Board of Trustees, who headed the Survey Team; T. Scott Sutton, associate dean, College of Agriculture and Home Economics, whose years of experience as OSU's project leader in India had provided him with essential background to assist in initiation and implementation of overseas projects; Raymond E. Cray, campus coordinator at OSU, whose six years as project leader for OSU's India project had given him valuable insights for launching the ESALQ project; and John M. Sitterley, who had been designated as OSU's first chief of party at Piracicaba, and whose laudable patience and tireless effort during the first four years of the Brazilian project helped mightily to ensure its success. Subsequent developments bore out the wisdom of the initial project plans made by this four-man Survey Team. Clearly, their efforts were key element's in the subsequent rapid development of agricultural productivity in the state of Sao Paulo as well as throughout the nation of Brazil.

Dr. John Sitterley, an agricultural economist (farm management specialist), OSU's first chief of party in Brazil was succeeded on 1 April 1968 by Dr. A. L. Moxon, an animal scientist (nutritionist), who had served as assistant chief of party during 1964–66. Professor Ferdinando Galli, head of the Department of Forestry, succeeded Dean Malavolta as dean of

ESALQ on 1 January 1971. Shortly thereafter, in February 1971, Professor Almiro Blumentschein, head of the Department of Genetics at ESALQ, was named co-director of the OSU/AID/ESALQ contract. Dr. Trevor Arscott, soils scientist from the Department of Agronomy at Ohio State, became OSU's chief of party at ESALQ in October 1971. That year marked the attainment of maturity for the OSU/AID/ESALQ contract in that it brought in additional emphasis to plant physiology, agronomy, horticulture, agricultural engineering, and forestry. At the same time, OSU was asked to maintain continuing effort toward the development of a School of Home Economics and the enhancement of graduate programs in animal science as well as in agricultural economics and rural sociology.

As the original OSU/AID/ESALQ contract drew to a close on 31 December 1973, it was evident that the supplementary contract for "Advising Brazil on Agricultural Economics Research" (which had extended from 15 August 1972 to 15 October 1973 and which was country-wide despite it being headquartered at Piracicaba) had clearly demonstrated the wisdom of maintaining relationships between OSU and ESALQ. Thus, a project entitled "Planning Higher Agricultural Education in Brazil," which ran from 26 November 1973 to 30 June 1975, set the stage for a sizable project on "The Development of Graduate Education in Agriculture in Brazil," which ran from 2 December 1975 through 30 June 1978. These contracts provided a logical sequence of assistance to ESALQ, as did several other contracts that succeeded the 1964–73 major OSU/AID/ESALQ contract, and in some cases overlapped with it. Among these projects of lesser duration was one entitled "An Analysis of Farm Capital Formation and Technological Innovation in the LDC's," and another entitled "Advising Brazil on Agricultural Economic Research." More recently, a project entitled "Research Linkages in Agriculture, Home Economics, and Natural Resources—OSU/OARDC and the School of Agriculture (ESALQ) in Brazil" was begun on 1 September 1980 and continued through 30 June 1985.

In total more than $4,000,000 in USAID support was expended on the several contracts to which reference has been made. Over the twenty-one year time span from 1964 to 1985, more than 75 faculty members from ESALQ studied at the Ohio State University. This number included 55 who completed their M.S. or Ph.D. degrees. Some 85 OSU faculty members were directly involved in the contract, including 75 who were on the campus at ESALQ for periods ranging from one month to 4 years.

Although credit for success of the OSU/AID/ESALQ project in Brazil must go to all participants from both the United States and Brazil, the individuals previously named played an especially significant role, as did also Dr. Mervin G. Smith, assistant dean and director of international affairs in the College of Agriculture at Ohio State, who succeeded Professor Raymond R. Cray as campus coorinator for OSU's efforts in Brazil. Special credit is due also to Dr. Dorothy D. Scott and Dr. Francille M. Firebaugh, both of whom served as directors of the School of Home Economics at Ohio State during the period of OSU's involvement in Brazil; to Dr. W. E. Krauss, associate director of the Ohio Agricultural Research and Development Center (OAEDC); to Dr. J. M. Beattie, and to Dr. Clive W. Donoho, Jr., each of whom served as the associate director of OAEDC during several years of the Brazilian contract.

Although the Ohio Cooperative Extension Service was not as heavily involved as was the Ohio Agricultural Research and Development Center, the leadership and support of the associate director of the Ohio Cooperative Extension Service (OCES), Edwin L. Kirby, and of his successors Orlo L. Musgrave and George R. Gist, proved noteworthy. Special recognition is due for the leadership and strong support provided by Richard H. Hohning, associate dean of the College of Agriculture and Home Economics, and by Dr. Bohning's successors Dr. Robert W. Teater and Dr. Kenneth W. Reisch.

On the Brazilian side, in addition to those administrators whose names have been cited, the leadership and efforts of the following individuals were noteworthy: Dr. Salim Simao, who was named dean of ESALQ on May 16, 1978; Professor Renato Amilcare Catani, who served as associate dean at ESALQ in the early days of the contract; and Dr. Aristeu Peixoto, who served as associate dean of ESALQ beginning in 1978 and was named dean of ESALQ following the term of Dr. Salim Simao. Two other individuals who made major contributions were Dr. Ibriam Abrao, associate dean at ESALQ and head of the Soils Department; and Dr. Joaquim Engeler, chairman of the graduate program and assistant to the dean, who also served for a number of years as head of agricultural economics and rural sociology at ESALQ.

Throughout the period of the several USAID contracts, many OSU faculty as well as ESALQ faculty developed very close personal as well as professional relationships. They shared in both undergraduate and graduate teaching programs and research projects, and they coauthored many publications.

In the early years of the contract, major emphasis was devoted to converting educational practices and procedures from the "classical European model" to the academic practices and procedures followed in colleges of agriculture within the United States land-grant university system. Those efforts entailed conversion from external to internal examinations; greater emphasis on laboratory instruction; increased emphasis on the basic biological and physical sciences; more reliance on individual student initiative and reponsibility; and much greater attention to the purchase and maintenance of current library acquisitions.

There can be little doubt but what the goal of planning and establishing graduate-level programs, including initiation of administrative procedures for certification of graduate-level faculty, development of graduate-level courses and graduate research programs, were among the most significant contributions of the contract. A follow-up contract, entitled "The Development of Graduate Education in Agriculture in Brazil," which extended from 2 December 1975 to 30 June 1978, was shared by Ohio State with nineteen other U.S. institutions. OSU provided 10% of the effort for that project, including one long-term and eleven short-term advisers. Eight Ph.D. graduate students completed their work at OSU under this program. Among the Ohio State University faculty members who served on this capsheaf effort were Dr. W. R. (Rod) Sharp, from the Department of Botany at OSU; Dr. Clarence R. Cole, Regents Professor of Veterinary Pathobiology and former dean of the College of Veterinary Medicine at OSU; and Robert H. Miller, professor of soils microbiology in the Department of Agronomy at Ohio State, and currently chairman of the Department of Soil Science at North Carolina State University.

The most recent collaborative effort involving ESALQ and the College of Agriculture at Ohio State bore the title "Research Linkages in Agriculture, Home Economics, and Natural Resources." This effort, which was launched on 1 September 1980, was continued through 30 June 1985. Some 15 OSU College of Agriculture faculty members were involved in this extension of earlier programs on the campus at ESALQ.

Overall, seventy-five OSU faculty members spent from one month to four years working on the campus of ESALQ at Piracicaba in Brazil. An equal number of faculty members from ESALQ studied at Ohio State. Financial support for this monumental effort was provided by USAID to ESALQ for enhancement of twenty-five subject-matter areas. The crowing achievement for those Ohio State University faculty members who participated in the Brazilian effort was the establishment of graduate programs in fifteen academic areas at ESALQ. Both undergraduate and graduate courses and curricula within twenty-five subject-matter areas were either initiated or upgraded. Of no less significance were:(1) establishment of a Department of Home Economics; (2) development of a large and modern library; (3) dramatic improvement in laboratory instructional endeavors; (4) installation of state-of-the-art equipment, which generally preceded the improvement of laboratory practices; and (5) initiation of internal examination procedures plus greatly expanded student services, all of which are a source of great pride both OSU and ESALQ.

The twenty-one year partnership and the significant achievement that it attained stand as a living memorial to Dr. Roy M. Kottman, whose confidence in the wisdom of international cooperation lead to the establishment of contractual relationships between OSU and ESALQ. It is to Dr. Roy Kottman, dean of the College of Agriculture and Home Economics, director of the Ohio Agricultural Research and Development Center, and director of the Ohio Cooperative Extension Service throughout almost two decades of the partnership between OSU and ESALQ that these papers are dedicated. His vision and tireless effort bear testimony to the concept that research and education can provide the "miracles" that will satisfy the "rising tide of expectations" of a rapidly expanding world population!

1. Dr. C. W. Donoho, Jr., became director of OARDC in April 1982.

Biotechnology of Plants and Microorganisms

A. CLAUSI

Opportunities for Genetic Engineering in the Food Industry

1

The challenge to be met in the future relationship between food processing companies and companies in the technical forefront of biotechnology is quite unique. The case for a close relationship is easy to make. All human and animal food is a product of bioprocesses. To understand better these processes—more importantly yet, to be able to manipulate these processes towards specific needs—is obviously a worthy objective.

Still, general agreement on the merits of marrige alone is no guarantee for harmonious bliss through "sickness and health" in every specific instance. To realize the large potential of cooperative programs between the food industry and companies on the leading edge of biotechnology, a conscious effort must be made toward better understanding of the irrespective needs and capabilities. This is not an easy task because we come from different "cultures." Let me try to explain what I mean by that statement.

The food industry (FI) cannot be described as technology intense; this is reflected in research expenditures as a percentage of net sales (fig. 1). It is obvious that food research is on the low end of the scale.

But research expenditures tell only part of the story. The FI got started by assuming gradually a bigger and bigger role in those food preparation "chores" which were originally executed in the home. This implies that traditional recipes, or methods of food preparation, formed the basis of this industry. Technology, based on scientific breakthroughs, was only pulled in on an *as needed* basis to afford the quality cost effectiveness and/ or distribution tolerance which is today associated with FI products.

It is true that FI leaders, particularly General Foods, Inc, (GF), have established significant technical capabilities over time in response to ever increasing appetites for business growth. The *range* of skills represented

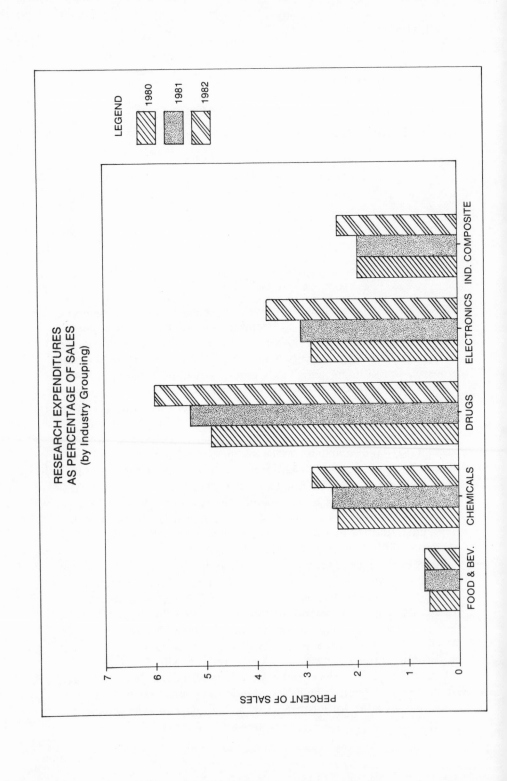

today in the central laboratory of a major food company such as GF rivals—if not exceeds—the *range* of skills in the typical industrial chemical or pharmaceutical laboratory. Not only are the skills represented that are normally found in the laboratories of, say, a chemical company, such as organic chemists, analytical chemists, and chemical engineers, there are microbiologists, nutritionists, and food scientists, to mention only a few. There is also a significant difference in the role that these laboratories play within the total organization. The technical capabilities of the average food company laboratory are used to *support* marketing concepts and strategies and so are very product development-oriented. FI laboratories are not traditionally used as a spawning ground for technologies that in turn could become the driving force for new product and process categories. I hasten to add that we at GF are in the process of modifying this traditional approach. We have identified a few technologies that are recognized and commensurately funded by the corporation as potential seedbeds for major new categories or businesses. We believe these efforts will complement our more traditional product development approach. But by and large, my statement still holds—FI research is conducted primarily to support defined marketing concepts.

Take in contrast the typical biotechnology company: staffed by scientists often transplanted from academic environments with in-depth knowledge in specific academic subspecialties they, understandably, want to do what they know how to do best. Being at the leading edge of a selected specialty is highly prized in such an environment as compared with the environment in industry, where the integration and profitable application of *available* technology is the standard mark of success.

Our most difficult challenge, therefore, is to bridge the gap that exists between the different research "cultures". I would like to share with you my thoughts on how, over time, it may be accomplished. The charges as I see them are as follows:

1. Biotechnology companies *and* food processors such as GF must proactively seek opportunities for dialogue about mutually beneficial opportunities. That is easier said than done, particularly for the smaller biotechnolgy boutiques with limited resources. However, it is vital. I know of more than one instance where a biotechnology company has refused contacts, or severely constrained contacts, with FI factors, either because of limited resources or, worse, because they generally felt they did not require a commercial input. "We haven't the time to teach the new biotechnology to industry," they said. They should have *taken* the time since some of these

companies are either not around today or are in difficult financial straits. The FI needs to know more about what biotechnology can do; conversely, the FI needs to tell Biotech more about what FI needs.

2. The dialogue must initially focus on areas in which there is the largest knowledge base—from both points of view. Speaking for GF, that means focusing first on opportunities in established GF product categories such as coffee, cereals, rice, and vegetables for our "Bird's Eye" brand. Focus particularly on those that are large volume/profit contributors.

Fortunately, these are the very products about which GF is likely to know exactly what functional properties should be improved—in other words, what is meaningful to consumers and worth the trip. Also, in all probability, GF will know a great deal about the physical and chemical attributes of these products, which would provide an essential contribution to a joint program.

Food processors may know what they want in their heartland areas, but they usually do not know what is technically feasible, how much it could cost to get there, and what is feasible short-term as well as long-term.

A good question for biotechnology to ask a potential FI partner is, "What attributes would you prefer your current product line to have if you had the power to change it at will?" From that may come more high-solids tomato projects. Maybe this strategy will not deliver the home runs, but it should surely keep biotechnology in the game.

Once this dialogue has begun, opportunities in less defined areas can be explored. Major food companies are extremely interested in identifying leverage areas that will allow them to participate in *new* food and beverage categories. Here, the information base is less extensive. However, GF will have a fairly good idea of what is needed in several categories of interest to it, and the need will be substantial.

In order to participate in a new category, especially one in which there is an appreciable competitive situation, a food company must provide substantially superior consumer benefits and/or restructure the cost dimensions so that it becomes the low-cost producer. The questions will be, "What can the new biotechnology bring to these opportunities?" and "What *additional* new biotechnology developments will be required to insulate us from competition?" Again, dialogue is essential—perhaps a more difficult, more extensive dialogue since the areas are less well defined.

3. The charge in this process must be to cultivate relationships with *selected* companies with bona fide interests in specific new-category development. Building relationships over extended periods of time will be essen-

tial to achieving an *open* dialogue between parties. Discussions with hidden agendas cannot be productive.

There are new biotechnology-driven concepts that admittedly are dreams of some of the country's more futuristic thinkers. Agriculture divorced from soil and phytoproduction independent from whole plants are examples of this. Perhaps these would be very important break-throughs, but the science usually is not even close to the stage that would encourage food processors to participate in some financial way. These may be the home runs we all dream about, but, frankly, the FI is not ready for such long-term thinking. It loses interest quickly in programs that extend well beyond its planning horizons. That is reality, be aware of it! I believe GF is one of the more forward thinking companies in the FI, but even GF finds it hard to stretch beyond a ten-year timeframe. To the degree that biotechnology firms can see their way clear to devote some resources to these long-range opportunities, serve as a bridge to academic activities in this terrain, and more clearly define benefits and risks to food processors, I believe it would be very valuable. Clearly, it is up to the biotechnology industry to decide to what extent it will rely on the development of major scientific breakthroughs.

4. This charge is essential to closing the gap of understanding and fostering closer ties with food processors. It has to do with first defining, then establishing a *company profile* that is understandable to potential FI partners. Because so many new biotechnology companies have formed recently, it is a very confusing competitive framework in which they operate. Hence, it is important to establish and communicate a corporate profile that flags its strengths and differentiates it from the competition. This will facilitate establishment of the essential framework within which a dialogue can take place with food processors.

5. In all business-technical pursuits, there must be a payoff that ultimately translates to profits. The payoff has got to be a primary part of the upfront dialogue with food processor partners. In this dialogue be flexible, practical, and equitable. Too many programs have failed to get off the launching pad because of unreasonable expectations about payoff.

There are times when *cost-plus contractual arrangements*, particularly at the early stages of a relationship with a food processor, are most appropriate. When dealing with an improvement in a heartland business where the business *knows* what it wants, knows the chemical pathways that are essential, *and* the application of existing, albeit new, biotechnology is all that is required, a negotiator would be flogged by his company if he consid-

8   *Biotechnology of Plants and Microorganisms*

ered fat royalty arrangements or joint ventures. Of course, when one is dealing in areas where mutual and more equal capabilities are requirements or where scientific breakthroughs are expected from a biotechnology company partner, then a more lucrative and equitable arrangement is in order.

To summarize, the appreciable gap of understanding between the FI and biotechnology companies must be closed if the relationship wanted is to be fully exploited.

1. Seeking opportunities to dialogue together is the only way to close that gap. The dialogue should be entered into in a way that would be productive.

2. Establish dialogue with companies on opportunities to apply the new biotechnology in areas in which the information base is the greatest— what can be done together short-term.

3. Move on to discussion with selected companies on *new* categories that are likely to require even newer technologies. These are more difficult and long-term discussions, and it is important to select the right FI partner—one that has a genuine interest and understands long-term commitments.

4. Establish a profile that is right for your company and communicate it to the FI industry.

5. Be flexible and equitable in defining the payment terms of agreements.

All of this really suggests the need to establish a strategic plan, a plan that defines clear goals and pathways to establish these goals, resource/skill needs, and preferred FI partners. Too many biotechnology companies start off with a good idea but fail to have a plan on how best to capitalize upon that idea. Some try to do too many things and prove that if you do not know where you are going, any road will get you there.

C. BALL

Opportunities for Genetic Engineering in the Pharmaceutical Industry

2

ABSTRACT

In this chapter genetic engineering has been considered to be synonymous with recombinant DNA technology. Detailed discussion of the contribution of classical genetics to the pharmaceutical industry has been omitted since it has been previously discussed, particularly in the Symposium series "Genetics of Industrial Microorganisms" (Murray, 1976). The impacts of recombinant DNA technology on the cheaper production of existing pharmaceutical products and the discovery of new ones are discussed. The technology exists for transferring a limited number of genes from one organism (the donor) into another organism (the host). Production of a metabolite is mainly dependent on the development of cloning vectors that work in the producing host, but for production of specific polypeptides, it often is possible to choose a preferred host.

INTRODUCTION

Microbial participation in the formation of pharmaceutical products can be divided into three categories (Aharonowitz and Cohen, 1981). Production of compounds in all three categories has been or will be influenced by recombinant DNA technology. The three categories are founded on the basis of how much of the key genetic information needed to make the actual chemical structure of the product is present in the microorganism. The first category consists of those products that are natural metabolites, such as antibiotics, all the information for their synthesis being native to the cell. The second category includes those products that are not natural metabolites of microorganisms but the microorganism contributes information for only a few biological steps used in the chemical conversion of chemicals from one form to another. This is termed bioconversion (for example, the bioconversion of steroids). The third category is characterized by the fact that none of the information that defines the structure of the product molecule is initially found in the genome of the microorganism. In this case, the information is inserted into the cell. As an example, microorganisms can be made to produce mammalian polypeptides.

PHARMACEUTICAL PRODUCTS

The types of pharmaceutical products that are likely to be influenced by recombinant DNA technology in any of the three general ways detailed in the introduction can be divided, according to the American Government Report of the Office of Technology Assessment, Washington, D.C. (OTA Report), into two groups, namely, small molecules and large molecules. Small molecules include amino acids, vitamins, antibiotics, alkaloids, steroid hormones, very short peptides, and aromatic compounds such as p-acetaminophenol, this last presently made only by chemical means. Large molecules include peptide hormones (table 1), enzymes, viral antigens (vaccines), preparations such as hemophilia clotting factors and miscellaneous proteins such as blood serum albumin. The medical conditions currently being addressed by production of these compounds include the use of amino acids in intravenous solutions, steroids as anti-inflammatory agents, antibiotics to control infectious disease, human growth hormone to treat dwarfism, insulin to treat diabetes, enzymes such as glucose oxidase for use as diagnostic agents and urokinase as a therapeutic antithrombotic agent, serum albumin currently used to treat shock, and blood clotting factors such as factors VIII and IX to treat hemophilia. Many of these products are currently obtained from animal cells, but in the future they could be produced by microorganisms.

In addition, there are a number of products not used clinically at the moment that might be used if produced in large quantities by microorganisms. These include certain hormonal polypeptides shown in table 1 and miscellaneous proteins such as interferons and monoclonal antibodies that could have a variety of uses both therapeutically and as diagnostics. Also, there are several human vaccines that might be produced safely by using recombinant DNA since only part of the viral antigen (viral coat protein) need be produced. Furthermore, there are possibilities of another use of certain compounds if available in larger quantities, for example, the speculative use of human growth hormones for treatment of wounds and burns.

TABLE 1

HUMAN LARGE POLYPEPTIDES*

POTENTIALLY ATTRACTIVE FOR BIOSYNTHESIS

	Amino Acid Residues	Molecular Weight
Prolactin	198	
Placental lactogen	192	
Growth hormone†	191	22,005
Nerve growth factor	118	13,000
Parathyroid homone (PTH)...............	84	9,562 bovine
Proinsulin	82	
Insulin-like growth factors (IGF-1 & IGF-2)	70,67	7,649; 7,471
Epidermal growth factor		6,100
Insulin†	51	5,734
Thymopoietin............................	49	
Gastric inhibitory polypeptide (GIP)................................	43	5,104 porcine
Corticotropin (ACTH)†	39	4,567 porcine
Cholecystokinin (CCK-39).................	39	
Big gastrin (BG).........................	34	
Active fragment of PTH..................	34	4,109 bovine
Cholecystokinin (CCK-33)	33	3,918 porcine
Calcitonin†	32	3,421 human
		3,435 salmon
Endorphins.............................	31	3,465
Glucagon†	29	3,483 porcine
Thymosin-α^1	28	3,108
Vasoactive intestinal peptide (VIP).........	28	3,326 porcine
Secretin†	27	
Active fragment of ACTH†	24	
Motilin	22	2,698

SOURCE: OTA Report

* Mainly hormones.

† Currently used in medical practice.

It is impossible to consider the pharmaceutical industry without considering the relative potential size of markets for various products. In the OTA Report, these aspects are considered but problems exist with such an appraisal. For example, market value data are given for a number of compounds already produced, but it is not clear how profitable various products are. Also, there are no good market data available for a number of newer compounds such as interferon and vaccines. Overall, however, there are some definite optimistic indications of strong potential market needs particularly in areas where the current source is animal cells and where a microbial alternative now or in the future would be desirable. Such products already discussed include insulin, serum albumin, urokinase, and human growth hormone where the potential sales in the United States of each product is of the order of $100 million per year. Another interesting aspect of the OTA Report is that it highlights the large sales volume of antibiotics (greater than $4 billion), and it might be assumed that this is a very profitable market judging by the relatively large and traditional research efforts in established pharmaceutical companies. It should be noted that the array of antibiotic substances made by each microbial genus does not bear much relationship to their clinical or commercial importance (Aharonowitz and Cohen, 1981). Of the almost 5,000 antibiotics known, only about 100 have been marketed, of which about 70% were derived from streptomycetes. However, the fungal-produced β-lactam antibiotics, penicillins, and cephalosporins contribute about 25% of the world market sales. Nevertheless, β-lactam antibiotics produced by eubacteria have been discovered recently (Imada et al., 1981; Sykes et al., 1981), and derivatives of them could establish a major market share along with β-lactams produced by actinomycetes as indicated in a recent review (O'Sullivan and Ball, 1982).

DEFINITION OF RECOMBINANT DNA TECHNOLOGY

Recombinant DNA technology (Miller and Baxter, 1981) often involves cutting DNA that codes a polypeptide of interest with one of several restriction endonucleases. A restriction endonuclease is a highly specific enzyme that cuts double-stranded DNA at a uniquely specified sequence of four to six nucleotides. These enzymes recognize mirror image sequences of DNA (palindromes) and cut them along an axis of symmetry, commonly generating short segments of single-stranded DNA that are complimentary to each other. Such complimentary ends of DNA strands are termed cohesive termini, or "sticky ends," because they have the ability to adhere to complimentary sticky ends by base pairing. A cloning vehicle or vector, such as a plasmid (table 2) is chosen with a single restriction site for the enzyme used to cut the foreign DNA. The plasmid is cut with the enzyme and mixed with the foreign DNA. The complimentary sticky ends align by hydrogen bonding, inserting the foreign DNA into the plasmid. The ends are then cemented by DNA ligase. The resulting new recombinant plasmid, carrying the gene for the desired protein, is then put into a microorganism. Plasmids that have entered the microorganism now behave as benign obligate intracellular parasites, replicating independently of the host chromosomal DNA and having full access to its protein synthetic machinery. Successfully transformed microorganisms can be selected on agar plates if the plasmid-cloning vehicle also confers antibiotic resistance to the microorganisms.

TABLE 2

DESIRED PROPERTIES OF A DNA MOLECULE THAT IS TO SERVE AS
A VECTOR OF DNA FRAGMENTS IN RECOMBINATION EXPERIMENTS (27) IN VITRO

1. The vector must be capable of autonomous replication in an appropriate host cell.
2. The site of insertion of a DNA fragment must be such that the insertion does not destroy an essential function of the vector.
3. The vector must carry a means for selection of transformed cells, such as a drug resistance determinant, the ability to confer immunity (e.g., to a colicin or to a phage), or the production of a phage plaque in a lawn of bacterial cells.
4. A means for distinguishing, or preferably selection of, recombinant DNA molecules from the parent vector DNA.

FOREIGN PROTEIN PRODUCTION

According to authorities in this field (Miller and Baxter, 1981), DNA from a variety of sources may be inserted into the DNA of bacteria (cloned) by means of bacteriophage or plasmid vectors. The DNA to be cloned may originate from the genome of another organism (genomic DNA), it may be copied enzymatically from messenger RNA (complimentary DNA or cDNA), or it may be synthesized by purely chemical means. In general, cloning of genomic DNA from higher organisms is useful for studying gene structure, but not for inducing bacteria to synthesize the proteins coded by the DNA. This is because the genes of higher organisms usually consist of stretches of DNA coding parts of a protein that are separated from one another by other DNA (intervening sequences) that does not code protein (Breathnach et al., 1977; Chow et al., 1977). Bacterial genes are not interrupted by intervening sequences, hence bacteria cannot make protein accurately from genes containing such interruptions. The use of complimentary DNA or chemically synthesized DNA circumvents this problem. These sources of DNA have been used successfully to produce an increasing number of pharmacologically useful proteins.

Chemically synthesized DNA offers advantages for many recombinant DNA experiments (Miller and Baxter, 1981). Chemical synthesis can yield a product of great purity, without the possibility that the DNA sequences coding other genes have been coisolated with the DNA sequences to be cloned. Furthermore, as the DNA is being chemically synthesized, its precise nucleotide sequence must be designed by the experimenter. This permits selection of triplet codons favored by the recipient host microorganism and permits the creating of designed proteins that do not exist in nature. However, chemical synthesis is complicated and may not produce long segments of DNA efficiently. Furthermore, as short polypeptides are rapidly degraded by the host bacteria, successful production of these peptides has had to be achieved by synthesizing them as a hybrid protein covalently linked to other large proteins and then cleaving the short polypeptide from the hybrid protein. A good example of this is the production of human insulin (Goeddel et al., 1979).

The cDNA technique is based on the ability to make DNA copies of RNA molecules. The RNA is copied into DNA by an enzyme called RNA-dependent DNA polymerase, usually referred to as a reverse transcriptase because it transcribes RNA back into DNA (Baltimore, 1970; Temin and Mizutari, 1970). The identification, isolation, and purification of this enzyme has permitted the copying of essentially any mRNA molecule into DNA. Such DNA is termed complementary DNA or cDNA because it is mirror image copy of the mRNA. Since mammalian mRNA has been processed free of RNA corresponding to intervening sequences, cDNA may be used by bacteria as genes.

An example of the interplay of the chemical synthesis and the cDNA approaches is the production of human growth hormone. Human growth hormone is a polypeptide 191 amino acid units long elaborated by tissues of the pituitary gland. The segment of the gene that codes the first 24 amino acids of the peptide was synthesized chemically from blocks of nucleotides. To obtain the rest of the gene, a series of enzymes was used. Reverse transcriptase was employed to copy the gene for the hormone from messenger RNA obtained from human pituitary tissues (Martial, et al., 1979). Restriction endonucleases cut out the needed fragment. The DNA ligase was then used to join the natural and the synthetic fragments. The complete gene was inserted into a modified version of plasmid pBR322 carrying the lac operon. The synthetic part of the growth-hormone gene had been constructed with its own initiation codon (ATG), the group of three bases that provides the signal to start the process of transcription. The hormone, therefore, could be produced independently in bacterial cells, without the need for attachment to a bacterial protein.

Within the technology described is a requirement (table 2) of methods for detecting the desired recombinant clones usually exploiting cDNA or mRNA hybridization probes. However, if one does not have a probe for the cDNA or mRNA of interest (Miller and Baxter, 1981), other techniques may be developed for the particular clone of interest. The final solution is to generate enough clones so as to make it likely not only that one of them contains the DNA of interest but that it also synthesizes the desired protein.

Clones can then be assayed in batches and the proper clone factored out in repeated assays of smaller and smaller batches. Such an approach was used for the cloning of human leukocyte interferon (Nagata et al., 1980). Messenger RNA prepared from 10^{11} virally induced leukocytes was used to prepare cDNA that yielded 20,000 clones. These were assayed in groups, from which one clone that synthesized interferon was finally identified by bioassay techniques. Thus, in principle, any gene producing an identifiable protein can be cloned, isolated, and characterized.

BIOCONVERSION

The use of microbial enzymes or microbial whole cells in either isolated or immobilized form to modify certain chemicals could play an increasing future role in the pharmaceutical industry (Kieslich, 1978). The potential role of genetic engineering to increase yields is great because of the common one-to-one relationship between gene and enzyme. Hitherto, some of the outstanding examples have been in the area of modification of steroids and β-lactam antibiotics. The essential point is that the enzyme operating at normal temperature and pressure and with great specificity is more efficient than a chemist working with abnormal temperatures and/or pressures and/or anhydrous conditions.

The manufacture of steroids includes several important microbiological steps in a process made up primarily of chemical syntheses (Murray, 1976; Aharonowitz and Cohen, 1981). The raw materials of production are the complex alcohols called sterols. Stigmasterol is a by-product of the soybean-oil industry; diosgenin is extracted from the roots of the Mexican barbasco plant. The sterols are converted in a series of chemical steps into one of two intermediates: compound S and progesterone. A fungal mold (either *Rhizopus nigricans* or *Curvularia lunata*, depending on the intermediate compound) is then used to add a hydroxyl group (-OH) to the four-ring steroid nucleus. This microbial step was critical in the development of a commercially feasible method of manufacturing steroids. The chemical method of hydroxylation is complex and difficult. Finding a biological method reduced the complete synthesis from 37 to 11 steps and greatly reduced the unit cost of manufacture. The other important microbial step is the removal of two hydrogen atoms from the steroid nucleus. Processes resembling this generalized scheme are employed in producing cortisone, hydrocortisone, prednisolone, and prednisone, which are all medically important synthetic molecules having pharmacological properties different from those of the natural steroids (Murray, 1976; Aharonowitz and Cohen, 1981).

The microbiological conversion of penicillin G into 6-aminopenicillanic acid (6-APA) has fewer steps and is cheaper than the chemical conversion (Aharonowitz and Cohen, 1981; Vandamme, 1977). In the production of semisynthetic penicillins, 6-APA forms a chemical nucleus to which side chains can be attached, yielding new antibiotics. In performing the chemical conversion, three steps must be carried out at low temperature and under strictly anhydrous conditions with a number of chemical solvents. The biological process relies on bacteria that make enzymes called acylases; the enzymes cleave away the benzyl group, leaving 6-APA. The fermentation can be carried out in water at 37° C. Recently, recombinant DNA methods have been used by Collins and colleagues (personal communication) to increase the yield of *E.coli* penicillin acylase by a factor of 40-fold.

Cephalosporin β-lactam acylases that efficiently yield 7-ACA upon hydrolysis have not been found. However, a two-step removal of the α-aminoadipyl side chain by oxidation to glutaric acid (Gilbert et al., 1972) followed by cleavage of the side chain (Shibuya et al., 1981) has been reported. Whether the isopenicillin N acylase (transacylase) of *Penicillium chrysogenum* can remove the cephalosporin C side chain or can be modified to do so, is yet unclear; nevertheless, considerations such as this indicate that there could be recombinant DNA possibilities in transferring the genes involved in bioconversions into an antibiotic producing organism thus avoiding a number of laborious chemical extraction and / or modification stages downstream in the production process.

In addition, it has been found (Shimizu et al., 1975) that most penicillin acylase-producing bacteria can efficiently synthesize penicillins from 6-APA. In recent years there have been some examples of very efficient synthesis described. Thus, amoxicillin has been produced with a non-immobilized whole-cell system of *Xanthomonas citri* (Kato et al., 1980). More than 90% of the added 6-APA was converted to amoxicillin under optimum conditions. Ampicillin synthesis has been studied with the immobilized enzyme from *Escherichia coli* (Marconi et al., 1975), and cephalexin synthesis recently has been studied using immobilized *X. citri* cells (Choi et al., 1981). Presumably, this technology could find application in the synthesis of a variety of cephalosporins and penicillins.

An interesting aspect of bioconversion is discussed in the OTA Report. This is a production of p-acetaminophenol (trade name Tylenol). This is prepared normally by chemical means from nitro benzene, but a proposed microbiological route from analine is suggested. The reaction sequence is analine, to p-aminophenol and then its reaction with acetic acid to p-acetaminophenol. Certain fungi can carry out these reactions so one could speculate about the possibilities of a biological mode of production. Detailed calculations, although very hypothetical, were carried out in the OTA Report supporting the biological method of production.

METABOLITE PRODUCTION

Among the metabolites of interest to the pharmaceutical industry, the antibiotics are the most important and the most challenging from the point of view of applying recombinant DNA methods. It is likely that vector methods for cloning in the original organism as host will have to be developed. A review of the vector development question (Queener and Baltz, 1979) reinforces this conclusion when it is claimed that *Streptomyces* are different from *E. coli* or *Bacillus subtilis* in that they contain a guanine plus cytosine content of 69–73% (Tewfits and Bradley, 1967), approaching the genetic code upper limit, and might be expected to have controlling sequences differing substantially from those—in *E.coli* or *B. subtilis* and to possess rather atypical distributions of specific tRNAs. The lower fungi may possess special transcriptional features (Queener and Baltz, 1979), e.g., mRNAs with polyadenylic acids at the their 3′ ends (Darnell, 1978) and 7-methyl-guanine caps at their 5′ ends (Filipowicz, 1978), genes-containing introns, and specific post-transcriptional mRNA splicing mechanisms, not found in prokaryotes. Also, little is known about the very subtle genetic interactions that control precursor and antibiotic enzyme levels simultaneously in highly evolved and highly productive industrial strains (Queener and Baltz, 1979). In addition, most antibiotic-producing microorganisms undergo a complex differentiation process and possess poorly understood mechanisms that avoid "suicide" while producing very high levels of antibiotics. Therefore, it appears that the only reasonable approach is to apply recombinant DNA techniques in *Streptomyces* and lower fungi to produce yield improvement and for new-product formation. Successful transfer and expression of many structural genes from one organism into another might be anticipated—particularly when those genes are known either to be part of the same operon or at least clustered in the native strain (Queener and Baltz, 1979).

Suitable vectors have not yet been developed in fungi; but they have been produced in *Streptomyces* (Chater et al., 1982), and it has been concluded that a recombinant DNA approach peculiar to the antibiotics industry would be the use of random cloning for strain improvement and for new products (Hopwood, 1981). In principle, a few thousand (table 3) colonies from a "self-cloning" experiment using either a production strain or an

TABLE 3

NUMBER* OF CLONES IN A COMPLETE GENOMIC LIBRARY FOR A
STREPTOMYCETE OF GENOME SIZE 10^4 KB (26)

LENGTH OF CLONED FRAGMENTS (KB)	PERCENTAGE PROBABILITY OF COMPLETE LIBRARY			
	80	90	95	99.99
5	3,218	4,604	5,988	18,414
10	1,609	2,302	2,994	9,206
15	1,072	1,534	1,998	6,136
20	804	1,150	1,496	4,601
25	643	920	1,148	3,679
30	536	766	997	3,065

* Basis of the claim (see text) that only about a thousand clones need to be examined to have a good chance of detecting transformation with a DNA fragment containing the gene sequence of interest.

undeveloped wild strain should yield clones in which increased copy numbers of genes for rate-limiting steps results in a substantial yield increase. The availability of clones of "relevant" genes would facilitate their further manipulation (e.g., by mutagenesis *in vitro*) and introduction into other independently improved lines. If the cloning experiment is interspecific, then one may hope that genes for enzymes able to introduce variations into the resident antibiotic biosynthetic pathways will be cloned, resulting in the formation of new products (Chater et al., 1982; Hopwood, 1981).

Some interesting speculations on the reasons for attempting gene transfer between different genera are made by Queener and Baltz (1979). They indicate that although the cell wall structures of *Penicillium* and *Cephalosporium* are naturally resistant to the action of β-lactam antibiotics, they are inherently suitable for the production of this class of compounds. Highly productive strains already exist, and significant improvements in existing manufacturing procedures for producing cephalosporin antibiotics could be achieved through the expression of one or two foreign genes in *C. acremonium*. The discovery of a transacylase capable of efficiently replacing the 7-α-aminoadipyl side chain of cephalosporin C with a variety of substituted acetic acids would make very attractive the cloning of the corresponding gene from its source in nature into *C. acremonium*. Recombinant DNA technology might also be extended by cloning other genes from streptomycetes such as those that are capable of producing 7-o-methyl-cephalosporins or cephalosporins with a carbamoyl group at position 3 (fig. 1). Expression of the streptomycete genes in *C. acremonium* might provide a means of producing the cephamycin antibiotics in high yield by directly employing highly developed *C. acremonium* strains.

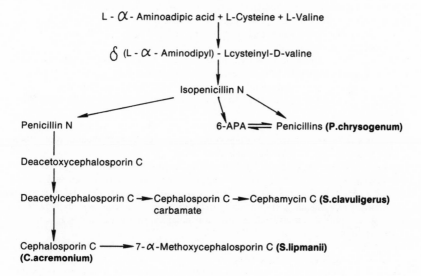

Possible Pathways to Penicillins and Cephalosporins

L - α - Aminoadipic acid + L-Cysteine + L-Valine

δ (L - α - Aminodipyl) - Lcysteinyl-D-valine

Isopenicillin N

Penicillin N

6-APA ⇌ Penicillins **(P.chrysogenum)**

Deacetoxycephalosporin C

Deacetylcephalosporin C →Cephalosporin C →Cephamycin C **(S.clavuligerus)**
carbamate

Cephalosporin C ——→ 7-α-Methoxycephalosporin C **(S.lipmanii)**
(C.acremonium)

REFERENCES

Aharonowitz, Y., and G. Cohen. 1981. The microbial production of pharmaceuticals. Sci. Amer. 245:140–52.

Baltimore, D. 1970. Viral RNA-dependent polymerase. Nature 226:1209–11.

Breathnach, R., J. L. Mandel, and P. Chambon. 1977. Ovalbumin gene is split in chicken DNA. Nature 270:314–19.

Chater, K. F., D. A. Hopwood, T. Kieser, and C. J. Thompson. 1982. Gene cloning in streptomyces. Current Topics in Microbiology & Immunology 96:69–95.

Choi, W. G., S. B. Lee, and D. D. Y. Ruy. 1981. Cephalexin synthesis by partially purified immobilized enzymes. Biotech. Bioeng. 23:361–71.

Chow, L. T., R. E. Gelinas, T. R. Broker, and R. J. Roberts. 1977. An amazing sequence arrangement at the 5' ends of adenovirus 2 messenger RNA. Cell 12:1–8.

Darnell, J. E., Jr. 1978. Implications of RNA-RNA splicing in evolution of eukaryotic cells. Science 202:1257–60.

Filipowicz, W. 1978. Functions of the 5'-terminal m7G cap in eukaryotic mRNA. FEBS Letters 96:1–11.

Gilbert, D. A., B. H. Arnold, and R. A. Fildes. 1972. Improvements in or relating to cephalosporin derivatives. Brit. Pat. 1, 272:769.

Goeddel, D. V., D. G. Klein, F. Bolivar, H. L. Neyneker, D. G. Yansura, R. Crea, T. Hirose, A. Kraszewski, K. Itakura, and A. D. Riggs. 1979. Expression in *Escherichia coli* of chemically synthesized genes for human insulin. Proc. Nat. Acad. Sci. USA 76:106–10.

Hopwood, D. A. 1981. Future possibilities for the discovery of new antibiotics by genetic engineering. *In* M. R. J. Salton and G. D. Shockman (eds), β-latam antibiotics, pp. 585–98. Academic Press, New York.

Imada, A., K. Kitano, K. Kintaka, M. Muroi, and M. Asai. 1981. Sulfazecin and Isosulfazecin, Novel β-lactam antibiotics of bacterial origin. Nature 289:590–91.

Kato, K., K. Kawahara, T. Takahasai, and S. Igarashi. 1980. Enzymatic synthesis of Amoxillin by the cell-bound amino ester hydrolase of *Xanthomonas citri.* Agric. Biol. Chem. 44:821–25.

Kieslich, K. 1978. Microbial transformations-type reactions. *In* R. Hutter, T. Leisinger, J. Nuesch, and W. Wehrli (eds.), Antibiotics and other secondary metabolites: Biosynthesis and production, pp. 57–100. Academic Press, New York.

Marconi, W., F. Bartoli, F. Cecere, G. Galli, and F. Morise. 1975. Synthesis of penicillin and cephalosporins by penicillin acylase entrapped in fibers. Agric. Biol. Chem. 39:277–79.

Martial, J. A., R. A. Hallewell, J. D. Baxter, and H. M. Goodman. 1979. Human growth hormone: Complementary DNA cloning and expression in bacteria. Science 205:602–7.

Miller, W. L., and J. D. Baxter. 1981. Synthesis of biologically active proteins by recombinant DNA technology. Drug Dev. Res. 1:435–54.

Murray, H. C. 1976. Steroids. *In* B. M. Miller and W. Litsky (eds.), Industrial microbiology, pp. 79–105. McGraw-Hill, New York.

Nagata, S., H. Taira, A. Hall, L. Johnsrud, M. Streuli, J. Ecsodi, W. Boll, K. Cantell, and C. Weissmann. 1980. Synthesis in *E. coli* of a polypeptide with human Leukocyte interferon activity. Nature 284:316–20.

O'Sullivan, J., and C. Bell. 1983. The biochemistry and genetic regulation of β-lactam antibiotic. *In* L. Vinning (ed.), Biochemistry and genetic regulation of commercially important antibiotics, pp. 73–94. Addison-Wesley, Reading, Mass.

O'Sullivan, J., and E. P. Abraham. 1971. Biosynthesis of β-lactam antibiotics. *In* J. W. Corcoran (ed.), Antibiotics IV Biosynthesis, pp. 101–22. Springer Verlag, Berlin.

Queener, S. W., and R. H. Baltz. 1979. Genetics of industrial microorganisms. Ann. Rept. Ferm. Proc. 3:5–45.

Shibuya, Y., K. Matsumoto, and T. Fujii. 1981. Isolation and properties of 7B-(4 carboxybutanamido) cephalosporic acid acylase–producing bacteria. Agric. Biol. Chem. 45:1561–67.

Shimizu, M., T. Mauike, H. Fujita, K. Kimura, R. Okachi, and T. Nara. 1975. Search for microorganisms producing cephalosporin acylase and enzymatic synthesis of cephalosporins. Agric. Biol. Chem. 39:1225–32.

Sykes, R. B., C. M. Cimarusti, D. P. Bonner, K. Bush, D. M. Floyd, N. H. Georgopapadakou, W. H. Koster, W. C. Liu, W. L. Parker, P. A. Principe, M. L. Rathnum, W. A. Slusarchyk, W. H. Trejo, and J. S. Wells. 1971. Monocyclic β-lactam antibiotics produced by bacteria. Nature 291:489–91.

Temin, H. M., and S. G. Bradley. 1967. Characterization of deoxyribonucleic acids from streptomycetes and nocardiae. Nature 226:1211–13.

Tewfik, E. M., and S. G. Bradley. 1967. Characterization of deoxyribonucleic acids from streptomycetes and nocardiae. J. Bact. 94:1994–200.

Vandamme, E. J. 1977. Enzymes involved in β-lactam antibiotic production. Adv. Appl. Microbiol. 21:89–123.

CARLOS E. WINTER, MARIA HELOIZA T. AFFONSO,
LUCILLE M. FLOETER-WINTER, AND WILLY BECAK

Perspectives of Applications of Biotechnology to Human Health

3

INTRODUCTION

Man as a social being is characterized by association in communities that provide protection and other advantages. A very important attitude of the community is the promotion of human health and the treatment of disease. The medical and scientific approaches to human health have two main aspects: the curative and the preventive. Infectious diseases have been approached through these two aspects; yet the abusive use of drugs leads to an increasing inefficiency of their therapeutic action. Moreover, this approach is too expensive for developing countries such as Brazil.

Prevention of infectious diseases seeks the reduction of their occurrence, through immunization, isolation, disinfection, and disinfestation of the population. In the case of genetically determined diseases, the prevention is made through genetic screening or antenatal counseling.

New biotechnology achievements can be applied toward preventing, and in some cases even curing, human diseases. The first such applications were done with the aim of treating human diseases. The pharmaceutical industry, with its infrastructure, was the first to use the techniques of recombinant DNA. Those applications were mainly done to improve the existing processes or the elaboration of substances that were not usually commercialized due to their limited production. The examples that are best known in this field are the human peptide hormones (Craig and Hall, 1983). However, many other biological products are being produced by bacteria or other simple organisms. Hypothetically many of them can be used in the treatment of diseases; coagulation factor VIII in hemophilia, vasopressin to control arterial pressure, and many others (Hale, 1979).

In this paper we emphasize the application of biotechnology in the production of more efficient and inexpensive vaccines to be used in immunization programs. Regarding preventive aspects, we also will discuss the utilization of recombinant DNA techniques in diagnosis of infectious and genetically determined diseases.

In developing countries the development of substances related to prevention of human diseases through the new biotechnological methodology is of great social importance. In this report we will discuss the programs being developed in several Brazilian laboratories aiming at improvement of human sanitary programs.

VACCINES

The control of infectious diseases is one of the main objectives of current public health programs. Many of these diseases, mainly those caused by bacteria, can be rapidly controlled through chemotherapeutic agents (e.g., sulfas and antibiotics). However, it should be noted that not all pathogens can be eliminated by such drugs. Moreover, the use of antibiotics has contributed after many years of indiscriminate prescription to an increase in the number of resistant strains of microorganisms. Thus, we face the ever increasing necessity to develop more efficient methods of pathogen control. This can be done by preventing the diseases or treatment after the infection. No doubt the most efficient way to prevent infectious diseases is active immunization through vaccination programs. The basic principle of immunological protection through vaccination is the modified antigen that loses its pathogenic characteristics but still retains the capacity to induce antibodies that neutralize the native antigen.

The current vaccines used have three types of antigens:

1. Dead microorganisms—vaccines with bacterial or viral agents inactivated by various methods in order to preserve their antigenic capacity.

2. Live or attenuated organisms—living vaccines developed through exhaustive selection of mutant strains. Although there are restrictions about the introduction of living organisms in a person, this king of vaccine gives a more efficient and enduring protection.

3. Toxoids—some bacteria exert their pathogenic effects through the

secretion of toxic proteins that act in various tissues of the host. These proteins are transformed into toxoids in order to make them lose their toxic activity but still retain their antigenic specificity. These toxoids induce the production of antibodies that are able to neutralize the native toxins by blocking their active sites.

Current vaccines have some problems related to the stability of the immunizing antigen and to contamination with harmful organisms or substances. Maintenance and control systems used to solve those problems are very expensive. Vaccines containing dead or attenuated microorganisms must be refrigerated through cold chains from the time of production up to the time of administration. In developing countries this can be an almost insoluble problem due to the lack of support and specialized personnel. Another problem is that protection by vaccines is limited to the vaccinating substance utilized, and very few of them are polyvalent. There are no efficient and cheap vaccines against several of the diseases present in three-fourths of the world. Among them the most important are malaria, trypanosomiasis, and hepatitis.

New techiques are being applied in the development of some new vaccines and improvement of existing ones. Of paramount importance are the techniques of genetic engineering and chemical synthesis of antigens as well as the interaction of both.

Antigens Produced by Living Microorganisms

Many methods have been developed to permit the introduction of specific DNA sequences in host cells: microinjection (Mueller et. al., 1978), cell fusion mediated by liposome vesicles containing DNA (Dimitradis, 1978; Ostro et al., 1978), and gene transfer mediated by erythrocytes (Rechsteiner, 1978). With the emergence of recombinant DNA techniques, methods were developed to introduce DNA using bacterial vectors (phages and plasmids [Graham and Van der Eb, 1973; Mantei et al., 1979; Wigler et al, 1977; Wold et al., 1979] or viral vectors for eukaryotic cells (Gruss and Khoury, 1981; Shimotohno and Temin, 1981).

Through recombinant DNA techniques, it is possible to identify the DNA sequences that code pathogenicity determinants, e.g., production of toxins (Mosely et al., 1983). By removing those specific sequences, we can transform the microorganism into an attenuated variant that retains all other characteristics and can be used as an immunizing agent. That is the approach in the production of vaccines against typhoid fever (Wahdan et al., 1982) and diarrheas caused by enterotoxigenic strains of *Escherichia coli* (Levine et al., 1983; Klipstein et al., 1983).

Another approach is the utilization of monoclonal antibodies to identify a protein antigen in a given organism. The ascertainment of the amino acid sequence of the antigen allows the synthesis of the corresponding DNA. Cloning of this DNA sequence in *E. coli* may result in production of a relatively great amount of the antigen. That is the case of the malaria vaccine being developed by the Nussenzweigs and colleagues at the New York University Medical Center (Ellis et al., 1983).

Another type of vaccine can be elaborated against the surface structures of the microorganism. Those structures, usually proteins, are very important in the first stages of infection, and are specific for each microorganism. High quantities of those proteins can be produced by DNA fragments cloned in *E. coli* containing the sequence coding adhesion structures. Those proteins with a high degree of purity can be used as a cheap and effective vaccine.

We will discuss further some of the problems involved in the utilization of this kind of technology. We emphasize the problems of regulation and expression of the cloned sequences in the host cell and the purification of synthesized antigen for use in vaccines. It is also important to consider the new technologies for construction of recombinant viral vectors to be used as live vaccines.

Bacterial Cells as Hosts

Genetic manipulation *in vitro* requires essentially the introduction of a recombinant DNA molecule in a host cell. The DNA sequence must be capable of autonomous replication in the host cell and of expression of the characteristics encoded by it.

In experiments of molecular cloning, the most widely used cell systems are bacterial organisms. Plasmids and temperate phages are normally used in these systems as molecular vectors. *Escherichia coli* is the choice bacterium for the expression of foreign genes. This is because its control mechanisms of gene expression are the best known of all bacteria so far studied. The successful expression of cloned genes in *E. coli* depends on the organization of the celluar machinery. All this must ensure that the cloned genes are as efficient as, or better expressed than, the resident genes.

The first step in cloning and expressing the sequences that contain the antigenic determinants of a given organism is the identification of these sequences in the genome of the organism. Once these sequences are identified and isolated, the corresponding DNA molecules can be enzymatically inserted into a vector (e.g., a bacterial plasmid). This recombinant plasmid is then introduced in *E. coli*, and the following step is to study the expression of the polypeptides coded by the DNA molecule. The identification of those polypeptides can be done by various methods.

One of the critical points in a cloning experiment is the expression of the introduced DNA sequence. Furthermore the synthesis of a functional protein depends upon a series of factors such as the transcription of the right gene, efficient translation of its mRNA, and in many cases the post-translational processing and compartmentalization of the growing peptide. A failure in any one of these steps may result in non-expression of the cloned gene. Therefore it is important to develop expression vehicles containing "strong" promoters that can react with DNA-dependent RNA polymerase of the host (Remaut et al., 1981). Another problem is that the synthesized polypeptide must be exported from the cell.

A very interesting example of the utilization of that technology is the molecular cloning of the polio virus genome (Racaniello and Baltimore, 1981a; Van der Werf et al., 1981). A complementary DNA molecule was sythesized from the viral RNA. The sequences that coded various viral polypeptides were then inserted into the plasmid pBR322. With that approach the cloning of almost the whole viral genome was possible, and study of the clones obtained indicated the complete sequence of the viral RNA. The transfection of one of the plasmids that contained the complete viral genome to CV-1 or HeLa cells led to the production of infectious viral particles (Rancaniello and Baltimore, 1981b). As we can see, the expression of specific proteins of the polio virus in *E. coli* may result in the development of a new type of polio vaccine safer and more efficient than those in use.

As with poliomyelitis, the utilization of genetic engineering techniques can lead to the production of new vaccines and provide an improvement of the vaccines now in use.

Viruses as Hosts

The administration of live attenuated microorganisms in immunization programs has been used since Jenner developed in 1789 a vaccination procedure against smallpox, using a natural mutant of smallpox virus, a nonvirulent vaccinia virus.

Success of living vaccines can be evaluated by the eradication of smallpox disease according to the World Health Organization. This knowledge, when associated with the new recombinant DNA technology, opens up the possibility of construction of a recombinant vaccinia virus containing antigen genes from other organisms. Those expressing recombinant viruses may then be used as live vaccines.

To consider the vaccinia virus as a suitable cloning and expressing vector, some biological characteristics were taken into account.

1. The vaccinia virus is not considered an oncogenic organism like other potential vectors such as the simian virus 40; herpes virus, adenovirus, retrovirus, and so on. This low risk point should be taken into consideration when one intends to use vaccinia virus as a vaccinal vector (Panicali and Paoletti, 1982).

2. The purified vaccinia DNA is not itself infectious, but it was found that the vaccinia virus genome can cause homologous recombination "*in vivo*" between DNA sequences of intact virus or calcium phosphaste-precipitated viral DNA (Sam and Dumbell, 1981; Nakano et al., 1982) and between recombinant plasmids containing both foreign and vaccinia DNA (Panicali and Paoletti, 1982; Weir et al., 1982; Mackett et al., 1982). This ability of making homologous recombination with the sequence of constructed recombinant plasmids, probably in the infected cell cytoplasm, provides a suitable way to introduce genome modifications into established vaccinia virus strains (Weir et al., 1982).

3. A large genome (approximately 180 kilobases) is found in vaccinia virus. This fact allows the insertion in the genome of relatively large foreign DNA sequences. Panicali and Paoletti (1982) succeeded in introducing a 13kb length DNA fragment. Therefore it is possible to construct a single recombinant virus expressing many antigens. Such a virus strain can then be used as a polyvalent vaccine (Mackett et al., 1982).

4. The transcription enzymes are packaged within the virus particle; this implies the existence of regulatory sequences inside the vaccinia genome (Kates and McAustin, 1967; Munyon et al., 1967). These appropriate transcriptional signals have not been identified, but no homology was found between them and prokaryotic or eukaryotic transcriptional initiation sequences (Venkatesan et al., 1981). Therefore the efficient expression of foreign genes will be possible only if the exogeneous DNA sequences have been inserted into the correct orientation and near or beside the proper promoter site of some vaccinia DNA gene (Mackett et al., 1982).

Some recent progress has been made on this problem with the studies describing a vaccinia thymidine kinase (TK) gene. These studies provide important information regarding the location and mode of action of vaccinia virus transcriptional signals, and also demonstrate the use of vaccinia virus as a selectable cloning and expression vector (Panicali and Paoletti, 1982; Weir et al., 1982; Mackett et al., 1982).

Synthetically Made Antigens

The feasibility of a completely synthetic vaccine (produced without the use of living organisms) emerged from studies about the immunochemistry of protein antigens. A great part of antigenic determinants used in a vaccination against viruses, bacteria, or parasites is transported through proteins and linked to their spatial structure. Therefore relatively short fragments of the peptide chain that are exposed at the surface can be analyzed as isolated antigenic determinants. One antigen can induce antibody synthesis against many different determinants present in its molecule, and some of them can be overlapping. Landsteiner and Tample (1920) showed that some of the protein antigenic determinants could be reduced to a single chemical radical or a maximum of a few amino acids in their sequence. More recently Sela and colleagues (1967) proposed two types of antigenic determinants in proteins: sequential and conformational. Sequential determinants are due to the amino acid sequence in a random coil, and conformational determinants are those due to the steric conformation of the protein. Antibodies directed to sequential determinants are able to react with peptides that possess the amino acid sequence of those determinants. This does not normally occur with those directed to conformational ones.

The detection of these two types of determinants in protein was mainly possible through the use of techniques developed for the chemical synthesis of proteins. The results of Sela and colleagues (1967) were obtained through the synthesis in liquid phase of a tripeptide and its subsequent condensation on different conformations. Today it is possible to synthesize with a high yield larger peptides with an defined sequence. This is due mainly to technical innovations developed since Merrifield (1965) published on automated peptide synthesis.

Synthesis of peptides with defined sequences for the immunochemist enables the identification of specific determinants (Sela, 1983). It was also possible to localize and identify proteins whose existence was known only from the sequence of their mRNA or from the DNA that codes them (Sutcliffe et al., 1980; Lerner et al., 1981).

Having obtained the peptide against which we intend to make the antibodies, we then must make it immunogenic. It is known (Dresser, 1962) that molecular weight is an important factor that influences immunogenicity. A peptide of a few amino acids has little or no chance of eliciting an immune response. What is normally done is to couple the synthetic peptide to a carrier macromolecule, usually albumin or gelatin (Sela, 1983; Arnon and Sela, 1983). More recently, synthetic carriers made from the polymerization of simple peptides have been utilized (Sela, 1983; Arnon and Sela, 1983). Yet, in spite of these modifications, the antigens are not necessarily immunogenic. What is then done is to use an adjuvant, a substance or a mixture of substances that enhances the immune response. The adjuvant normally used with experimental animals is the complete Freund adjuvant (Freund, 1947). This adjuvant is very toxic and could not be used in a vaccine for human use. Many other substances have been tested as to their ability to enhance the immune response (Stewart-Tull, 1980). One of these compounds that has a very low toxicity, derived from studies of the bacterial wall, is N-acetylmuramyl-L-alanyl-D-isoglutamine (or muranyl dipeptide [MDP]). It has been used with success as a substitute for Mycobacteria Freund complete adjuvant (Ellouz et al., 1974; Kotani et al., 1975).

The vaccines currently used against diseases caused by either viruses or bacteria have a number of disadvantages. These could be avoided with the use of completely synthetic vaccines. Recently many reviews have been published about synthetic vaccines and their possible importance in the future prevention and control of infectious diseases (Arnon, 1980; Lerner, 1983). The first step in the direction of sythetic vaccines was the study of viral antigenic determinants. The study of these determinants was initially made using peptides obtained by the clevage of the protein from the viral capsid as was the case of TMV (Anderer, 1963). The most extensive studies, however, were made with MS2, a bacteriophage from *E. coli* (Langbeheim et al., 1976). The amino acid sequence of the protein from the viral capsid of MS2 was already known (Lin et al., 1967). This fact facilitated making possible the synthesis of peptides containing part of the structure of the capsid protein. Langbeheim and colleagues (1976) showed, for the first time, that a synthetic peptide could be used to induce neutralizing antibodies against the intact virus. After that, Arnon and his colleagues (1980) linked the synthetic peptide to MDP and thus produced a completely sythetic antigen. This antigen could induce antibodies against the virus in the absence of adjuvants.

The first results with synthetic antigens with a possible application in human health problems were obtained by Audibert and colleagues (1981). They were able to synthesize a peptide with a part of the diphtheria toxin amino acid sequence that induced active antitoxic immunization against diphteria in laboratory animals. They obtained similar results with an antigen produced by binding the synthetic peptide to MDP (Audibert et al.,

1982). Although antidiphtheric vaccine is one of the most easily made vaccines, it involves growing large quantities of a highly dangerous organism. In this case a synthetic vaccine would eliminate this potential risk.

Until a few years ago, the synthesis of a specific peptide depended upon the sequencing of antigens that could be obtained in relatively high amounts. This was not the case of some viruses that could not be cultivated outside their hosts (e.g., hepatitis B virus) or that grew very little *in vitro* (as occurs with some strains of foot and mouth disease virus). With recombinant DNA techniques and more precise nucleic acid sequencing methods, it was possible to obtain the amino acid sequence of proteins without the need to purify them. Viral proteins previously available in minimal quantities could then have their amino acid sequence documented. Having cloned and sequenced a part of the genome of hepatitis B virus (Valenzuela et al., 1979), three groups of investigators (Lerner et al., 1981; Dressman et al., 1982; Prince et al., 1982) developed synthetic determinants of the viral capsid. These peptides were able to induce the production of antibodies that reacted with the intact virus.

Another disease in which this approach may be profitable in the future is influenza. The influenza viruses have an enormous instability in their antigenic determinants (Webster et al., 1982). A vaccine that is effective today may not be next year. A peptide that is part of one of the proteins from the surface of the influenza virus was synthesized (Muller et al., 1982). This peptide, which is common to at least nine strains of virus, was bound to a carrier protein and induced antibodies in guinea pigs that reacted not only with the original peptide but with the intact virus. Those antibodies interfered also with the growing of the virus *in vitro*. Furthermore, guinea pigs immunized with this conjugate are partially protected against infections from other viral strains. Similar results (Green et al., 1982) increased the probability of obtaining an effective influenza vaccine.

Bearing in mind reported developments in the fields of peptide immunochemistry and recombinant DNA, synthetic vaccines may be a reality in the medium or long term. Yet, we should recall the advertence (Sela, 1983) in a recent review on synthetic vaccines extensible to other non-conventional approaches to vaccine production: "It is important to remember that while such successful vaccines should contain the right antigenic specificity and adequate built-in adjuvanticity, the investigators should also consider the genetic background of the recipients and the best methods of causing a long-lasting protection, as well as antigenic competition and how to cope with it in order to prepare molecules possessing several protective specificities."

LABORATORIAL DIAGNOSIS

Infectious Disease Diagnosis

One of the most serious aspects of the infectious diseases is their ability to disseminate in the community. Human populations have been challenged frequently by outbreaks of infectious illnesses such as plague, cholera, and smallpox. Actually outbreaks of epidemics such as measles, meningitis, hepatitis, and diarrheas are not uncommon, especially in children. These diseases lead to serious damage to human health.

Effective control of such diseases must occur to avoid dissemination risks as well as to provide suitable treatment of the patients. The first step in the direction of such a procedure, in many infectious diseases, is the isolation and identification of the pathogen. The accumulated basic knowledge in some areas such as microbiology, parasitiology, immunology, and molecular biology enhances the understanding of some characteristics of the microorganisms, mainly structural, genetic and metabolic.

The experimental exploration of these characteristics improved the development of new techniques that allow the rapid identification of pathogenic organisms. One of the characteristics being used for the identification of viruses, bacteria, or other agents is their surface structure. Antisera prepared against surface components can be utilized in the serological typing of such organisms.

In a great number of cases, especially with regard to the epidemiologically important diseases, it is fundamental to identify the serological type of the pathogenic agent. In the case of poliomyelitis, for instance, the proportion of viruses from types 1, 2, and 3 used in the vaccine, depends on the serotype that is predominant in a given region. The serological identification is also important in some parasitary illnesses such as schistosomiasis, malaria, and trypanosomiasis.

With regard to bacterial diseases, the identification procedure requires investigation of a great number of characteristics: colony morphology; biochemical tests such as sugar fermentation; amino acids and enzyme production; serological typing; and production of virulence factors, for instance, toxins. Thus, from the clinical specimen, it is necessary to isolate and identify the bacteria by means of different tests. Only then the patient can be properly medicated.

Nevertheless, in a great number of cases, these identification tests are not performed for several reasons: lack of specialized personnel for the interpretation of results; difficulties in the execution of a series of tests and, mainly, their high costs. Obviously, the non-identification of the microorganism can bring several serious consequences, such as the administration of wrong antibacterial drugs. The main efforts in this field, therefore, are concentrated in the development of new techniques that allow a rapid, efficient, and inexpensive clinical diagnosis. A decisive contribution in this area has been given by genetic engineering techniques.

The production of monoclonal antibodies by hybridoma technique (Yelton and Scharff, 1981) against specific antigenic determinants of the cellular surface provides an efficient method for the accurate identification of several microorganisms. These monoclonal antibodies are effective in recognizing several parasitary infections such as schistosomiasis, malaria, and trypanosomiasis. They may be used also for identification of some viral infections such as poliomyelitis (Yewdell and Gerhard, 1981).

Among the bacterial diseases, the infectious diarrheas are those more extensively studied with regard to the development of new identification techniques, mainly those caused by enterotoxigenic *E. coli* (ETEC). This bacterium is a major cause of diarrhea illness in adults and children in the developing countries. Its virulence is mainly due to the production of two enterotoxins (LT and ST) (Sack, 1975) that are active in the small intestine, promoting alterations in hydrosaline metabolism (Kantor et al., 1974; Newsome et al., 1978) with consequent diarrhea.

It is very difficult to identify ETEC strains in clinical laboratories because current methods detect enterotoxin production by animal (Dean et al., 1972) or cell culture assays (Donta et al., 1974), which are very complicated in the routine work. Moreover, the cost and inconvenience of these methods make them unsuitable for large scale examination.

Genetic engineering techniques have contributed to the development of alternative assays for the detection of ETEC. These assays employ a simple method that can be used routinely by the majority of clinical laboratories. Genes that encode both LT and ST enterotoxins have been isolated and characterized using recombinant DNA techniques (So et al., 1979). These sequences, when labeled *in vitro* with ^{32}P, can be used as a probe for detection of homologous DNA sequences, by colony hybridization. Unlike biological and immunological assays, which detect the enterotoxins themselves, these new methods detect the genes that encode the enterotoxins.

Obviously, this method still presents some limitations for use in the routine of a clinical laboratory, and some alterations must be made. For instance, the necessity of ^{32}P-labeled probes requires their frequent preparation, which is inconvenient for many laboratories. This probem can be solved hereafter by the use of a nonradioactive labeled probe (Malcolm, 1982).

Besides in the identification of ETEC, the possibility to employ this method for the identification of other pathogenic gram-negative bacilli such as *Shigella, Yersinia,* and *Salmonella* has been glimpsed. By using recombinant DNA techniques, the isolation of the DNA sequences that encode their virulence determinants will be possible. The isolation of these sequences can be enhanced by the fact that, in most of these bacteria, the pathogenic factors are plasmid-mediated (Elwell and Shipley, 1980). These sequences utilized as probes will allow a rapid and efficient identification of these microorganisms directly from the isolation culture medium where the clinical specimen was inoculated.

It should still be observed that these techniques will be extended, not only for the identification of other pathogenic bacteria, but also for virus and other pathogens.

Antenatal Diagnosis

The number of different human diseases described as genetically determined is about 2,500 (McKusick, 1974). For most of these diseases, only the symptoms are known, and the genetic condition is not understood. About 1-2% of all live births are carriers of genetically determined diseases (Carter, 1977). For some of these diseases such as phenylketonuria, hemophilia, and diabetes, only a palliative treatment is applied. In other cases the carrier will never have a normal life, and then diagnosis and therapeutic abortion remains the only option (Little, 1981).

Elucidation of the molecular structure of human genes has been possible by the application of recombinant DNA technology. The isolation of DNA from fetal cells (fibroblasts), obtained by amniocentesis, opens the possibility to make a comparative analysis, taking adult normal genes as standard. This fact may help enormously in antenatal diagnosis. These studies will increase our understanding of genetic disease as well as open the possibilities to decrease the incidence of such diseases by the application of therapeutic abortion.

Human hemoglobins have been intensively studied in recent years because they can be isolated and biochemically analyzed (Bunn et al., 1977). Four different globin chains were described; two pairs of these chains, named alpha, beta, gamma and delta, combine with a prosthetic group and form the hemoglobin molecule (Muirhead and Perutz, 1963). Chain combinations vary with the developmental stage of the individual (Bunn et al., 1977).

The molecular structure of globin genes is understood well because the messenger RNAs coding the globin polypeptide chains are isolated easily. This makes possible the construction of appropriate probes, necessary to detect the position of restriction sites in globin gene DNA sequences. DNA analysis of hemoglobinopathies was then possible, and this is the best example of clinical applications of recombinant DNA technology applied to antenatal diagnosis.

Defects of hemoglobins that cause hemoglobinopathies are divided into two classes: structural globin chain alterations, caused by amino acid substitution, and frequency variations between different globin chains (alpha-thalassemias; beta-thalassemias). The clinical consequences of structural variants are complex (Carter, 1977). In thalassemic individuals, however, the treatment is expensive, time-consuming, and painful for the patient. The clinical effects of hemoglobinopathies may be detected only after birth, and the diagnosis is made by direct analysis of hemoglobin isolated from peripheral blood.

It is possible to do antenatal diagnosis of hemoglobinopathies on 3-month-old fetuses when already a small amount of alpha- and/or beta-globin chains are found in peripheral circulation (Wood et al., 1977; Heywood et al., 1966). The antenatal diagnosis involves direct examination of fetal blood. Acquiring the sample, however, is very complicated and expensive because it requires sophisticated equipment and also involves a high fetal death rate. Only 0.05 ml of fetal blood can be obtained by sampling from a placental vein, after fetus/placenta localization with the use of ultrasound devices, and observing the vein with an optic fiber visualization system called a fetoscope (Alter, 1979). This blood sample may then be analyzed biochemically to detect and/or establish the ratio of alpha/beta chains.

The globin genes (normal or altered) may be directly analyzed (Southern, 1975). This method can be done easily and has been used currently in antenatal diagnosis. Radioactive probes, generally complementary DNA made by copying the messenger RNA's that code globin chains, and fetal DNA are the biological materials required for application of this Southern blotting method (Little et al., 1978; Wilson et al., 1978). Fetal DNA is isolated easily from fetal fibroblasts collected by amniocentesis, a relatively simple technique that has only a 3% fetal death rate, the same rate as spontaneous abortion in the same gestational stages (Alter, 1979). From 20 ml of amniotic fluid, about $5\mu g$ of fetal DNA can be isolated. This quantity is more than sufficient to perform the analysis by genetic engineering methods. Southern (1975) described a method to detect specific sequences among many DNA fragments discriminated by agarose gel electrophoresis. The DNA to be analyzed, therefore, must first be cleaved by a restriction enzyme to generate specific fragments. The combined use of various restriction enzymes provides the necessary information to construct a gene map of the restriction sites on the DNA fragment that one is studying. About 400 different restriction enzymes have been described, and these enzymes cleave about 90 different specific nucleotide sequences (Roberts, 1983). Therefore, it is possible to make a detailed screening of a gene and its neighboring sequences. This was the way used to establish the molecular structure of normal alpha- and beta-globin loci as well as their neighboring sequences (Maniatis et al., 1980).

Structural changes can be detected by the same approach. Three kinds of structural alterations may be detected by Southern blotting. The deletion of a DNA sequence within a gene will modify the restriction pattern by decreasing the size of some specific restriction fragment. The localization of the deletion within the gene may be narrowed if some enzymes are used together. If a point mutation occurs in the restriction site of a given enzyme the restriction pattern will change only for this particular enzyme. A third kind of detectable alteration is a small deletion involving a restriction site, causing a discrete modification of restriction pattern for that and other enzymes. Little (1981) provides an excellent review with many examples of all three kinds of alterations.

With increasing knowledge of the molecular structure of globin genes, it was possible to verify a great number of alterations. These observations made it possible to say that some homozygous individuals are really heterozygous for different molecular forms (Orkin et al., 1979; Flavell et al., 1979).

With a more sophisticated technology, involving genomic genes and messenger RNAs, it has been possible to obtain details about splicing of the recently transcribed RNAs. These studies have shown some splice defects associated with hemoglobinopathies that could not be detected by the technology just described. The development of more refined techniques, such as translation *in vitro*, will provide much information about the regulation of hemoglobin synthesis in developing organisms. An excellent review of those problems was made by Weatherall and Clegg (1982).

Many papers have been published, reflecting the initial investment in the application of recombinant DNA technology, in the study of the globin gene structure. Success was possible because it was easy to obtain appropriate probes. As previously shown, the obtention of probes is the sole limitation for a generalization of this technology in antenatal diagnosis for other human genes. As more human genes are identified and isolated, more abnormal genes will be diagnosed precociously.

PERSPECTIVES

For a long time, Brazilian scientific research in the field of human health was directed mainly to basic problems. The traditional research institutions, founded by Oswaldo Cruz, Vital Brazil, Adolfo Lutz, and others, produced an important number of active scientists working today in many different parts of the country. A variety of problems related to public health has emerged because of the high rate of population growth. Consequently, many of those scientists began to direct their attention to the application of acquired knowledge during recent years to solve the new health challenges.

This new approach led to the development of a series of fields such as production of therapeutically important drugs as hormones, enzymes, and antibiotics; improvement in the production of vaccines; more effective control of infectious diseases; new methods for identification of pathogens, and such. However, a series of barriers in the current techniques imposed limitations to the development of new products, such as vaccines, as well as the development of new methods for the control of pathogens and diseases.

In developing countries such as Brazil, there are several health problems that elicit a particular interest. They affect a great proportion of people, and the available methods to solve them are very expensive. For example, there is a very high incidence of parasitary diseases (schistosomiasis, Chagas's disease, and malaria) that cause dramatic damage to the population. In the same way, infectious diarrheas are the major mortality cause among children. Other infant illnesses (poliomyelitis and measles) are controlled by expensive imported vaccination programs because the application of technology for production of such vaccines is still not available in Brazil.

The emergence of biotechnology techniques in Brazil, stimulated by grants from government financial agencies, opened a new field for the investigation of health problems. Thus many research groups began to employ such techniques in trying to optimize existing and new products that could not normally be obtained by traditional methods. Among those groups working with various aspects of various organisms, which could lead eventually to an applications to human health, are the groups led by:

1. Professor S. B. Henriques, from Universidade Federal de Minas Gerais (MG) surface antigenic determinants of *Schistosoma mansonii*).

2. Professor C. M. Morel, from Fundacao Oswaldo Cruz (RJ); Professor W. Colli, from Universidade de São Paulo (SP); and Professor E. F. P. Camargo, from Escola Paulista de Medicina (SP) (different aspects of antigenic determinants of *Trypanosomoa cruzi* and its interaction with host cells).

3. Professor W. Becak, from Instituto Butantan (SP) (molecular cloning and gene expression of the poliomyelitis virus RNA).

In laboratorial diagnosis and development of new products are groups led by:

1. Professor R. R. Brentani, from Universidade de São Paulo (SP) (cloning of insulin genes and diseases related to collagen metabolism).

2. R. R. Golgher, from Universidade Federal de Minas Gerais (MG) (characterization and purification of human interferon from various sources).

3. Professor A. C. M. Paiva, from Escola Paulista de Medicina (SP; and Professor M. Tominaga, from Universidade de São Paulo (SP) (synthetic peptides)

4. Professor H. G. Pereira, from Fundacao Oswaldo Cruz (RJ) (rapid methods for detection of human virus)

5. Professor F. Salzano, from Universidade Federal do Rio Grande do Sul (RS) (diagnosis of genetic determined diseases, in particular, hemoglobinopathies)

6. Professor L. R. Trabulsi, from Escola Paulista de Medicina (SP) (identification of pathogenic enterobacteria from human stools using radioactive probes)

In comparison with population size and the number of problems to be solved, there are still too few and too scattered groups employing genetic engineering techniques in the human health field in Brazil. However, recognition of the importance of this new technology and governmental support are contributing to the establishment of new groups in several Brazilian research institutions. With the joint efforts of scientists and government, we hope that soon Brazil will be well prepared in terms of biotechnology to attend to some of its important problems related to human health.

ACKNOWLEDGMENTS

We would like to thank Rosely M. Aniceto Teixeira for typing the manuscript of this paper.

REFERENCES

Alter, B. P. 1979. Prenatal diagnosis of hemoglobinopathies and other hematologic diseases, J. Pediatr. 95:501–13.

Anderer, F. A. 1963. Preparation and properties of an artificial antigen immunologically related to tobacco mosaic virus. Biochem. Biophys. Acta 71:246–48.

Arnon, R. 1980. Chemically defined antiviral vaccines. Ann. Rev. Mircobiol. 34:593–618.

Arnon, R., M. Sela, M. Parant, and L. Chedid. 1980. Antiviral response elicited by a completely synthetic antigen with built-in adjuvanticity. Proc. Natl. Acad. Sci. USA 77:6769–72.

Arnon, R., and M. Sela. 1983. Les Antigènes et vaccins synthétiques: La Recherche, J. Freund (1947). Some aspects of active immunization. Ann. Rev. Microbiol. 1:291–308.

Audibert, F., M. Jolivet, L. Chedid, J. E. Alouf, P. Boquet, P. Rivailler, and O. Siffers. 1981. Active antitoxic immunization by a diphtheria toxin synthetic oligopeptide. Nature (London) 289:593–94.

Audibert, F., M. Jolivet, L. Chedid, R. Arnon, and M. Sela. 1982. Successful immunization with a totally synthetic diphtheria vaccine. Proc. Natl. Acad. Sci. USA 79:5042–46.

Bunn, H. F., B. G. Forget, and H. M. Ranney. 1977. Human hemoglobins. Saunders Co., Philadelphia.

Carter, C. O. 1977. Monogenic disorders. J. Med. Genet. 14:316–20.

Craig, R. K., and L. Hall. 1983. Recombinant DNA technology: application to the characterization and expression of polypeptide hormones. *In* R. Williamson, ed., Genetic engineering, 4:57–125. Academic Press, London.

Dean, A. G., Y. C. Ching, R. G. Williams, and L. B. Harden. 1972. Test for *Escherichia coli* enterotoxin using infant mice: application in study of diarrhea in children in Honolulu. J. Infect. Dis. 125:407–11.

Dimitriadis, G. J. 1978. Translation of rabbit globin mRNA introduced by liposomes into mouse lymphocytes. Nature (London) 274:923–24.

Dressmann, G. R., Y. Sanchez, F. Ionescu-Matiu, J. T. Sparrow, H. R. Six, D. L. Peterson, F. B. Hollinger, and J. L. Melnick. 1982. Antibody to hepatitis B surface antigen after a single inoculation of uncoupled synthetic HBsAg peptides. Nature (London) 295:158–60.

Dresser, D. W. 1962. Specific inhibition of antibody production. I. Protein-overloading paralyses. Immunology 5:378.

Donta, S. T., H. W. Moon, and S. C. Whipp. 1974. Detection of heat labile *Escherichia coli* enterotoxin with the use of adrenal cells in tissue culture. Science 183:334–36.

Ellis, J., L. S. Ozaki, R. W. Gwadz, A. H. Cochrane, V. Nussenzweig, R. Nussenzweig, and G. N. Godson. 1983. Cloning and expression in *E. coli* of the malarial sporozoite surface antigen gene from *Plasmodium knowlesi*. Nature (London) 302:536–38.

Ellouz, F., A. Adam, R. Ciorbaru, and E. Ledered. 1974. Minimal structural requirements for adjuvant activity of bacterial peptidoglycan derivatives. Biochem. Biophys. Res. Comm. 59:1317–25.

Elwell, L. P., and P. L. Shipley. 1980. Plasmid-mediated factors associated with virulence of bacteria to animals. Ann. Rev. Microbiol. 34:456–96.

Flavell, R. A., R. Bernards, J. M. Kooter. E. De Boer, P. F. R. Little, G. Annison, and R. Williamson. 1979. The structure of the human beta-globin gene in beta-thalassemia. Nucleic Acids Res. 6:2749–60.

Freund, J. 1947. Some aspects of active immunization. Ann. Rev. Microbiol. 1:291–308.

Graham, F. L., and A. J. van der Eb. 1973. New technique for the assay of infectivity of human adenovirus 5 DNA. Virology 52:456–57.

Green, N., H. Alexander, A. Olson, S. Alexander, T. M. Shinnick, J. G. Sutcliffe, and R. A. Lerner. 1982. Immunogenic structure of the influenza virus hemagglutinin. Cell 28:477–87.

Gruss, P., and G. Khoury. 1981. Expression of simian virus 40-rat preproinsulin recombinants in monkey kidney cells: use of preproinsulin RNA processing signals. Proc. Natl. Acad. Sci. USA 78:133–37.

Hale, A. J. 1979. Genetic engineering: do we need it? how would we do it? An approach to therapeutics. Biochem. Soc. Symp. 44:123–31.

Heywood, J. D., M. Karon, and S. Weissman. 1966. Amino acids: incorporation in α and β-chains of hemoglobin by normal and thallassemic reticulocytes. Science 146:530–31.

Kantor, H. S., P. Tao, and S. L. Borbach. 1974. Stimulation of intestinal adenyl cyclase by *Escherichia coli* enterotoxin: comparison of strains from an infant and an adult with diarrhea. J. Infect. Dis. 129:1–9.

Kates, J. R., and B. R. McAustan. 1967. Poxvirus DNA-dependent RNA polymerase. Proc. Natl. Acad. Sci. USA 58:134–41.

Klipstein, F. A., F. Engert, J. D. Clements, and R. A. Houghton. 1983. Vaccine for entertoxigenic *Escherichia coli* based on synthetic heat-stable toxin cross-linked to the B subunit of heat-labile toxin. J. Infect. Dis. 147:318–26.

Kotani, S., Y. Watanabe, F. Kinoshita, T. Shimono, I. Morisaki, T. Shiba, S. Kusomoto, T. Tarumi, and K. Ikenaka. 1975. Immunoadjuvant activities of synthetic N-acetyl-muramyl peptides or -amino acids. Biken J. 18:105–11.

Landsteiner, K., and A. Tample (1920) cit. in Landsteiner, K. 1945. The specificity of serological reactions. Harvard University Press, Cambridge, Mass.

Langbeheim, H., R. Arnon, and M. Sela. 1976. Antiviral effect on MS-2 coliphage obtained with a synthetic antigen. Proc. Natl. Acad. Sci. USA 73:4636–40.

Lerner, R. A., N. Green, H. Alexander, F. T. Liu, J. G. Sutcliffe, and T. M. Shinnick. 1981. Chemically synthesized peptides predicted from the nucleotide sequence of the hepatitis B virus genome elicit antibodies reactive with the native envelope protein of Dane particles. Proc. Natl. Acad. Sci. USA 78:3403–7.

Lerner, R. A. 1983. Synthetic vaccines. Scientific American 248:48–56.

Levine, M. M., J. B. Kaper, R. E. Black, and M. L. Clements. 1983. New knowledge on pathogenesis of bacterial enteric infections as applied to vaccine development. Microbiol. Rev. 47:510–50.

Lin, U. Y., C. M. Tsung, and H. Fraenkel-Conrat. 1967. The coat protein of the RNA bacteriophage MS2. J. Mol. Biol. 24:1–14.

Little, F. R. 1981. DNA analysis and the antenatal diagnosis of hemglobinopathies. *In* R. Williamson, ed., Genetic engineering, 1:61–102. Academic Press, London.

Little, P., P. Curtis, C. Contelli, J. Van den Berg, R. Dalgelish, S. Malcolm, M. Courtney, D.

Westaway, C. Andueller, A. Graessmann, and M. Graessmann. 1978. Ma translation of rabbit globin mRNA introduced by liposomes into mouse lymphocytes. Nature (London) 274:923–24.

Mackett, M., G. L. Smith, and B. Moss. 1982. Vaccinia virus: a selectable eukaryotic cloning and expression vector. Proc. Natl. Acad. Sci. USA 79:7415–19.

Malcolm, S. 1982. Lighting up time. Nature (London) 300:108–9.

Maniatis, T., E. F. Fritsch, J. Lauer, and R. M. Lawne. 1980. The molecular genetics of human hemoglobins. Ann. Rev. Genetics 14:145–78.

Mantei, N., B. Werner, and C. Weissmann. 1979. Rabbit betaglobin mRNA production in mouse L cells transformed with cloned rabbit betaglobin chromosomal DNA. Nature (London) 281:40–46.

McKusick, V. A. 1974. Mendelian inheritance in man. Catalogues of autosomal dominant, autosomal recessive, and S-linked phenotypes. Johns Hopkins University Press, Baltimore.

Merrifield, R. B. 1965. Automated synthesis of peptides. Science 150:178–85.

Mosely, S. L., J. W. Hardy, M. I. Huq, P. Echeverria, and S. Falkow. 1983. Isolation and nucleotide sequence determination of a gene encoding a heat-stable enterotoxin of *Escherichia coli*. Infect. Immun. 39:1167–74.

Mueller, C., A. Graessmann, and M. Graessmann. 1978. Ma translation of rabbit globin mRNA introduced by liposomes into mouse lymphocytes. Nature (London) 274:923–24.

Muirhead, H., and M. F. Perutz. 1963. Structure of hemoglobin. Nature (London) 199:633–39.

Mueller, G., M. Shapira, and R. Arnon. 1982. Anti-influenza response achieved by immunization with a synthetic conjugate. Proc. Natl. Acad. Sci. USA 79:569–73.

Munyon, W., E. Paoletti, and J. T. Grace, Jr. 1967. RNA polymerase activity in purified infectious vaccinia virus. Proc. Natl. Acad. Sci. USA 58:2280–87.

Nakano, E., D. Panicali, and E. Paoletti. 1982. Molecular gentics of vaccinia virus: demonstration of marker rescue. Proc. Natl. Acad. Sci. USA 79:1593–96.

Newsome, P. M., M. N. Burgess, and N. A. Mullan. 1978. Effect of *Escherichia coli* heat-stable enterotoxin on cyclic GMP levels in mouse intestine. Infect. Immun. 22:290–91.

Orkin, S. H., J. M. Old, H. Lazarus, C. Altay, A. Gurgey, D. J. Weatherall, and D. Nathan. 1979. The molecular basis of alpha-thalassemias: frequent occurrence of dysfunctional alpha loci among non-Asians with H6 disease. Cell 17:33–42.

Kstro, M.J., D. Giacomoni, D. Lavelle, W. Paxton, and S. Dray. 1978. Evidence for translation of rabbit globin mRNA after liposome-mediated insertion into a human cell line. Nature (London) 274:921–23.

Panicali, D., and E. Paoletti. 1982. Construction of poxviruses as cloning vectors: insertion of the thymidine kinase gene from herpes simplex virus into the DNA of infectious vaccinia virus. Proc. Natl. Acad. Sci. USA 79:4927–31.

Prince, A. M., H. Ikram, and T. P. Hop. 1982. Hepatitis B virus vaccine: identification of HBsAg/a and HBsAg/d but not HBsAg/y subtype antigenic determinants on a synthetic immunogenic peptide. Proc. Natl. Acad. Sci. USA 79:579–82.

Rancaniello, V. R., and D. Baltimore. 1981a. Cloned poliovirus complementary DNA is infectious in mammalian cells. Science 214:916–18.

Rancaniello, V. R., and D. Baltimore. 1981b. Molecular cloning of poliovirus cDNA and determination of the complete nucleotide sequence of the viral genome. Proc. Natl. Acad. Sci. USA 78:4887–91.

Rechsteiner, M. C. 1978. Red cell-mediated microinjection. Natl. Cancer Inst. Monogr. 48:57–64.

40   *Biotechnology of Plants and Microorganisms*

Remaut, E., P. Stanssens, and W. Fiers. 1981. Plasmid vectors for high-efficiency expression controlled by the pL promoter of coliphage lambda. Gene 15:81–93.

Roberts, R. J. 1983. Restriction and modification enzymes and their recognition sequences. Nucleic Acids Res. 11:135–67.

Sack, R. B. 1975. Human diarrheal disease caused by enterotoxigenic *Escherichia coli*. Ann. Rev. Microbiol. 29:333–53.

Sam, C. K., and K. R. Dumbell. 1981. Expression of poxvirus DNA in coinfected cells and marker rescue of thermosensitive mutants by subgenomic fragments of DNA. Ann. Virol. Inst. Pasteur. 132E:135–50.

Sela, M. 1983. From synthetic antigens to synthetic vaccines. Biopolymers 22:415–24.

Sela, M., B. Schechter, I. Schechter, and F. Boreck. 1967. Antibodies to sequential and conformational determinants. Cold Spring Harbor Symp. Quant. Biol. 32:537–45.

Shimotohno, K., and H. M. Temin. 1981. Formation of infectious progeny virus after insertion of herpes simplex thymidine kinase gene into DNA of an avian retrovirus. Cell 26:67–77.

So, M., F. Heffron, and B. J. McCarthy. 1979. The *E. coli* gene encoding heat stable toxin is a bacterial transposon flanked by inverted repeats of IS 1. Nature (London) 277:453–56.

Southern, E. M. 1975. Detection of specific sequences among DNA fragments separted by gel electrophoresis. J. Mol. Biol. 98:503–17.

Stewart-Tull, D. E. S. 1980. The immunological activities of bacterial peptidoglycans. Ann. Rev. Microbiol. 34:311–40.

Sutcliffe, J. G., T. M. Shinnick, N. Green, F. T. Lie, H. L. Niman, and R. A. Lerner. 1980. Chemical synthesis of a polypeptide predicted from nucleotide sequences allows detection of a new retroviral gene product. Nature (London) 287:801–5.

Valenzuela, P., P. Gray, M. Quiroga, J. Zaldivar, H. M. Goodman, and W. J. Rutter. 1979. Nucleotide sequence of the gene coding for the major protein of hepatitis B virus surface antigen. Nature (London) 280:815–19.

Van der Werf, S., F. Gregegere, H. Kopecka, N. Kitamura, P. G. Rothberg, P. Kourilsky, E. Wimmer, and M. Girard. 1981. Molecular cloning of the genome of poliovirus type 1. Proc. Natl. Acad. Sci. USA 78:5983–87.

Venkatesan, S., B. M. Baroudy, and B. Moss. 1981. Distinctive nucleotide sequences adjacent to multiple initiation and termination sites of an early vaccinia virus gene. Cell 25:805–13.

Wahdan, M. H., C. Serie, Y. Cerister, S. Sallam, and R. Germanier. 1982. A controlled field trial of live *Salmonella typhi* strain Ty21a oral vaccine against typhoid: 3-year results. J. Infect. Dis. 145:292–95.

Weatherall, D. J., and J. B. Clegg. 1982. Thalassemia revisited. Cell 29:7–9.

Webster, R. G., W. G. Laver, G. M. Air, and G. C. Schild. 1982. Molecular mechanism of variation in influenza viruses. Nature (London): 296:115–21.

Weir, J. P., G. Bajszar, and B. Moss. 1982. Mapping of the vaccinia virus thymidine kinase gene by marker rescue and by cell-free translation of selected mRNA. Proc. Natl. Acad. Sci. USA 79:1210–14.

Wigler, M. S. Silverstein, L. S. Lee, A. Pellicer, Y. C. Cheng, and R. Axel. 1977. Transfer of purified herpes virus thymidine kinase gene to cultured mouse cells. Cell 11:223–32.

Wilson, J. T., L. B. Wilson, J. K. de Ried, L. Villa-Komaroff, A. Efstratiadis, B. G. Forget, and S. Weissman. 1978. Insertion of synthetic copies of human globin genes into bacterial plasmids. Nucleic Acids. Res. 5:563–81.

Wold, B., M. Wigler, E. Lacy, T. Maniatis, S. Silverstein, and R. Axel. 1979. Introduction and expression of a rabbit beta-globin gene in mouse fibroblasts. Proc. Natl. Acad. Sci. USA 76:5684–88.

Wood, W. G., J. B. Clegg, and D. J. Weatherall. 1977. Developmental biology of human hemoglobins. *In* E. B. Brown, ed., Progress in hematology, 10:43–90. Green and Stratton, New York.

Yelton, D. E., and M. D. Scharff. 1981. Monoclonal antibodies: a powerful new tool in biology and medicine. Ann. Rev. Biochem. 50:657–80.

Yewdell, J. W., and W. Gerhard. 1981. Antigenic characterization of viruses by monoclonal antibodies. Ann. Rev. Microbiol. 35:185–206.

ROY M. KOTTMAN

Impact of High Technology on Agriculture

4

Prior to the 1970s, discussions on improving agriculture through heredity involved topics of Mendelian genetics—selection, hybridization, and subsequent selection of those phenotypes that, in various stages of homozygosity or heterozygosity, provided the potential for increased yields, improved resistance to diseases and pests, or more desirable color, nutritional composition, or flavor. How vastly different are our discussions today! Geneticists are now dealing with the identification and characterization of genes, along with exciting concepts about gene splicing, replication, regulation, and transfer. We are presently concerned about cell and protoplast culture, somatic hybridization, protoplast fusion, embryo transfer, and a number of other techniques, all of which underlie those magical buzzwords "Genetic Engineering!" Truly it is wizardry of sorts. The techniques of "Genetic Engineering" are a mystery to laymen, and are only imperfectly understood by many who call themselves biological scientists.

The science and practice of biotechnology are on center stage! They are at the cutting edge of the biological and agricultural sciences. What I have seen of the work in laboratories, greenhouses, and field plots at various biotechnology companies suggests to me that biotechnology is very much a part of the new frontier in genetics. It is, however, not the process or the technique, but the impact of high technology genetics on agriculture here in the United States and in the rest of the world, that is captivating in terms of both social and economic outcomes. The intriguing question for mankind is whether the new thrusts in plant breeding are two years, *or* two decades, away from having cataclysmic impact on production of the world's major crops. Are we one year, *or* one decade, from transferring disease and insect resistance from one organism to another? Can we incorporate at will those genes with fertilizer-sparing capabilities, or water-conserving attributes, or yield-enhancing qualities into any plant communities we choose to transfer them through protoplast fusion and subsequent regeneration of whole plants that can then be cloned using tissue culture techniques to produce genetically identical plants suitable for commerical production? The answer is, of course, neither yes nor no, but maybe. The scientific advances that will facilitate our doing so are just now being made. The pathways to success will be strewn with temporary setbacks, but the future looks very bright indeed.

Many crop improvements made over the past half-century have come about as a result of scientific breakthroughs. To gain an appreciation of what genetics and plant breeding have already accomplished in the battle to overcome world hunger, we need look only to the work of our distinguished colleague Dr. Norman E. Borlaug, whose plant breeding efforts with wheat gave rise to the so-called "Green Revolution."

As a result of Dr. Borlaug's work, we find that some of the most dramatic increases in wheat yields have occurred in one or another of the so-called developing countries. But even as we recognize those achievements we should not lose sight of the fact that wheat yields in the United States have increased by 30% since 1965. And though our domestic increase of 30% is dwarfed by the 450% increase that has been achieved in Punjab State in north India, the impact of our increase in wheat yields looms large in world markets because of the fact that we harvest 80 million acres of wheat. In consequence, total production of wheat in the United States has increased from 1.3 billion bushels to 2.8 billion bushels (Agric. Statis., 1982) in less than two decades. However, neither foreign nor domestic yield increases are totally attributable to genetic improvement. Improved farm equipment, fertilizers, pesticides, and, in many instances, improved irrigation practices have played a major role. The point to be remembered, however, is that genetic manipulation is the foundation upon which future yield increases, as well as quality enhancement and consumer acceptance, must be built.

To some of us whose careers have been spent working at one or another university within the corn belt of our great nation, it is difficult to talk about plant breeding without making reference to hybrid maize. The development of hybrid maize has enabled our national production of that versatile crop to double over the past two decades. National production has increased from 4.1 billion to 8.2 billion bushels annually (Agric. Statis., 1982), and yield per acre has increased by 48% over that same period.

Soybean breeding, which thus far has been limited to laborious cross-pollination and phenotypic selection, has lagged behind both wheat and maize with only a 24% yield increase over the past seveteen years. (Agric. Statis., 1982)

The foregoing reference to yield increases prompts me to state that as we look to the future we need not be ashamed of the past. Nor is there need to restrict our review of past accomplishment to the cereal grains and oilseeds. Potato breeding has resulted in yield gains of 29% (Agric. Statis., 1982), and although of lesser magnitude, many fruit and vegetable crops, as well as ornamental and floral crops, and even tobacco, have been dramatically enhanced in terms of yield, utility, and overall contribution to the economy. Tomato yields have increased by 48% (Agric. Statis., 1982) as a result of genetic and environmental manipulation utilizing what we have come to look upon as conventional plant breeding.

The outer limits of what can be accomplished through genetic engineering truly boggles the mind. What has taken ten or twenty years to accomplish by way of conventional plant breeding methods is now being telescoped into a matter of months of laboratory effort followed by only one or two years of field evaluation utilizing field locations in both the temperate and tropical zones.

In the late 1960s, I asked Dr. R. Bruce Curry, professor of agricultural engineering at the Ohio Agricultural Research and Development Center, to undertake computer simulation studies of maize production in Ohio. When Dr. Curry completed his work on that project, his report indicated that with the solar energy conditions that prevail in Ohio, along with all the other environmental factors that on the average prevail during the summer growing season in Ohio, it should be possible to produce 428 bushels of maize per acre. He made it clear, however, that such yield levels could be reached only if the genetic capability of maize plants could be enhanced to the point that the plants could take full advantage of Ohio's summertime environment. Since that time, I have read reports from agronomists at the University of Illinois which suggest that maximum maize yields of 600 bushels per acre are possible if, indeed, the genetic makeup of the maize plant can be so altered as to enable it to take advantage of environmental conditions as they exist in central Illinois.

Consider, if you will, the impact of what is now possible through genetic engineering as scientists work with a crop the average yield of which is 110 bushels per acre but whose potential is 600! Given a short-run goal of achieving just one-fourth of that projected difference between maximum potential yields and present-day average yields across the corn belt, we are talking about an increase of 125 bushels per acre on 80 million acres. This equates to a whopping 10-billion-bushel increase in total United States maize production, which at $3.00 per bushel will result in $30 billion dollars of new wealth for our nation. That $30 billion bonanza is just waiting for those biotechnologists who will unlock the secrets that can be unlocked best through the evolving techniques of genetic engineering.

Potential as dramatic as that cited for improvement of maize exists with respect to wheat, rice, soybeans, potatoes, tomatoes, sugar cane, sugar beets, many of the fruits and vegetables, and some of the floral and ornamental crops, as well as coffee, tea, tobacco, herbs and spices.

The number and expertise of scientists now and in the future, along with what is done to support them, will largely determine how fast and how far genetic engineering will enable crop production to progress over the next two decades. We know, for example, that within the agricultural experiment stations of our nation's land-grant universities there are 250 full-time equivalents (some 350 individual scientists) engaged in biotechnological research (Natl. Assoc. State Univ., 1982). That number is expected to expand by nearly 50% over the next two years (Natl. Assoc. State Univ., 1982).

During 1981-82 the state agricultural experiment stations were committed to the expenditure of nearly $35 million on biotechnological research, with 14% of that amount being contributed through private funding (Natl. Assoc. State Univ., 1982). Although I have not seen comparable data on the Agricultural Research Service, USDA, I am of the opinion that a nearly equal commitment of personnel and dollars is being made.

As would be expected, similar effort is under way in other nations (Bentley, 1983). Japan, West Germany, and France are reported to be giving high priority to the development of their biotechnology industries. In a survey of more than 600 leading industries in Japan, biotechnology was rated as having the greatest growth potential of all newly emerging industries. More than 150 Japanese companies are reported to be spending more than $217 million annually on research and development in biotechnology. And it has been reported that the government of Japan is spending $200 million per year in converting biotechnology from the laboratory into commercial ventures.

Clearly the race is on to exploit the potential of genetic engineering, biotechnology, or call it what you will. I happen to believe that biotechnology companies are in a very favorable position to capitalize on practical applications of what has thus far been largely a laboratory exercise involving basic research in biotechnology.

In an annual report for 1983, the chairman of a new biotechnology corporation's board of directors highlighted the opportunities before us in this succinct statement: "The technological revolution which we are currently witnessing is expected to have a major economic impact on agribusiness, the world's largest industry, with over 500 billion dollars in it's annual sales worldwide!" (Connor, 1983). To this I would add a predictive postscript that by the year 2020, agriculture's $500 billion annual sales worldwide will exceed $1 trillion. I firmly believe that the majority of the gargantuan increase in production will have been achieved through "miracles" wrought in biotechnology laboratories both here in the United States and in the rest of the world. Developing and maintaining leadership in that highly competitive arena will require high levels of vision, dedication, and inspired leadership.

I am intrigued with a thought expressed by Smith (1955) in her monograph entitled, "I Captured the Castle," wherein she wrote: "I have found that sitting in a place where you have never sat before can be inspiring!" It seems to me that all of us are sitting in a place where we have never sat before, and I find our situation in biotechnology to be very inspiring!

This paper is condensed from remarks on the occasion of the first Scientific Advisory Board meeting, DNA Plant Technology, Inc., Walt Disney Conference Center, Lake Buena Vista, Florida, 8 September 1983.

REFERENCES

Agricultural Statistics, USDA. 1982. P. 2.
National Association of State Universities and Land-Grant Colleges. 1982. Emerging technologies in agriculture: issues and policies.
Progress Report of the Committee on Biotechnology, Division of Agriculture.
Bentley, O. L. 1983. Research and education. USDA Secretary's Summit Meeting, Washington, D.C.
Smith, D. 1955. *In* J. Bartlett, ed., Familiar quotations. Little Brown and Co., Centennial Edition, Boston.
Connor, J. 1983. DNA Plant Technology annual report.

PAULO F. C. DE ARAUJO, JOAQUIM J. C. ENGLER,
AND RICARDO SHIROTA

Biotechnology in Agriculture of Developing Countries: An Economic Perspective

5

Beginning with the end of World War II, several writers have concerned themselves with the disparities in national income between developed and developing countries. Consequently, we have many theories as to the process of economic development. Some of these theories emphasize the role of agriculture in stimulating this economic process. In the large majority, they are based on the evidence that the income generated in developing economies comes mainly from the agricultural sector (de Araujo, 1975; Johnston and Mellor, 1961; Hayami and Ruttan, 1971). However, it is in this sector that a considerable percentage of the total population is concentrated.

As a general rule, however, agriculture in developing countries is characteristically dualistic in nature and constitutes a sector in which a traditional segment predominates. For this reason there are only limited possibilities for generating an agricultural surplus that would allow a rapid transfer of capital to the urban-industrial area. Even the transfer of labor—almost entirely unskilled—takes place in a disorderly manner, with the end result that it generates mounting social tensions in the urban environment.

In this paper we outline, first of all, some considerations with respect to the sources of agricultural growth. In particular, the growth of Brazilian agriculture will be analyzed on the basis of contributions of various authors. Included in this topic are some considerations regarding the potential sources for stimulating the growth of agricultural production in Brazil.

Second, we discuss succinctly some of the experiences related to the Green Revolution in Asia, a subject that was prominently discussed in the literature during the 1960s and 70s. Third, in light of the positive and negative aspects of the Green Revolution, we analyze the prospects for biotechnology in fomenting Brazilian agricultural production. In this part of our exposition, we will try to suggest a possible frame of reference, together with the precautions to be taken, in order that the results of this new technology may prove to be socially beneficial through the provision of a new springboard for agricultural growth. Some final considerations will refer to alternative policies for the agricultural sectors of countries undergoing development.

AGRICULTURAL GROWTH

Sources of Agricultural Growth

Agricultural production can be raised, broadly speaking, by two basic methods: increased use of the fundamental means of production and increased productivity, partial or total.

Because of the natural limitations of productive resources, the first alternative is not feasible beyond certain limits. A typical case is that of the land input itself; this element is incorporated into the productive system, initially in the more fertile regions whose topography is best suited for cultivation. Thus, if the agricultural frontiers have not already reached their absolute limit, the incorporation of additional areas tends towards less fertile regions. This necessarily entails higher production costs for a given level of productivity. According to this hypothesis, the frontier will be pushed back in places more distant from the consuming centers. This means that if technical efficiency is to be maintained economic efficiency will certainly be reduced in proportion to the increased cost of transporting both inputs and products. At the level of the production unit, consideration must also be given to the existence of doubts as to the economies of scale. This still further impedes agricultural growth obtained through increased use of the various means of production.

The second alternative depends on increased productivity, to be obtained in two basic ways: improvements in the quality of the means utilized, and scientific and technological developments leading to technological innovations (Thompson, 1976; Engler, 1978; Vera and Tollini, 1979).

Technological Development in Agriculture

The study of partial indices of productivity—production per unit area and per worker—between different countries, carried out by Hayami and Ruttman (1971), contains appreciable differences in these values, as demonstrated in table 1 (the same data are shown graphically in figure 1).

TABLE 1

ESTIMATED LAND AND LABOR PRODUCTIVITIES IN AGRICULTURE,
43 COUNTRIES, 1960, IN WHEAT UNITS

COUNTRY	OUTPUT	
	Per Hectare (Y/A)	Per Male Worker (Y/L)
Argentina	0.37	39.9
Australia	0.09	106.4
Austria	2.33	31.7
Belgium (and Luxemburg)	6.12	52.7
Brazil.......................	0.60	9.4
Canada	0.58	75.8
Ceylon......................	2.85	3.9
Chile	0.48	12.9
Colombia	0.84	10.3
Denmark	4.60	47.4
Finland	2.02	30.9
France......................	2.49	35.9
Germany, Fed. Rep.	4.00	38.6
Greece......................	1.22	9.9
India	1.06	2.1
Ireland	1.58	21.0
Israel.......................	1.84	28.9
Italy........................	3.00	16.1
Japan	7.47	10.7
Libya.......................	0.04	n.a.
Mauritius	5.33	11.6
Mexico	0.27	5.2
Netherlands	7.21	43.1
New Zealand	1.19	141.8
Norway.....................	3.09	31.1
Pakistan	n.a.	2.4
Paraguay	0.94	5.0
Peru........................	0.56	10.2
Philippines	1.88	3.8
Portugal	n.a.	7.4
South Africa	0.16	11.7
Spain.......................	1.08	12.2
Surinam	4.46	17.1
Sweden	2.33	44.3
Switzerland	3.16	29.3
Syria	0.36	9.4
Taiwan	10.24	8.1
Turkey	0.59	7.1
U.A.R.	6.90	4.4
U.K.	1.94	44.0
U.S.A.......................	0.80	99.5
Venezuela	0.28	8.4
Yugoslavia	1.14	n.a.

SOURCE: Hayami and Ruttan, 1971.

(Y/A: yield per unit area; Y/L: yield per unit labor)

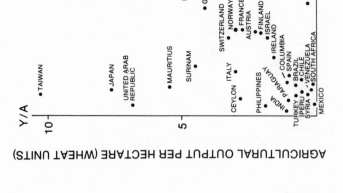

Fig. 1. International comparison of agricultural output per male worker and per hectare of agricultural land. Output data are 1957–62 averages, and labor and land data are of year closest to 1960. (Data from table 1.) (Hayami and Ruttan, 1971)

Figure 1 enables us to visualize the concentrated grouping of the developing countries, generally clustered around the point of origin of the graph. In contrast, the developed countries with modern agriculture are found well above and/or to the right of the point of origin.

The authors use these data to develop the well-known theory of induced innovation, according to which the technology variable is endogenous to the process of agricultural development. Countries such as Japan that have only a limited quantity of the land factor but an abundance of the labor factor have achieved the growth of their agriculture through the development of chemical and biological technologies, which compensate for the scarcity of the land factor.

Reversing the situation in Japan, the United States had relatively abundant land and the labor factor was scarce. Thus, the main effort was devoted to developing labor-saving innovations that rendered this factor highly productive. In practice the solution was the adoption of mechanical innovations that led to a high ratio of land area cultivated per worker.

Hence, the development of agriculture in third world countries would be conditioned on the efficiency with which the relative limitations of the various factors might be detected. Once the limiting factors are identified and defined, efforts would be channeled toward the development of technological innovations that might permit the substitution of these factors by others more abundant (and/or cheaper).

It still remains to consider that the relative limitation of the several factors can be—and generally is—substantially different among the various regions of the same country. In the following, we will exemplify this case with a rapid analysis of the relevant characteristics of Brazilian agriculture and its recent evolution.

Brazilian Agriculture: Characteristics of Its Recent Evolution-Interregional Differences

In Brazilian agriculture, by analysis of the data provided by partial indices of productivity Vera and Tollini (1979) show differences in the productivity of the land and labor factors similar to those found by Hayami and Ruttman (1971) among the 43 countries included in their analysis (table 2 and figure 2). Several recently published works have tried to explain the reasons for these variations. Among the hypotheses advanced is one recognizing that in São Paulo there have been significant investments in the infrastructure of research, rural extension, and education (Schuh, 1975).

52   *Biotechnology of Plants and Microorganisms*

TABLE 2

ESTIMATED PRODUCTIVITY OF LAND AND LABOR IN BRAZILIAN AGRICULTURE BY
STATE AND REGION, 1970

REGION AND STATE	PRODUCTION IN CRUZEIROS AT 1970 VALUE	
	Per hectare	Per Worker
North..........................	33.20	824.06
Rondônia	23.23	1,843.31
Acre..........................	22.95	1,523.83
Amazonas.....................	56.82	897.70
Roraima	12.65	2,436.51
Pará..........................	32.32	632.83
Amapá	24.97	1,439.32
Northeast	60.94	598.18
Maranhão.....................	48.87	446.07
Piauí	19.56	362.25
Ceará.........................	40.03	474.28
Rio Grande do Norte	40.17	596.49
Paraíba	74.74	585.87
Pernambuco...................	125.55	711.45
Alagoas.......................	161.74	841.47
Fernando do Noronha	40.00	3,164.71
Sergipe	95.15	617.12
Bahia.........................	68.26	714.82
Southeast	134.04	2,352.82
Minas Gerais	71.38	1,514.67
Espírito Santo	121.33	1,522.19
Rio de Janeiro	198.69	2,535.65
São Paulo.....................	254.80	3,663.21
South.........................	185.72	2,012.65
Paraná........................	210.19	1,555.85
Santa Catarina................	173.94	1,600.55
Rio Grande do Sul	173.80	2,859.89
West Central	22.80	2,007.60
Mato Grosso	17.66	2,166.57
Goias........................	29.02	1,896.15
Distrito Federal	96.21	2,246.02
Brazil........................	84.88	1,420.06

SOURCE: Vera and Tollini, 1979.

An interesting comparison of the foregoing statements may be made by reference to table 3, which contains the rates of growth in productivity achieved by labor and land factors in various groups of countries and in Brazil—in the latter case, by region and at the aggregate level. In the specific case of the State of São Paulo, the rates are obviously higher than those of the developed countries. This can be explained by the rapid transfer of labor to the urban sector arising from the process of industriali-

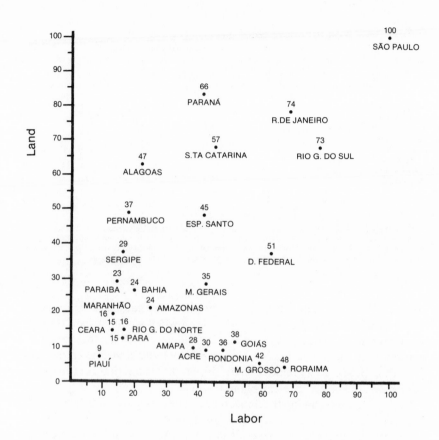

Fig. 2. Partial and total productivity of land and labor, per state, Brazil, 1970. (Vera and Tollini, 1979)

zation. Land was becoming scarcer as the state's agricultural frontiers approached their limit. Supplementing these factors was an accentuated investment in chemical-biological research on fertilizers and insecticides that made possible a rapid increase in agricultural productivity.

The evidence of these influences becomes still stronger when we analyze the contributions made to the growth of agricultural production in Brazil through additions to the area under cultivation and to the increased yields

TABLE 3

ANNUAL RATES OF GROWTH IN PRODUCTION PER WORKER
AND PRODUCTION PER CULTIVATED AREA (1955-65)

Category	Growth Rate of Labor Productivity	Growth Rate of Land Productivity
Developed countries*	4.7	2.1
Intermediate countries	4.4	2.0
Developing countries	1.4	2.1
Brazil......................	4.0	2.0
South Central	4.1	2.1
São Paulo.....................	5.4	4.8
Northeast	3.8	0.6

SOURCE: Hayami and Ruttan (1971), p. 74.

* Developed countries: Australia, Belgium, Canada, Denmark, France, Germany, Netherlands, New Zealand, Sweden, Switzerland, United Kingdom and United States. Developing countries: Brazil, Ceylon, Colombia, India, Mexico, Peru, Philippines, Syria, Taiwan, Turkey, United Arab Republic. Intermediate countries: Argentina, Austria, Chile, Finland, Greece, Ireland, Israel, Japan, Portugal, South Africa, Spain and Venezuela. (Pastore et al., 1976).

obtained (table 4). The differences in productivity gains between regions are appreciable, and are even more striking when a comparison is made between states. In the results obtained by Thompson (1976), there are indications that the states having higher productivity and higher rates of growth in productivity relative to the growth in production are those that have made the larger investments in research.

Still further reinforcing the importance of research in agricultural development, an interesting exercise is to measure the social returns of the investments in research. Various authors, both with reference to Brazil as well as to other countries, have arrived at notably high internal rates of return from these investments (Ayer and Schuh, 1972; Fonseca, 1979; Griliches, 1958; Peterson, 1967; Arino and Hayami, 1975).

We may infer from these results the need for a strong stimulus to such programs of investment in agricultural research. It must be recognized, nevertheless, that these programs—generally maturing over a long period—have received little attention from policy-makers.

THE GREEN REVOLUTION: A LESSON TO BE CONSIDERED

The history of the process known as the Green Revolution began in 1944 with the creation of the Centro Internacional de Mejoramiento de Maiz e Trigo (CIMMYT) in Mexico.

TABLE 4

CONTRIBUTION OF AREA (CA) AND YIELD (CY) TO THE GROWTH OF
BRAZILIAN AGRICULTURE, BY REGION AND BY STATE, 1947-75

Region and state or Territory	CA	CY	Weight
North	99.94	0.06	0.01*
Rondônia	97.80	2.20	0.06
Acre.........................	104.46	- 4.46	0.09
Amazonas.....................	94.97	5.03	0.15
Roraima	98.34	1.66	0.02
Pará.........................	99.23	0.77	0.67
Amapá	196.96	-96.96	0.01
Northeast	101.80	- 1.80	0.27
Maranhão....................	64.37	35.63	0.09
Piauí	116.46	-16.46	0.04
Ceará........................	121.71	-21.71	0.15
Rio Grande do Norte	116.07	-16.07	0.06
Paraíba	115.79	-15.79	0.08
Pernambuco..................	83.44	16.56	0.19
Alagoas.....................	120.51	-20.51	0.08
Sergipe	127.07	-27.07	0.04
Bahia.......................	97.35	2.65	0.27
Southeast	80.26	19.74	0.32
Minas Gerais	115.00	-15.00	0.30
Espírito Santo	103.81	- 3.81	0.03
Rio de Janeiro	95.04	4.96	0.08
São Paulo....................	59.39	40.61	0.59
South........................	83.81	16.19	0.32
Paraná.......................	81.94	18.06	0.37
Santa Catarina................	81.28	18.72	0.13
Rio Grande do Sul	85.85	14.15	0.50
West Central	110.66	-10.66	0.08
Mato Grosso	96.75	3.25	0.26
Goiás........................	115.55	-15.55	0.74
Brazil........................	89.84	10.16	1.00

SOURCE: Vera and Tollini (1979).

* The states are weighted with reference to their respective regions; the regions are weighted with reference to the country as a whole.

The program, sponsored by the Rockefeller Foundation, was distinguished by three unique characteristics: (1) the number one priority was to build up the production of wheat in order to feed the world's hungry populations; (2) no distinction would be made between basic and applied science, and the emphasis would be placed wherever it might be necessary in order to advance the program toward its objectives; (3) training and guidance would be provided to local personnel in order to prepare them to direct the program.

Wheat was the crop chosen because of its important place in human alimentation, and if Mexican wheat could be improved, perhaps consuming populations in the rest of the world would also be benefited. The research workers had in view, in addition to climatic adaptation to tropical regions, the obtaining of varieties resistant to rust. Later on the research efforts were directed to adjusting the new varieties to other factors, such as soil fertility, the use of insecticides, weed control, and irrigation (Borlaug, 1972; Sprague, 1967; Brown, 1970).

First Results: High Production Levels

Once adequate agricultural practices were identified, phenomenal wheat productions were obtained with these new varieties. The impact was so great that in 1957, thirteen years after beginning the research, 70% of the area devoted to wheat growing in Mexico was sown with the new high-production varieties (Paarlberg, 1971).

In effect, when CIMMYT started its work, Mexico was importing one-half of the wheat consumed by its population. As a result of the program, this panorama was drastically altered through a substantial growth in population. The country became self-sufficient and began to produce a surplus for export. Moreover, in 1963 it became, in fact, a net exporter of grains (Williamsen, 1976).

These first results were so promising that another international organization was formed in 1961 to carry out research on rice in the Philippines: the International Rice Research Institute (IRRI). The success of this program was truly important for many reasons: it was set up in a specially famished part of the world and not in one of those countries already producing surplus foodstuffs; it was located in the tropics, where agriculture is technically behind that of temperate climates; and, finally, it was established in countries having high birth rates and, therefore, in need of greater food production.

Under these circumstances, the Green Revolution made it possible to solve three crucial problems. In the first place, there was the problem of the agricultural frontiers: nearly all the land having characteristics favorable to agriculture was already under cultivation. Thus, increased production could only be obtained by pushing the agricultural frontiers into areas whose soil was ill-suited to agriculture or, alternatively, by improving the productivity of the land factor.

Second, there was the spector of hunger in conjunction with a fatalistic outlook regarding the incompatibility of the rate of population growth and the rate of increase in the food supply. In effect, population growth was far outpacing the increase in agricultural production. Moreover, a large part of the food supply was being provided by the rich countries, whose low rate of population increase gave them no cause for worry. Finally, there was the problem of placing the so-called underdeveloped countries, which had little hope of economic progress, on a sounder economic footing. The growth of their agricultural production would enable their people to attain better levels of nutrition, and the exportation of surplus products could be expected to generate resources to finance their economic growth.

In addition to Mexico, many Asiatic countries would also be benefited by CIMMYT's new varieties of wheat and, later on, by those of IRRI's rice. In Pakistan 20% of the total area planted to wheat is given over to the new varieties which are responsible for 42% of the total production. In that country, 1967 saw the planting of 600,000 acres of the new varieties, but the estimated area in 1968 was close to 3.5 million acres. In India the corresponding figures were 18% and 36%, respectively, and the planted area jumped from 700,000 acres to 6 million acres. With this, Pakistan produced a group of about 6.5 million metric tons of wheat and India about 16.5 million metric tons—both new records (Gaud, 1968).

In this way the majority of the technological innovations in food production have helped to fend off the specter of hunger in the immediate future. The image of agriculture in the developing countries has changed from a simple economic prop supporting other sectors of their economies to one of an effective and potentially growing contribution to the process of development. However, this rapid expansion of certain cereal crops in the developing countries has led to many problems and distortions, in parallel with the valuable contributions made by the highly productive varieties.

Some Economic Implications of the Green Revolution

If, during the first stage, the anticipated increase in production is in fact obtained, this automatically gives rise to a new set of adjustments and second generation problems that must be solved in order to sustain and accelerate development.

The process received its major impulse in the 1960s. By the end of that decade, however, some analyses indicated that the Green Revolution was generating more problems than solutions (Falcon, 1970; Warton, 1969; Munthe-Kaas, 1970). Some analysts affirm also that the positive contributions of the process were appreciably diminished by the resulting problems, such as: (1) soil erosion; (2) the cost (and risk) of adopting the new technologies (improved varieties of seed generally demand irrigation and the intensive application of fertilizers and insecticides); (3) high costs of storage, processing, distribution, and marketing; and (4) the high cost of transferring the new technology to low-income producers, more averse to risk and reluctant to adopt new technologies.

Additionally it has been pointed out that because of the adoption of labor-saving technologies the Green Revolution would affect the level of employment in agriculture. On the other hand, Shaw (1970), cited by Falcon (1970), shows that in some areas the shorter vegetative cycle might well enable the planting of several crops during the year, thus smoothing out the demand for labor. At the same time, Shaw recommends that the analysis should be made in its specific regional context. On the whole, the evidence seems to indicate that the new varieties create employment.

Finally, there is the problem of access to the new technologies by small producers, which would result in expelling them from the agricultural sector toward the cities.

Besides all these problems, there is also the need to consider the difficulty of reconciling the increased production arising from the new technologies with the useful functions performed by many millions of agricultural workers and small farmers. And it is not a simple task, at short and medium term, to create and adapt a whole infrastructure, involving storage and processing facilities and traditional marketing customs, to deal with new, large-volume production.

Thus, as usually happens in the process of development, the experience of the Green Revolution has been characterized by a history in which technological development, in its second and third stages, has caused serious distortions in the production and distribution structures. Even so, by opening the way to new possibilities of production, and by providing successful examples to promote a new international system of agricultural research, it has made a contribution that cannot be ignored.

BIOTECHNOLOGY AS A POTENTIAL INSTRUMENT FOR AGRICULTURAL DEVELOPMENT

In scientific circles today, there is a growing expectation that biotechnology may prove to have an economic and social impact fully as important as that attributed to electronics in recent times. This expectation is founded, basically, on the recent laboratory advances made in the fields of molecular biology, immunology, and the culture of cells and tissues. The fact is, as Baltimore (1982) emphasizes, that biotechnology is not in any way a new assemblage of knowledge concerning biology. What is "new" consists of the methodologies and techniques developed in the laboratory, which hold an immense potential for application in the areas of health, energy, and agriculture. The "old" biotechnology, also according to Baltimore, concentrated largely on the manipulation of microorganisms or food plants, and selections were made primarily with a view to higher levels of production.

What the "new" biotechnology does today is to use the available knowledge of the cell interior to guide investigation and control the products being sought. In other words, it involves the separation of parts of the cell that have desirable characteristics and, at the same time, are capable of preserving those characteristics, and the reinsertion of these parts in the cells or plants to be reproduced.

Another aspect to be stressed is that the principal agents utilized in biotechnology (DNA and its derivatives) are found in nature and represent the result of year after year of evolution in the several natural DNA recombinant forms. A further point is that, although we are discussing a perhaps "far-out" technology, biotechnology can contribute in a significant way to the solution of crucial problems faced by third world countries.

This does not imply, however, that the economic potentialities of these techniques will be realized easily or that their possible results should be overestimated. Not insignificant investments must be made in research and, above all, in highly specialized human capital before we can expect to arrive at solutions for, by way of example, plant and animal diseases, and even human ills.

In agriculture especially, the effort to obtain varieties with favorable characteristics is very old (Goodman, 1978).

As we have already seen, the Green Revolution of recent years was the result of efforts that led to the improvement and better selection of species of agronomic interest. Today, some specialists regard biotechnology as a second green revolution.

The idea of a revolutionary development is acceptable in the sense that biotechnology allows penetration into the interior of the cell and the manipulation of its genetic material in the form of specific genes, with the results forthcoming much faster than ever before. Over and above this, the feasibility of transferring genes from one species to another opens the possibility of obtaining results unimaginable in advance.

Studies are already in progress looking to obtain vegetable species having higher photosynthetic efficiency, improved fixation of atmospheric nitrogen, and greater tolerance of soil problems such as salinity, alkalinity, acidity, soluble aluminum, and deficiency of nutrients. Another research objective is to increase protein and vitamin contents, an undertaking already beginning to show positive, and to a certain extent surprisng, results. Some varieties of rice developed by IRRI have very high productivities and are also tolerant of soil problems (see table 5).

TABLE 5

PERFORMANCE OF RICE VARIETIES/LINES TOLERANT OF PROBLEM SOILS

Soil Problem	Variety or line	Yield (t/ha)
Salinity	IR50, IR5657-44	3.6
Alkalinity	IR36, IR4595-4	3.6
Strong acidity	IR42, IR4683-54	4.2
Peat	IR42, IR8192-31	3.1
Aluminum toxicity	IR43, IR6115-1	3.8
Phosphorus deficiency	IR52, IR8192-200	4.4
Zinc deficiency	IR36, IR8192-31	2.9
Iron deficiency	IR36, IR52	2.8

SOURCE: Swaminathan, 1982.

In animal improvement, some of the objectives of research now in progress are: improvement in food conversion efficiency; increased rate of gain in weight; higher average weight at time of slaughter; and better rates of reproduction. The evident potential of biotechnology, already reinforced by many positive results, is a strong stimulus to encourage the developing countries to control and apply these techniques. Such countries will thus be able to fortify their agricultural sectors, making possible the transfer of increasing volumes of surplus production to finance the growth of other sectors of their economies.

Nevertheless, though the potential of biotechnology cannot be denied, it must be borne in mind that its development and the targets to be aimed at should be compatible with the economic needs of each country. There is no doubt that many of the problems in connection with the Green Revolution could have been avoided by a more careful diffusion of the high-yield varieties. Even so, the experience gained therefrom can be extremely useful in the preparation and execution of biotechnological policies in third world countries.

Agriculture and Biotechnology in Brazil: A Major Challenge

The makeup of some of the growth indicators pertaining to Brazilian agricultural production over the period from 1934 to 1980 shows that the greater outputs attained by some essential crops have depended heavily on increases in the area cultivated (see figures 3–6). An extreme case is that of beans, where productivity actually declined from 0.87 t/ha in the 1934/38 period to 0.47 to t/ha in 1976/80, thus cancelling out a major part of the benefit normally to be expected from the larger area under cultivation. In rice, corn, and wheat crops, productivity did not vary to any appreciable extent during the period analyzed.

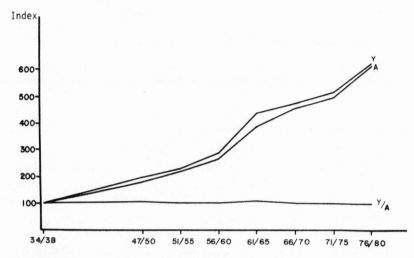

Fig. 3. Evolution of the harvested area (A), production (Y), and yield (YA) of rice, Brazil, 1934–38 to1976–80. (Hayami and Ruttan, 1971)

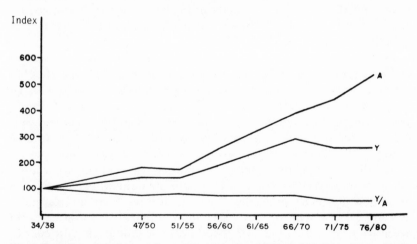

Fig. 4. Evolution of the harvested area (A), production (Y), and yield (YA) of bean, Brazil, 1934–38 to 1976–80. (Hayami and Ruttan, 1971)

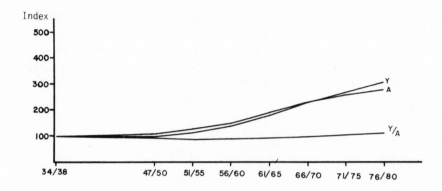

Fig. 5. Evolution of the harvested area (A), production (Y), and yield (YA) of corn, Brazil, 1934–38 to 1976–80. (Pastore et al., 1976)

The facts quoted give cause for concern when we remember the economic and social importance of these foodstuffs. In Brazil it is corn that accounts for the largest area under cultivation (about 12 million ha). Corn likewise is responsible for employing the country's largest contingent of agricultural workers during a significant part of the year. Its importance is all the greater because it provides a basic source of food for livestock in general, as well as for human consumption by low-income families in the form of its derivatives and by-products. Beans are also of prime importance, since they constitute one of the people's basic foodstuffs. Some 24% of the protein consumed by the country as a whole is provided by this leguminous plant.

Apart from its importance as an article of food, wheat is the second commodity on the list of Brazilian imports, coming immediately after petroleum. Per capita consumption of wheat has been rising steadily with the passing years, thus unfavorably affecting the balance of trade. Rice, together with beans, is the country's basic food. Outside these, perhaps with the exception of wheat, most other products originate in the traditional sectors of Brazilian agriculture, where the predominating technologies are those calling for minimum capital investment. Aggravating this situation, the land frontiers in the traditional regions are virtually exhausted. The expansion of production to other regions such as the "Cerrado" (savannas and stunted brushlands), Amazonia, and the Brazilian Northeast could accentuate still further the overall decline in the production efficiency of these crops.

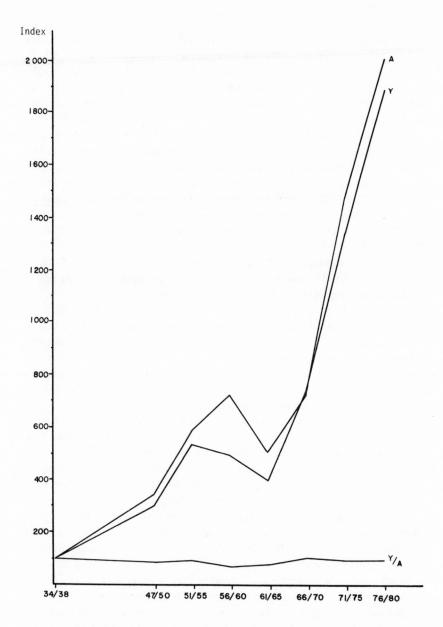

Fig. 6. Evolution of the harvested area (A), production (Y), and yield (YA) of wheat, Brazil, 1934–38 to 1976–80.

Covering an area of some 183 million ha (20% of Brazilian territory), the "Cerrado" is a region possessing an enormous agricultural potential. Its topography is suitable for mechanization, and geographically it lies in relatively close proximity to the consuming centers. In addition, it already has a reasonable highway infrastructure. At the same time, the region presents problems of soil and climate that affect agricultural production. Toxicity of aluminum and manganese, high acidity, low nutrient content as to both macro- and microelements, and low nutrient retaining capacity, among others, are problems to be solved. The introduction of agricultural activities into this region, based on the varieties known today, is only possible by means of heavy investments to improve the chemical characteristics of the soil. This being so, the obtaining of varieties tolerant of these adversities would make a very great contribution to more efficient agriculture in the "Cerrado."

In contrast to the conditions typical of the "Cerrado," the ecosystem of Amazonia is decidedly unstable, which perhaps constitutes the principal problem standing in the way of opening up the agricultural frontiers of this region. Amazonian soils in general are low in fertility, and even the most fertile soils are rapidly exhausted when cultivated. An idea of the conditions existing may be had from the fact that 70% of the available mineral nutrients are found in the biosphere. Forest clearing operations must be conducted with maximum care, and the crops to be introduced must have a highly developed root system capable of retaining the nutrients. Otherwise, they will be leached out rapidly by the heavy rains.

The Northeast, with its own peculiar climate and soil characteristics, requires other precautions. Its principal drawback is the lack of water. Irrigation is an extremely delicate matter, inasmuch as the available water is obtained from underground aquifers and has a high content of dissolved salts. For these conditions, varieties capable of tolerating the water shortage and the high salinity of the soil are a necessity.

In tables 6 and 7 we present the main problems in terms of the regions involved and the products selected, which are those judged to merit larger investments in research. Research scientists in biotechnology should be challenged to attack these problems.

Finally, we recognize the superficiality and timidity with which the research and financial institutions charged with the country's technological development have approached the question of agricultural biotechnology. Few institutions and extremely scanty human and material resources are involved in research projects in this area. We would estimate that Brazil has

TABLE 6

SELECTED CROPS, MAJOR PROBLEMS, AND RESEARCH ISSUES, BRAZIL, 1983

Crop	Problem/Research issue
Rice	
Under Irrigation	High yield varieties
	Varieties resistant to some important diseases such as *Pyricularia oryzae* and *Helminthosporium oryzae*
	Varieties resistant to *Lissoshoptrus* sp. and nematodes
	Varieties resistant to freezing
	Multiple crop varieties
Without Irrigation..........	High yield varieties and resistant to hydric deficiency in the dry season ("veranico")
	Varieties resistant to "brusone" (*Pyricularia oryzae*) and "mancha parda" (*Cochliobolus miyabeanus*)
	Varieties tolerant of soil problems
Beans......................	Fixation of genetic characteristics in more productive varieties
	Varieties resistant to major crop diseases: "mosaico dourado", "ferrugem" (rust), and "bacteroise"
	Varieties resistant to major pests, such as: *Empoasca* spp. and *Ellasmopalpus lignosellus*
	Varieties more efficient in symbiotic nitrogen fixation
	Varieties resistant to hydric deficiency
Corn	Varieties resistant to important diseases and pests
	Varieties suitable to mixed crop cultivation
	Varieties more suitable to regional climate characteristics
	Varieties more suitable to "in natura" consumption
Wheat	High yield varieties
	Varieties resistant to important diseases and pests in the traditional regions of the country
	Development of new and productive varieties for the "Cerrado" region

TABLE 7

SELECTED REGIONS OF BRAZIL AND RESEARCH ISSUES, 1983.

"Cerrado" region	Northeast region	Amazon region
Acidity	Hydric deficiency	Diseases and pests
Aluminium toxicity	Alkalinity	Multiple crop varieties
Manganese toxicity	Salinity	
Micro- and Macronutrients deficiency		Micro and Macronutrients deficiency
Hydric deficiency in the dry season ("veranico")	Mixed crop cultivation varieties	Varieties more suitable to local climate
Diseases and pests	Small-farm technology	

invested approximately US $1 million in the 1980/83 period for the development of agricultural biotechnology.

At present among the Brazilian institutions involved in biotechnological research projects pertaining to agricultural problems, the leading ones are: University of São Paulo, University of Campinas, Empresa Brasileira de Pesquisa Agropecuaria, and Instituto Agronomico de Campinas. These and other institutions are developing research projects mainly on sugar cane, beans, rice, corn, coffee, tomatoes, wheat, and grapes. And in most of these on-going projects there are promising results for the near future.

FINAL COMMENTS

Biotechnology applied to agriculture may represent a significant means to increase agricultural output in developing countries. Simultaneously, it may constitute a strategic and key variable to promote and accelerate economic development, on a sounder basis. However, we should not overestimate the potential contributions of such new and revolutionary technologies. There are many issues to be observed carefully, and solved, if biotechnology is to perform a positive role in this context. This also means that economic perspectives of biotechnology in agriculture of developing countries will be dependent largely on the political and financial support that policy makers recognize as needed by selected institutions and researchers.

In effect, one may accept as a general rule that agricultural research systems in developing countries are often plagued by sequences of progress and growth followed by erosion and losses of capacity, as planning (and budgetary) targets respond to political changes and interests.

The experience of Brazilian agriculture suggests that some crops and regions are undergoing technological advances. Thus, to such crops and regions the research efforts and priorities should be goal-oriented. By doing this, it is very likely that the social returns from investments in biotechnology will fit best the economic needs of the country. Also, two ways of increasing agricultural surplus could be simultaneously stimulated: (1) to promote technological development in favor of a more efficient agricultural production in the West-Central section of Brazil; and (2) to create a small-farm technology for the arid and poor regions of the Northeast.

In the short run, priority should be given to a research program to be implemented on the basis of the availability of human resources and institutional experience. Rather than implementing a nation wide program, specific projects with very particular objectives and methodologies would accelerate the achievement of positive net gains to society.

Additionally, a multidisciplinary group of national researchers has to be prepared for anticipating the social distortions and needed adjustments resulting from the adoption of the new technologies in the agricultural sector. This is to say that the interchange of ideas and know-how both with the research systems derived from the Green Revolution has to be promoted urgently.

One of the positive features of biotechnology is the potential for reducing the time span of the innovation process. In order to achieve this, biotechnology researchers of developing countries will have to be exposed continuously to the frontier techniques in the more advanced research centers.

REFERENCES

Araujo, P. F. C. de. 1975. Agricultura no processo de desenvolvimento economico. *In* Araujno, P. F. C. de, and G. E. Schuh (eds.), Desenvolvimento da agricultura: natureza do processo e modelos duelistas, pp. 83–97. Livraria Pioneira Editora.

Arino, M., and Y. Hayami. 1975. Efficiency and equity in public research: rice breeding in the Japanese economic development. Amer. J. Agric. Econ. 57:1–10.

Ayer, H. W., and G. E. Schuh. 1972. Social rates of return and other aspects of agricultural research: the case of cotton research in Sao Paulo, Brazil. Amer. J. Agric. Econ. 54:4.

Baltimore, D. 1982. Priorities in biotechnology. *In* Priorities in biotechnology research for international development: proceedings of a workshop, pp. 30–37. National Academy Press, Washington, D.C.

Borlaug, N. B. 1972. La Revolucion verde: paz y humanide. Serie Reimpresos y Traducciones CIMMYT No. 3.

Brown, L. 1970. Seeds of change–7th: Green revolutions and development in the 1970's. Praeger, New York.

Engler, J. J. de C. 1978. Analise da productividade agricola entre regioes do Estado de Sao Paulo. Piracicaba, ESALQ/USP. 132 pp.

Falcon, W. F. 1970. The green revolution: second generation problems. Amer. J. Agric. Econ. 52:698–710.

Fonseca, M. A. S. 1976. Retorno social aos investimentos em pesquisa na cultura do cafe. Piracicaba, ESALQ/USP, MS thesis.

Gaud, W. S. 1968. New seeds: the green revolution. Development Digest 6:3–6.

Goodman, M. M. 1978. Historia e origem do milho. *In* E. Paterniani (ed.), Melhoramento e producao de milho no Brasil, pp. 1-31. Piracicaba, ESALQ/USP, Fundacao Cargill.

Griliches, Z. 1958. Research costs and social return: hybrid corn and related innovations. J. Pol. Econ. 66:419–31.

Hayami, Y., and V. Ruttan. 1971. Agricultural development: an international perspective. Chapter 2. Johns Hopkins University Press, Baltimore.

Johnston, B. F., and J. W. Mellor. 1961. The role of agriculture in economic development. Amer. Econ. Review 51:563–93.

Monteiro, A. 1975. Avaliacao economica da pesquisa agricola: o cao do cacau no Braxil. Vicosa, MG, Federal University of Vicosa, MS thesis.

Munthe-Kaas, H. 1970. Green and red revolutions. Far East. Econ. Rev. 68:21–24.

Paarlberg, D. 1971. Norman Borlaug, pioneer in the green revolutions. Devel. Digest 9:6–14.

Pastore, A. C., E. R. de A. Alves, and J. A. B. Rizzieri. 1976. A inovacao e os limites a modernizacao na agricultura brasileiria. Revista de Economia Rural.

Peterson, W. 1967. Returns to poultry in the United States. J. Farm Econ. 49:1415–25.

Schuh, G. E. 1975. A modernizacao de agricultura brasileira: uma interpretacao. Technologia e Desenvolvimento Agricola, Instituto de Pesquisa Economico e Social, IPEA Monograph Series, No. 17, pp. 7-45.

Sprague, G. F. 1967. Agricultural production in the developing countries. Science 157:774–78.

Swaminathan, M. S. 1982. Perspectives in biotechnology research from the point of view of third world agriculture. *In* Priorities in biotechnology research for international development: proceedings of a workshop, pp. 38-63. National Academy Press, Washington, D.C.

Thompson, R. L. 1976. The metaproduction function for Brazilian agriculture: an analysis of productivity and other aspects of agricultural growth. Purdue University, Ph.D. diss. 177 pp.

Vera, F., and H. Tollini. 1979. Progresso technologico e desenvolvimento agricola. *In* A. Veiga (ed.), Ensaios sobre politica agricola brasileira, pp. 87-136. Secretaria da Agricultura, Sao Paulo.

Warton, C. R., Jr. 1969. The green revolution: cornucopia or Pandora's box? Foreign Affairs 47:464–76.

Williamsen, E. J. 1976. The agriculture of Mexico. Scientific American 235:128–50.

W. R. SHARP AND D. A. EVANS

Plant Cell Culture: The Foundation for Genetic Engineering in Higher Plants

6

ABSTRACT

Progress in the release of new varieties of crop plants using genetic engineering depends on the development of appropriate recombinant DNA technology and cellular genetics procedures. The latter includes the technologies of protoplast fusion, haploid plant production, embryo culture, and induction *in vitro* of cellular genetic variability (somaclonal variation). New cellular genotypes can be selected from cell cultures or through recombinant DNA technology that result in altered genotypes through modification of existing genetic material or the insertion of alien genes. The key step for the development of a new crop variety from a genetically modified cell rests on the ability to regenerate plants from the genetically modified cell using tissue culture procedures. It is important to note that the capacity for plant regeneration or *in vitro* propagation (IVP) from a given cell is dependent upon the existence of genetically determined regeneration. Breeding programs need to be established for certain crops that are recalcitrant, e.g., cereal, legumes, and so on, in order to develop parental genotypes suitable for use in genetic engineering experiments in which the goal is to obtain crop plants. Plantlets may be regenerated through either organogenesis or embryogenesis. Multiplication of genetically modified plants can be accomplished through tissue culture cloning.

INTRODUCTION

Plant breeders have been interested in transferring useful genes between plant species for years. The introduction of a useful genetic trait from a wild species into a cultivated crop is accomplished following production of interspecies hybrids. Because plants have approximately 30,000 genes, it is impossible to preselect the genetic traits to be transferred using conventional breeding, so hybrids contain a mixture of all genes of both species. Hybrids must be backcrossed several times to cultivated crops to eliminate undesirable genes and to retain desirable genes. Conventional breeding requires 6–8 generations to successfully transfer a gene from one species to a second species.

Recombinant DNA technology will have an important impact on agricultural markets within the next 20 years, or perhaps sooner. These breakthroughs will occur through a marriage of molecular biology with plant tissue culture in the development of new crops. During the immediate future, plant tissue culture and cellular genetics research alone will result in the development of breakthroughs in the production of new crop varieties not possible through conventional breeding programs.

Tissue culture methods have been developed that permit regeneration of plants from organized tissues or in some species from single cells. Tissue culture regeneration produces two categories of plants: clones and genetic variants. The conditions for the control of each category can be altered depending on the desired agricultural use of the regenerated plants.

Plants produced asexually through shoot apex/axillary bud culture *in vitro* are of uniform quality, and can be produced more rapidly than is possible through sexual reproduction or by conventional modes of propagation (Murashige, 1974). A millionfold increase per year in the ratio of clonal multiplication over conventional methods is not unrealistic. Clonally propagated homozygous superior plants produce uniformly superior seeds, improve progeny evaluation of breeding experiments and have improved vigor and quality. Moreover, clonal propagation technology can be used for the elimination of pathogens and in propagation of disease-free germplasm for international germplasm shipment. The advantages of tissue culture cloning over conventional propagation techniques derive from a reduction in time to achieve large-scale propagation. Meristematic explants are generally used for clonal propagation because they have a lesser degree of genetic variability as compared with non-meristematic explant material.

Potential agricultural applications of cloning include the increase of plants with poor seed set and low germinability (e.g., hybrid carrot seed), increase of rootstocks (e.g., M9 apple rootstock), increase of parental lines for breeding (e.g., asparagus), propagation of hybrids (e.g., broccoli), multiplication of virus-free plants (e.g., potato), and propagation of forest trees (e.g., Douglas fir). Tissue culture cloning is especially useful in propagation of hybrids in crops where male sterility is not commercially available since continuous commercial production of hybrid seed is expensive (e.g., greenhouse cucumber).

Tissue culture can be used for isolating somaclonal variants by using either non-meristematic tissue explants or single cells for plantlet regeneration. Factors that appear to be associated with the degree of somaclonal variability in plants obtained from tissue culture include: (1) explant source, (2) explant age, (3) culture environment, (4) culture duration, (5) nutrient additives, and (6) growth stimulants or regulators. Categories of somaclonal variability observed following regeneration consist of changes in: (1) chromosome number and structure, (2) simple nuclear gene mutation, (3) cytoplasmic gene mutation, and (4) epigenetic changes. Genetic analysis must be made to ascertain the basis for genetic variability. Exciting possibilities exist for taking advantage of the wide use of genetic variability associated in certain regenerated plants for development of new cultivars.

The techniques of recombinant DNA and gene cloning developed in the last decade have been used to demonstrate the ability to transfer genes between cells. The technology is sufficiently advanced so that potentially any gene can be inserted into any plant cell. The insertion of specific useful genes into otherwise productive crop species offers an opportunity to shorten the time requirement for the creation of new crop plants and signals the beginning of a second green revolution. Opportunities exist for the transfer or modification of genes in the plant for nitrogen fixation, herbicide resistance, salt tolerance, and pesticide production. Hand-in-hand integration of the techniques of plant tissue culture with recombinant DNA technology is crucial to the production of new crop plants. Although recombinant DNA can be used to transfer a specific gene from the cells of one species to cells of a second species, it is necessary to utilize tissue culture methodology to recover fertile, mature plants from the genetically modified cells to develop new varieties. Reliable regeneration protocols are necessary to apply recombinant DNA technology to crop improvement.

PATTERNS OF PLANT REGENERATION

High frequency plant regeneration is important for cloning, tissue culture breeding, and disease eradication. Regeneration can be accomplished through either the culture *in vitro* of meristematic explants or non-meristematic explants depending on the propagation or crop improvement goal.

Meristematic explants include isolated meristems, shoot tips, and axillary buds; non-meristematic explants include leaf, petiole, and root. The difference between the two explant sources is that the meristematic explants consist of cells with a developmental commitment to the plantlet pattern of development, whereas the non-meristematic explant cells are committed to alternative developmental pathways. Non-meristematic cells can redifferentiate in tissue culture and become committed to the plantlet pattern of development when subjected to an appropriate tissue culture regime.

Shoot Apex/Axillary Bud Propagation

Certain species or genetic variants are often difficult or time-consuming to propagate using established horticultural methods. The culture of organized structures for mass clonal propagation using shoot apex/axillary bud meristems to produce genetically identical plants is important on a commercial basis (Murashige, 1974). The apical shoot tip or axillary bud is surface-sterilized and placed onto a simplified culture medium for production of multiple shoots that are subsequently separated and subcultured onto a rooting medium. Rooted plantlets are then transplanted to soil or peat pots, and thereafter, transferred to the greenhouse or field. This procedure is widely used in the commercial tissue culture propagation industry because plants produced by this method are as a rule genetically uniform. This is because such plants result from preexisting or newly formed meristems without an intervening callus stage.

A general three-stage procedure has been described for production of plants from meristematic explants that normally requires alteration of culture medium or growth conditions between stages (Murashige, 1974). In stage 1, tissue is established *in vitro*. In stage 2, cultures are treated to produce multiple shoots. Stage 3, propagules undergo root formation and conditioning prior to transfer to the greenhouse. Although transfer to high light intensity is important in stage 3 (Murashige, 1978), the media and culture conditions are not altered for many species between stages 1 and 2.

The frequency of plant regeneration occurring in tissue culture using explants from apical meristems, axillary buds, or shoot tips can be manipulated by altering the types and concentrations of growth regulators used in the culture medium. The growth regulators benzyladenine (6BA), napthylene acetic acid (NAA), and gibberellic acid (GA) are especially important for high-frequency regeneration. Some mass propagation schemes require the reculture of shoots propagated *in vitro* to decrease the time for achieving production goals.

Adventitious Shoot Development from Either
Sporophytic or Gametophytic Tissues

Formation of localized meristems in cell cultures or induced organogenic-determined cells (IODC) leading to the organization of a well defined shoot and/or root meristem has been described (Evans et al., 1981). The primordia originate from small groups of cells and have a developmental commitment to either shoot and/or root organogenesis. The redetermination of cells comprising mature tissues to shoot/root formation is often dependent on the ratio of cytokinin to auxin in the nutrient medium (Evans et al., 1981; Skoog and Miller, 1957). High concentrations of cytokinin favor shoot formation, whereas high concentrations of auxin favor root formation. Factors important to determining the frequency of IODC development include explant age, seasonal variation, explant source, culture medium, growth regulators, and environmental conditions.

Induced Organogenic-Determined Cells (IODC)—One-Step. Shoot induction can occur from non-organogenic-determined cells cultured on a medium containing appropriate concentrations of specific growth regulators. The growth regulators promote cellular proliferation and redetermination of certain cellular populations to the shoot morphogenesis pattern of development.

The most frequently used growth regulator additives in the culture medium are indole acetic acid (IAA) and kinetin (KIN), 6BA, IAA + 6BA, or NAA + 6BA. Cytokinin is removed from the culture medium for root development.

Tobacco has been used as a model system for investigations of the IODC mode of adventitious shoot development. Totipotency was first demonstrated with *Nicotiana tabacum* by regeneration of mature plants from single cells (Vasil and Hildebrandt, 1965). Some of the basic factors regulating organogenesis have been elucidated using tobacco. Plant regeneration from isolated protoplasts also was accomplished first with *N. tabacum* (Takebe et al., 1971), and the first somatic hybrid plant was obtained between *N. glauca* and *N. langsdorffii* (Carlson, et al., 1972).

Induced Organogenic-Determined Cells (IODC)—Two-Step. Shoot induction in many of the monocotyledons requires a two-step culture medium regime that mimics the one used for somatic embryogenesis (see IEDC section). Species of numerous families of dicotyledonous plants can be induced to undergo plant regeneration readily *in vitro*, but monocotyledons in general have been more difficult to culture. This is unfortunate, since the *Gramineae* include many of the world's most important crops.

The first-step culture medium usually contains 2,4-D. In the second-step culture medium 2,4-D is either deleted, added at a reduced concentration, or replaced by a weaker auxin. The primary limitation to regeneration of monocots has been explant source, since only meristematic tissues have been used to initiate callus cultures capable of plant regeneration from most recalcitrant species cultured *in vitro*. In sugar cane a variety of explants may be used for plant regeneration including apical meristems, young leaves, and pith parenchyma. In addition, plants have been regenerated from long-term callus cultures in sugar cane (Nadar and Heinz, 1977). For each of these explants in sugar cane, the two-step procedure is used. More recently a similar two step regeneration scheme has also been accomplished for cell suspensions of pearl millet (Vasil and Vasil, 1980).

Somatic Embryogenesis

Two general patterns of embryogenic development *in vitro* are discernible: (1) direct embryogenesis from tissues marked by the absence of callus proliferation (e.g., nucellar cells of polyembryonic varieties of *Citrus*, epidermal cells of hypocotyl in wild carrot and *Ranunculus sceleratus*), and (2) indirect embryogenesis in which callus proliferation occurs prior to embryo development (e.g., secondary phloem of domestic carrot, inner hypocotyl tissues of wild carrot, leaf tissue explants of coffee, pollen of rice). An understanding of these two different patterns of development depends upon consideration of the determinative events of cytodifferentiation during the mitotic cell cycle. It has been documented that the fate of *determined* daughter cells following mitosis occurs at least one mitotic cycle prior to differentiation (Yeoman, 1970). In other words, cells that undergo embryogenesis directly are the daughters of a prior determinative cell division. Such determined cells may undergo a post-mitotic arrestment until environmental conditions are favorable for commencement of the mitotic developmental sequence characteristic of embryogenesis. The two general patterns of embryogenic development *in vitro*, direct and indirect, may be further characterized by the relative times of determination and differentiation of the cultured cells into embryogenic cells. Direct embryogenesis proceeds from preembryogenic-determined cells (PEDC) (Kato and Takeuchi, 1966), whereas indirect embryogenesis requires the redetermination of differentiated cells, callus proliferation, and differentiation of embryogenic-determined cells (IEDC).

Apparently, PEDCs require either synthesis of an inducer substance or removal of an inhibitory substance prior to resumption of mitotic activity and embryogenic development. Conversely, cells undergoing IEDC differentiation require a mitogenic substance to reenter the mitotic cell cycle and/or exposure to specific concentrations of growth regulators. Cytodifferentiation and the emergence of multicellular organization are multistep processes in which each step leads to the establishment of a particular pattern of gene activation, resulting in transition to the next essential state of development (Street, 1978). Arrestment may occur at any step in this process.

Preembryogenic-Determined Cells (PEDC). The natural occurrence of polyembryony in many species of *Citrus* (Bacchi, 1943) and its economic importance have long been realized. Thus, not only a zygotic embryo but also several adventive embryos may be found within a single seed. These adventive embryos have been shown to originate in single cells of the nucellus, near the micropyle of the ovule, and appear to be initiated after fertilization, before or soon after the first zygotic division (Bacchi, 1943).

In addition to the natural occurrence of nucellar polyembryony, cultures *in vitro* of nucellar explants may give rise to embryos and eventually to fully developed plants (Kochba et al., 1972). Nucellar tissues from both unfertilized as well as fertilized ovules may undergo embryogenesis *in vitro* (Button and Bornman, 1971). Cultures of monoembryonic cultivars such as the Shamouti orange also have been shown to be embryogenic (Button et al., 1974). Early characterization of embryogenesis in nucellar cultures consists of an initial proliferation of callus and subsequent development of "pseudobulbils" in the absence of exogenous plant growth regulators (Rangaswamy, 1958). Some of these pseudobulbils continue embryogenic development and eventually become entire plantlets. However, in other experiments with several cultivars of normally monoembryonic species, callus and pseudobulbil formation were not a prerequisite to embryo development, and embryos arose directly from nucellar tissue (Button and Bornman, 1971). Embryos may arise directly from the nucellus, or indirectly from nucellar callus (Mitra and Chaturvedi, 1972). Many published experiments include addition of growth regulators such as NAA and Kinetin, as well as addition of complex addenda such as malt, yeast extract, or coconut water. These additions may be beneficial in increasing the frequency of observed embryos. However, it must be stressed that embryogenesis may be observed in the absence of the components.

Embryogenesis has been characterized in habituated cell cultures, i.e., autonomous for exogenous growth regulators (Button et al., 1974). This callus was found to be composed not of unorganized parenchymatous tissue but solely of numerous proembryos. Embryogenesis has been observed from single cells in the periphery of the callus, as well as from existing proembryos. Some of these developing embryos may enlarge only to a globular stage commonly referred to as "pseudobulbils". These rarely develop into plants. Other proembryos follow the developmental sequence characteristic of zygotic embryogenesis and eventually develop into plants. The fact that this callus was habituated in no way decreased its embryogenic potential. That the presence of exogenous growth regulators actually depressed embryogenesis lends further support to the concept that exogenous growth regulators may be viewed best as inductive agents of determination.

It is our view that embryogenesis from nucellar tissues, both *in vivo* and *in vitro*, may be considered best as cases of PEDC-mediated embryogenesis. The cells of the nucellus are actually PEDCs, and their proliferation as a callus mass and subsequent embryogenesis may be viewed as the cloning of PEDCs. Thus, it appears that embryogenesis in the nucellus is autonomous and may be observed even in the absence of exogenous growth regulators. Although low concentrations of Kinetin and NAA have been shown to be beneficial, they are not absolute requirements for embryogenic determination. Exogenous growth regulators probably contribute to the cloning of these PEDCs, thus increasing the relative number of embryogenic cells. Of course, it is also possible that an additional population of PEDCs requiring exogenously supplied growth regulators for development further contribute to the relative number of embryos.

Induced Embryogenic-Determined Cells (IEDC). The concept of IEDC involves the redetermination of differentiated cells to the embryogenic pattern of development. Evidence exists that the auxin or auxin/cytokinin concentration in the primary culture medium or in nonmitotic-differentiated cells is critical not only to the onset of mitotic activity in non-mitotic differentiated cells but to the epigenetic redetermination of these cells to the embryogenic state of development.

A statement characterizing the role of growth regulators in gene expression as direct or indirect cannot be made. Regardless of how growth regulators control gene expression, evidence exists that the auxin, 2,4-D elicits a response at the transcriptional and translational levels that occurs shortly after subculture onto a secondary or induction medium (Sengupta and Raghavan, 1980a; 1980b).

Numerous examples of IEDC have been reported in which embryogenic cells resulting from cellular redetermination proceed through the various stages of embryo development and form plants. Development of the embryogenic-determined cells is usually restricted during culture on the primary or conditioning medium. Thereafter, cells need to be subcultured onto a secondary culture medium or induction medium for continued development in the embryogenic-determined cells. Documented examples of IEDC have been reported for the following crops: cacao (Pence et al., 1979), caraway (Ammirato, 1977), carrot (Reinert et al., 1967), celery (Williams and Collin, 1976), coffee (Sondahl et al., 1979), cotton (Price and Smith, 1979), eggplant (Matsuoka and Hinata, 1979), endive (Vasil and Hildebrandt, 1966), grape (Krul and Worley, 1977), palm (Reynolds and Murashige, 1979), and pumpkin (Jelaska, 1974).

Androgenesis and gynogenesis involve the development of IEDCs. In these instances either the microspore or megaspore must undergo a quantal mitotic division resulting in an embryogenic-determined cell. In the former, a quantal mitotic division results in a generative cell and a vegetative cell. The vegetative cell, or in some instances the generative cell (Raghavan, 1976), or a fusion cell is then determined as an embryogenic mother cell (Sunderland, 1974).

Other Modes of IVP

Protoplasts. The plant cell wall interferes with our ability to genetically modify plant cells. The wall can be digested with enzymatic treatment, but the resulting protoplasts have reduced viability. Nonetheless, it is possible to genetically alter protoplasts with mutagen treatment, protoplast fusion, or uptake of exogenous DNA. Genetically modified protoplasts must then be regenerated into intact plants. Application of protoplast technology for crop improvement is limited to the species in a small number of plant genera in which plants can be regenerated from protoplasts.

Protoplasts can be isolated following treatment of plant tissue with mixtures of cellulase, pectinase, Driselase, and Rhozyme (Gamborg and Wetter, 1975). Procedures have been devised for the release of protoplasts from leaf tissue, stem tissue, flower petals, or cell suspension or callus cultures. Isolation of protoplasts is usually achieved with enzymes dissolved in a solution containing osmotic stabilizers to prevent bursting and swelling of the protoplasts; the enzyme solution is removed by filtration and centrifugation. Isolated protoplasts are amenable to genetic modification following enzyme removal. After chemical or physical treatment, modified protoplasts are cultured in nutrient medium. In some cases protoplast culture medium may be extremely complex. Viable protoplasts regenerate a cell wall and undergo mitosis in the culture medium. First mitosis usually is observed in 2–7 days after removal of digestive enzymes. As the protoplast-derived cells continue to reproduce, new culture medium must be added, or culture medium sufficient for regeneration must be substituted for the protoplast culture medium.

Although protoplasts can be isolated from any plant tissue (Vasil and Vasil, 1980), the condition of the donor plant prior to protoplast isolation is extremely important if plants are to be regenerated (Shepard and Totten, 1977). In many cases greatest success in plant regeneration has been achieved with protoplasts isolated from cultural cells. Cultured cells are already grown under aseptic and controlled environmental conditions and in many cases are already programmed for rapid cell division. If leaf tissue is to be used for protoplast isolation, optimum growth conditions of donor plants must be ascertained.

Protoplast-derived cell colonies must usually be transferred to fresh medium for plant regeneration. In most cases where organogenesis occurs, a sequence of culture media must be used to achieve shoot, then root, formation. If embryogenesis is possible, only a single transfer of culture medium is required to recover plants from cultured cells. Consequently, regeneration of plants from cells derived from protoplasts is achieved using the same hormone requirements as plant regeneration from callus cells produced from intact plant tissue.

Plants have been regenerated from protoplasts isolated from at least 35 plant species (Vasil and Vasil, 1980; Evans, 1982). Most species capable of regeneration though, are Solanaceous plants, including species in the *Datura, Hyoscyamus, Lycopersicon, Nicotiana, Petunia*, and *Solanum* genera. Very few economically important crop species have been regenerated from protoplasts. Successful regeneration protocols have been published for asparagus (Bui-Dang-Ha and MacKenzie, 1973), carrot (Dudits et al., 1976), rapeseed (Kartha et al., 1974), tobacco (Takebe et al., 1971), potato (Shepard and Totten, 1977), millet (Vasil and Vasil 1980), and alfalfa (Kao and Michayluk, 1980). Extension of this list to other economically important crops, particularly cereals and legumes, is essential for application of protoplast technology for crop improvement.

Methodology for the isolation and fusion of protoplasts has been described (Gamborg and Wetter, 1975). Following fusion treatment, protoplasts regenerate cell walls and undergo mitosis. Hybrid cells must be identified in mixtures of parental, homokaryotic and hybrid cells. Identification of hybrids is usually accomplished using genetic markers (e.g., White and Vasil, 1979). Somatic hybrid plants have been recovered from interspecies combinations not possible using conventional genetics (Evans, 1982). In most cases the species combinations in interspecific hybridization represent Solanaceous crop plants and include only tomato, potato, tobacco, and carrot among cultivated crops.

Haploids. Haploid plants can be recovered using the IVP schemes of either androgenesis or plant regeneration from haploid callus cells obtained from cultured microspores. It has been reported that most species capable of haploid production proceed by direct androgenesis (Sharp et al., 1983). Taxonomic differences are apparent, though, between androgenesis and callus-mediated regeneration. The majority of Solanaceous species capable of producing haploid plants undergo direct embryogenesis, whereas the majority of Graminaceous species undergo callus-mediated haploid plant regeneration (Sharp et al., 1983). Crop species capable of haploid plant production from cultured anthers include: rapeseed, tobacco, rye, potato, corn, asparagus, pepper, strawberry, rubber, barley, sweet potato, wheat, grapes, and clover.

Embryos. Pollination and embryo rescue *in vitro* is an important mode of IVP for the development of interspecific and intergeneric hybridizations that are rare in nature. Success in this technique depends upon two basic considerations (Yeung et al., 1981). These are (1) using pollen grains and ovules at the proper stage, and (2) defining nutrient media that will support germination, pollen tube growth, and embryo development.

Embryo abortion, which occurs quite frequently as a result of unsuccessful crosses in breeding, can be avoided if it is caused by failure of endosperm development. This is accomplished through the aseptic culture of the embryo in a nutrient medium. This approach has been successful for interspecific crosses in cotton, barley, tomato, rice, and jute. Success has also been achieved with intergeneric hybrids, e.g., *Hordeum* and *Secale, Hordeum* and *Hordelymus, Triticum* and *Secale,* and *Tripsacum* and *Zea.* Another novel use of embryo culture has been in the production of monoploids and doubled monoploids of barley.

GENETICS

Genetic Basis of Plant Regeneration

Progress in the application to agriculture of embryogenesis and organogenesis *in vitro* using nonmeristematic explants is dependent upon an understanding of the genetic basis of these phenomena. Such a basis has been established for the frequency of organogenesis in alfalfa and tomato. Genetic lines of alfalfa with 67% regeneration were selected from hypocotyl tissue culture-regenerated plants with an initial plant regeneration fre-

quency of only 12% (Bingham et al., 1975). Shoot morphogenesis in tomato germplasm originating from cultured leaf discs with different regeneration potentials were examined (Frankenberger and Tigchelaar, 1980).
An analysis of diallel F_1 hybrids was clearly supportive of a genetic basis
for shoot morphogenesis. Work using selections obtained from potato
clones with low frequency androgenesis has resulted in identification of
recombinants with a 30-40% frequency of androgenesis (Jacobsen and
Sopory, 1978).

Application of Genetic Variability

Regenerated Plants. Phenotypic variation has been reported in a
number of plant species regenerated by adventitious shoot development.
Such genetic variability may be agriculturally useful when integrated into
existing breeding programs. The phenotypic variability recovered in regenerated plants reflects either preexisting cellular genetic differences or tissue
culture-induced variability. Variability has been observed in both asexually and sexually propagated crops. Plants regenerated from callus of sugar
cane have been examined in detail. A wider range of chromosome numbers
(2n = 71–300) has been observed in plants regenerated from sugar cane callus (Heinz et al., 1977) than in parent lines. These clones of regenerated
sugar cane were examined for disease resistance. Resistance to three diseases (eyespot disease, Fiji disease, and downy mildew) was observed in
some plants regenerated from a susceptible sugar cane clone (Heinz et al.,
1977). One regenerated line was simultaneously resistant to both Fiji disease and downy mildew. These lines are being incorporated into a conventional breeding program. More recently, plants regenerated from mesophyll protoplasts of potato have been analyzed. As in sugar cane, lines have
been identified that are resistant to diseases (late blight and early blight), to
which the parent line, Russet Burbank, is susceptible (Shepard et al., 1980).
Similarly, sexually transmitted variability in plants regenerated from lettuce has been observed (Sibi, 1978). Evans and Sharp (1983) documented
the ocurrence of single gene mutations in the progeny of tomato plants
regenerated from cell culture. In addition to single gene changes, cytoplasmic genetic changes have also been detected in plants regenerated from cell
culture (cf. Evans et al., 1984). It is hoped that new varieties will soon be
developed using these techniques.

Protoplast Technology. Phenotypic variability has been observed in
plants regenerated from mesophyll protoplasts (Shepard et al., 1980).

Experimenters attempting to apply protoplast regeneration to crop improvement should be cognizant of this potential variability. Separate clones should be monitored and screened for agriculturally useful variability.

Protoplast fusion offers the opportunity to introduce unique genetic variability into crop plants. Novel nuclear-cytoplasmic combinations can be accomplished that are not possible using conventional hybridization. In most plant species, the male cytoplasm is excluded during fertilization. During fusion of protoplasts, the two cytoplasms are mixed. Following fusion, mixed cytoplasms normally segregate to one of the two parental types, thereby permitting effective cytoplasmic transfer. More recently, interspecific recombination of both mitochondrion and chloroplast DNA has been observed in somatic hybrid plants. Agriculturally important traits known to be cytoplasmically encoded include male sterility and herbicide and disease resistance. Therefore, protoplast fusion may be uniquely applied to the transfer of cytoplasmic traits between breeding lines and to create unique cytoplasmic recombinants that incorporate multiple cytoplasmic traits in a single breeding line.

Incorporation of foreign DNA and treatment of protoplasts with chemical or physical mutagens can be used to isolate single genetic changes. As knowledge of plant molecular genetics involving isolation and synthesis of specific genes and manipulation of plasmid vectors accumulates, it will be possible to transfer single genes into plant protoplasts with subsequent plant regeneration.

Haploid Plants. Haploid plants are important in plant breeding because they greatly reduce the time for obtaining homozygous lines following diploidization in allogamous populations. Doubled haploids, when integrated into a conventional breeding program, have resulted in rapid development of new varieties. Diploidized haploid lines can be produced using colchicine treatment or alternatively may be produced using leaf midrib culture. Selection from a population of doubled haploids, particularly an F_2 population, can result in rapid fixation of desired characteristics. Kasha and Reinbergs (1980) have reported that use of doubled haploids have also been proposed for use in tobacco, *Asparagus, Brassica*, rice, corn, and potato.

Embryo Culture. Work accomplished by Zenkteller and colleagues (1975) pertains to the production of interspecific and intergeneric hybrids by pollination and embryo culture *in vitro*. Another application pertains to overcoming self-incompatibility. Self-incompatibility is usually controlled by a physiological barrier that prevents fusion of otherwise fertile gametes. Placental pollination *in vitro* has been especially successful for overcoming self-incompatibility. The entire ovule mass of an ovary that is still intact on the placenta is cultured following pollination to permit embryo development (Yeung et al., 1981).

LITERATURE CITED

Ammirato, P. V. 1977. Hormonal control of somatic embryo development from cultured cells of caraway. Plant Physiol. 59:579–86.
Bacchi, O. 1943. Cytological observations in Citrus. III. Megasporogenesis, fertilization, and polyembryony. Bot. Gaz. 105:221–25.
Bingham, E. T., L. V. Hurley, D. M. Kaatz, and J. W. Saunders. 1975. Breeding alfalfa which regenerates from callus tissue in culture. Crop Sci. 15:719–21.
Bui-Dang-Ha, D., and I. A. Mackenzie. 1973. The division of protoplasts from *Asparagus officinalis* L. and their growth and differentiation. Protoplasma 78:215–21.
Button, J., and C. H. Bornman. 1971. Development of nucellar plants from unpollinated and unfertilized ovules of the Washington navel orange *in vitro*. So. Afr. J. Bot. 37:127–34.
Button, J., J. Kochba, and C. H. Bornman. 1974. Fine structure of an embryoid development from embryogenic ovular callus of "Shamouti" orange (*Citrus sinensis* osb.). J. Exp. Bot. 25:446-57.
Carlson, P. S., H. H. Smith, and R. D. Dearing. 1972. Parasexual interspecific plant hybridization. Proc. Nat. Acad. Sci. USA 69:2292–94.
Dudits, D., K. N. Kao, F. Constabel, and O. L. Gamborg. 1976. Embryogenesis and formation of tetraploid and hexaploid plants from carrot protoplasts. Can. J. Bot. 54:1063–67.
Evans, D. A. 1979. Chromosome stability of plants regenerated from mesophyll protoplasts of *Nicotiana* species. Z. Pflanzenphysiol. 95:459–63.
Evans, D. A. 1982. Plant regeneration and genetic analysis of somatic hybrid plants. *In* E. D. Earle and Y. Demarly (eds.), Plant regeneration and genetic variability, chap. 3. Praeger, New York.
Evans, D. A., and W. R. Sharp. 1983. Single gene mutations in tomato plants regenerated from tissue culture. Science 221:949–51.
Evans, D. A., W. R. Sharp, and C. E. Flick. 1981. Plant regeneration from cell cultures. *In* J. Janick (ed.), Horticultural reviews, 3:214–314. AVI Press, Westport, Conn.
Evans, D. A., W. R. Sharp, and H. P. Medina-Filho. 1984. Somaclonal and gametoclonal variation. American J. Bot. 71:759–74.
Frankenberger, E. A., and E. C. Tigchelaar. 1980. Genetic analysis of shoot formation from excised leaf discs of tomato. In Vitro 16:217.
Gamborg, O. L., and L. R. Wetter. 1975. Plant tissue culture methods. Nat. Res. Council, Saskatoon, Canada.
Heinz, D. J., M. Krishnamurthi, L. G. Nickell, and A. Maretzki. 1977. Cell, tissue, and organ culture in sugarcane. *In* J. Reinert and Y. P. S. Bajaj (eds.), Plant cell tissue and organ culture, pp. 3-17. Springer-Verlag, Berlin.

Jacobsen, E., and S. K. Sopory. 1978. The influence and possible recombination of genotypes on the product of microspore embryoids in anther culture of *Solanum tuberosum* dihaploid hybrids. Theoret. Appl. Genet. 52:119–23.

Jelaska, S. 1974. Embryogenesis and organogenesis in pumpkin explants. Physiol. Plant. 31:257–61.

Kartha, K. K., M. R. Michayluk, K. N. Kao, O. L. Gamborg, and F. Constabel. 1974. Callus formation and plant regeneration from mesophyll protoplasts of rape plants (*Brassica napus* L. cv. Zephyr). Plant Sci. Lett. 3:265–71.

Kasha, K. J., and E. Reinbergs. 1980. Achievements with haploids in barley research and breeding. *In* D. R. Davies, D. A. Hopwood, eds. The plant genome, pp. 215-30. The John Innes Charity, Norwich.

Kasperbauer, M. J., and G. B. Collins. 1972. Reconstitution of diploids from anther derived haploids in tobacco. Crop Sci. 12:98–101.

Kochba, J., P. Spiegel-Roy, Y. Safran. 1972. Adventitious plants from ovules and nucelli in citrus. Planta 106:237–45.

Konar, R. N., and K. Natajara. 1965. Experimental studies in *Ranunculus sceleratus* L. Development of embryos from the stem epidermis. Phytomorphology 15:132–37.

Krul, W. R., and J. F. Worley. 1977. Formation of adventitious embryos in callus cultures of "Seyval", a French hybrid grape. J. Amer. Soc. Hort. Sci. 102:360–63.

Matsuoka, H., and K. Hinata. 1979. NAA-induced organogenesis and embryogenesis in hypocotyl callus of *Solanum melogena* L. J. Exp. Bot. 30:363–70.

Mirtra, G. C., and H. C. Chaturvedi. 1972. Embryoids and complete plants from unpollinated ovaries and from ovules of *in vivo*-grown emasculated flower buds of *Citrus* spp. Bull. Torr. Bot. Club 99:184–89.

Murashige, T. 1974. Plant propagation through tissue cultures. Ann. Rev. Plant Phys. 25:135–66.

Murashige, T. 1978. The impact of tissue culture on agriculture. *In* T. A. Thorpe (ed.), Frontiers of plant tissue culture, pp. 15-26. Univ. Calgary Press, Calgary.

Nadar, H. M., and D. J. Heinz. 1977. Root and shoot development from sugarcane callus tissue. Crop Sci. 17:814–16.

Pence, V. C., P. M. Hasegawa, and J. Janick. 1979. Asexual embryogenesis in *Theobroma cacao* L. J. Amer. Soc. Hort. Sci. 104:145–48.

Price, H. J., and R. H. Smith. 1979. Somatic embryogenesis in suspension cultures of *Gossypium klotzschianum* Anders. Planta 145:305–7.

Raghavan, V. 1976. Role of the generative cell in androgenesis in henbane. Science 191:388–89.

Rangaswamy, N. S. 1958. Culture of nuceller tissue of Citrus in vitro. Experientia 14:111–22.

Reinert, J., M. Tazawa, and S. Semenoff. 1967. Nitrogen compounds as factors of embryogenesis *in vitro*. Nature 216:1215–16.

Reynolds, J. F., and T. Murashige. 1979. Asexual embryogenesis in callus cultures of palms. In Vitro 15:383–87.

Sengupta, C., and V. Raghavan. 1980a. Somatic embryogenesis in carrot cell suspension. I. Pattern of protein and nucleic acid synthesis. J. Exp. Bot. 31:247–58.

Sengupta, C., and V. Raghavan. 1980b. Somatic embryogenesis in carrot cell suspension. II. Synthesis of ribosmal RNA and Poly (A) + RNA. J. Exp. Bot. 31:259–68.

Sharp, W. R., S. M. Reed, and D. A. Evans. 1983. Production and application of haploid plants. *In* P. B. Vose and S. Blixt (eds.), Contemporary Bases for Crop Breeding, chap. 13. Pergamon Press, Elmsford, N.Y.

Shepard, J. F., D. Bidney, and E. Shahin. 1980. Potato protoplasts in crop improvement. Science 208:17–24.

Shepard, J. F., and R. E. Totten. 1977. Mesophyll cell protoplasts of potato. Plant Physiol. 60:313–16.

Sibi, M. 1978. Multiplication conforme, non conforme. Le Selectionneur Francais 26:9–18.

Skoog, F., and C. O. Miller, 1957. Chemical regulation of growth and organ formation in plant tissue cultured in vitro. *In* Symposium of Society of Experimental Biology, vol. 11:118–31. Academic Press, New York.

Sondahl, M. R., D. A. Spahlinger, and W. R. Sharp. 1979. A histological study of high frequency and low frequency induction of somatic embyros in cultured leaf explants of *Coffea arabica* L. Z. Pflanzenphysiol. 94:101–8.

Street, H. E. 1978. Differentiation in cell and tissue cultures-regulation at the molecular level. *In* H. R. Schutte and D. Gross (eds.), Regulation of developmental processes in plants, pp. 192–218. VEB Kongress-und Werbedruck, Oberlungwitz.

Sunderland, N. 1974. Anther culture as a means of haploid production. *In* K. J. Kasha (ed.), Haploids in higher plants: advances and potential, pp. 91-122. Univ. of Guelph, Guelph, Ontario.

Takebe, I., G. Labib, and G. Melchers. 1971. Regeneration of whole plants from isolated mesophyll protoplasts of tobacco. Naturwissen. 58:318–20.

Vasil, I. K., and A. C. Hildebrandt. 1965. Differentiation of tobacco plants from single isolated cells in microcultures. Science 150:889–92.

Vasil, I. K., and A. C. Hildebrandt. 1966. Variation of morphogenetic behavior in plant tissue culture. I. *Cichorium endiva.* Amer. J. Bot. 53:860–69.

Vasil, I. K., and V. Vasil. 1980. Isolation and culture of protoplasts. *In* I. K. Vasil (ed.), International Review of Cytology, 11 B:1-19. Academic Press, New York.

White, D. W. R., and I. K. Vasil. 1979. Use of amino acid analogue-resistant cell lines for selection of *Nicotiana sylvestris* somatic cell hybrids. Theoret. Appl. Genet. 55:107–12.

Williams, L., and H. A. Collin. 1976. Embryogenesis and plantlet formation in tissue cultures of celery. Ann. Bot. 40:325–32.

Yeoman, M. M. 1970. Early development in callus cultures. Int. Rev. Cytol. 29:383–409.

Yeung, E. C., T. A. Thorpe, and C. J. Jensen. 1981. *In vitro* fertilization and embryo culture. *In* T. A. Thorpe (ed.), Plant tissue culture methods and application in agriculture, pp. 253–71. Academic Press, New York.

Zenkteller, M., E. Misiura, and I. Guzowska. 1975. *In* Form, structure, and function in plants, p. 180. B. M. Johri Commemoration Vol., Sarati Prakashan, Meerut.

OTTO J. CROCOMO AND
JOSE BARBOSA CABRAL

Interspecific Hybridization in
Phaseolus: Embryo Culture

7

INTRODUCTION

Approximately 180 species of the genus *Phaseolus* can be found in the hot regions of both hemispheres, with predominance in the Americas (Vieira, 1967; Burkart, 1951). The cultivated species come from two different origins, America and the Far East, and have morphological characteristics according to their origin. The main Far East cultivated species are: *Ph. acutifolius* Jacf., *Ph. calcaratus* Roxb., *Ph. angularis* (Willd) Wight., *Ph. mungo* L., and *Ph. aureus* (*Ph. radiata* L.). The main American species are: *Ph. coccineus* L., *Ph. vulgaris* L., *Ph. lunatus* L., *Ph. acutifolius* Freem., *Ph. specious* HBK., *Ph. candidus* Vell, and *Ph. caracolla* L. The three latter are ornamental species; the others are seed crops and/or forage crops (Burkart, 1952).

In Brazil, beans, as a source of protein, are a staple food. The most-used bean in the southern regions of Brazil is *Ph. vulgaris* (common or snap bean), and in the semi-arid region of the northeast, it is *Vigna unguiculata* (mascassar bean). The latter region contributes about 1/3 of the total national production of beans.

In that region, the intensity of rainfall is low and sometimes takes place in very short periods, contributing to the instability of crop production (Miranda et al., 1981a, b). Production is low also because of the high evaporation of rivers, dams, plants, and the soil itself (Mafra, 1981). To that one must add the physical characteristics of the majority of the soils of that region, which have very low effective depth with sometimes abrupt changes in the texture of the profiles, limiting water availability and the growth of root systems (Mafra, 1981). This means that a better knowledge of the climatic conditions, water availability, and bean genotypes adapted to water stress, high temperatures, salinity tolerance, disease resistance, and crop improvement is necessary to increase productivity in that region. Several varieties of *Phaseolus* species have been shown to suit some of these conditions.

A set of genes called osmoregulatory (*osm*) function in several microorganisms including *Escherichia coli*, *Salmonella*, and *Klebsiella*. When microorganisms are exposed to an increase in osmolarity in the growth media, they respond by an elevation of the osmotic tension in the cell interior. This is accomplished in bacteria by an increase of the internal concentration of proline and in some cases glutamate or gamma-amino butyrate (Measures, 1975). These substances prevent harmful effects of water stress.

Valentine's group in California isolated mutants of *S. typhimurium* that overproduce L-proline (Anderson et al., 1980).In the selection, the L-proline analogue, L-azetidine-2-carboxylate, was used knowing that mutants which excrete L-proline will dilute-out and thus antagonize the analogue. The results corroborate the hypothesis that L-proline plays a central role in osmoregulation in microorganisms. An analogous system may take place in higher plants (Flowers et al., 1977). Proline in that case would play an important role in the mechanism of salt tolerance in higher plants.

Interspecific crosses are valuable tools for modern plant-breeding technologies (Vieira, 1967). They can be utilized to transfer desirable genetic characters from a native species to a commercial one. The interspecific cross between *Ph. acutifolius* and *Ph. coccineus* and of these two with *Ph. vulgaris* will enable the breeder to combine drought resistance together with disease resistance due either to bacteria, fungi, or virus (Coyne, 1964). *Phaseolus coccineus* can be used to transfer common bean mosaic or yellow bean mosaic virus resistance (Coyne, 1964) together with low temperature tolerance (Hubbeling, 1957) to *Ph. vulgaris*. All this allows for the establishment of a bank of germplasm.

Interspecific and intergeneric crosses permit transference of desirable characteristics, but, as a rule, such crosses are not successful. The plants from these difficult hybridizations bear seeds that are unable to germinate under normal conditions as a result of embryo axis abortion. In such crosses, fertilization occurs normally, and the embryo axis develops relatively well; but a number of irregularities set in eventually, the embryo axis dies, and the seed collapses.

It is well known that embryo axes at different stages of development can be separated from the endosperm tissue in which they lie, cultured *in vitro* under aseptic conditions, and give rise to intact and fertile plants. The factors influencing nutrition, growth, and differentiation of embryo axes are several and should be know if growing them outside the environment of the ovule is the aim.

INTERSPECIFIC HYBRIDIZATION IN *PHASEOLUS*

Mendel (1866) was the first to report interspecific hybridization in the genus *Phaseolus*. He also described the hybrid *Ph. vulgaris* (pistillate) × *Ph. coccineus* (staminate) and segregations in F_2 Spontaneous hybrids between these two species were observed (Tscher Mark-Seysenegg, 1942). That cross was studied in detail by Lamprecht (Lamprecht, 1935, 1941, 1948a, 1948b).

There are several varieties of *Ph. vulgaris* that are sensitive to drought when grown in the semi-arid region of northeast Brazil (Magalhaes and Millar, 1978). On the other hand, *Ph. acutifolius*, *Ph. coccineus*, and *Ph. lunatus* have varieties that have drought tolerance. So there is great interest to obtain interspecific hybrids between these 3 species and *Ph. vulgaris* to transfer to the latter the drought-resistance character.

Phaseolus polystachyus (ticket bean), a wild species found in some regions of the United States that may possess resistance to disease and pests, was crossed with *Ph. lunatus* (lima bean). After several trials, 7 hybrid plants were obtained, but none of them produced flowers or seeds. However, all the hybrids expressed completely, or nearly completely, the hypogeal germination habit on the staminate *Ph. polystachyus* parent (Lorz, 1952), That was an attempt to solve the emergence problem in *Ph. lunatus*. The seedlings of the latter show a high mortality as a result of the breaking of the hypocotyl ("neck-breaking") as it elongates and pushes the extremely large cotyledons through an encrusted or impacted soil.

Using parent heterozygotes, *Ph. vulgaris* × *Ph. lunatus* crosses were made that produced 10 hybrid plants, 8 of which were similar to *Ph. vulgaris*. The other two had no similarity to any of the parents and were steriles (Honma and Heeckt, 1959).

Other workers (Al-Yasiri and Coyne, 1964) have shown that growth regulators have influence on the development of pods and the abortion of hybrids in *Ph. vulgaris* × *Ph. acutifolius* crosses. The growth regulators used were naphthalene acetamide and potassium gibberelate in the following treatments: (1) lanolin + 1% naphthalene acetamide; (2) lanolin + 2% potassium gibberelate; (3) lanolin + 1% naphthalene acetamide + 2% potassium gibberelate; (4) lanolin (control). The mixture as a paste was applied to scratches made with a dissecting needle on the base of the flowers immediately after pollination and followed by three or more applications at 5-day intervals. Two weeks later, an increase in pod size was observed as a result of growth regulator applications, the best combination being naphthalene + gibberelate. However, from this time on the growing of the pods in all the treatments occurred mainly in the period 15–20 days after pollination, and was much more intense in the control plants. Naphthalene alone slowed down the rate as well as the delayed pod abscission. Seeds collected from the remaining pods contained a small embryo axis and a seed coat. The excised embryo axes were viable and could be cultured *in vitro.*

Working with 7 *Phaseolus* species (*calcaratus, mungo, angularis, lunatus, coccineus, acutifolius,* and *vulgaris*), crossing them in all possible combinations and using the criteria of pod development, Al-Yasiri and Coyne (1966) classified the interspecific crosses among these species in three types:

1. cross is compatible and yields mature hybrid seeds;

2. cross is partially compatible, but pods collapse in the early stages of development; and

3. cross is incompatible and pods fail to develop.

Compatibility was observed in *Ph. vulgaris* × *Ph. coccineus* crosses. Twelve partial compatible crosses were *Ph. vulgaris* × *Ph. acutifolius*; *Ph. acutifolius* × *Ph. vulgaris*; *Ph. acutifolius* × *Ph. coccineus*; *Ph. coccineus* × *Ph. acutifolius*; Ph. vulgaris × *Ph. mungo*; *Ph. mungo* × *Ph. calcaratus*; *Ph. vulgaris* × *Ph. lunatus*; *Ph. angularis* × *Ph. acutifolius*; *Ph. angularis* × *Ph. vulgaris*; *Ph. angularis* × *Ph. calcaratus*; *Ph. vulgaris* × *Ph. calcaratus*; *Ph. angularis* × *Ph. mungo*; *Ph. mungo* × *Ph. vulgaris*. All eleven other combinations were incompatible. The success in interspecific crosses seems to depend upon the parent varietal lines utilized, the parental heterozygosity, and the environmental conditions at the time of pollination.

The genetic nature of incompatibility in certain varieties of *Ph. vulgaris* and *Ph. coccineus* could be related to the abnormal development of the seedlings (Kedar and Bemis, 1960). *Phaseolus coccineus* var. Scarlet Runner is characterized by hypogeal germination, is completely (or nearly) cross-pollinated, and is a source of disease resistance. On the other hand, *Ph. vulgaris* var. Blue Lake is epigean and is usually self-fertilized. Both species have 2n = 22. Both have been crossed with each other frequently, and an interspecific barrier between the above-cited varieties has been observed. Crosses of two varieties of *Ph. vulgaris* and *Ph. coccineus* respectively produced uniform F_1 populations of abnormal seedlings that were designated as B-dwarfs and T-dwarfs. Through genetic recombination, the F_1 of these two *Ph. vulgaris* varieties when crossed with *Ph. coccineus*, produced a population that segregated into four equal classes: B-dwarfs; T-dwarfs, BT-dwarfs, and Normals. In other work (Bemis and Kedar, 1961), the same authors discussed the inheritance of abnormal seedling development. These results open the possibility to enhance viability in interspecific crosses when offspring of intraspecific crosses of *Ph. vulgaris* varieties are used as one parent.

Fertilization in some interspecific crosses can be increased when obstacles that slow down pollen tube growth are overcome by special treatments. The obstacles may include deposits of callose on pollen tube walls, the presence of inhibiting substances in the style, and a deficiency of growth-promoting substances in the pollen (Ibrahim and Coyne, 1975). Application of White's nutrient solution to *Ph. coccineus* stigma surfaces prior to pollination by *Ph. vulgaris* overcame the obstacles to fertilization. Early seed development occurred, but did not prevent embryo abortion after 3 weeks. This suggests that inhibiting substances translocated from the leaves of *Ph. coccineus* to the pods may have caused late embryo abortion (Ibrahim and Coyne, 1975).

Crosses between *Ph. vulgaris* (Brazilian var. Carioca 1232) were made in an attempt to transfer genetic resistance to common bacterial blight (*Xanthomonas phaseoli*) from the latter to *Ph. vulgaris*. Pods formed, but they dropped two weeks after pollination (Mohan, 1982). On the other hand, the offspring of the cross *Ph. vulgaris* (vars. Carioca and Iguacu) × *Ph. coccineus* had low pollen fertility (30–35%). Backcrosses between that F_1 and commercial varieties of *Ph. vulgaris* had resistance to common bacterial blight and to virus; several morphological differences were also detected.

Attempted interspecific hybridizations between the species *Ph. vulgaris, Ph. coccineus, Ph. acutifolius*, and *Ph. lunatus* were made by Smartt (1970). These interspecific crosses indicated that a close genetic relationship exists between *Ph. vulgaris* and *Ph. coccineus*, that *Ph. acutifolius* is rather less closely related, and that *Ph. lunatus* is well separated from the other species genetically. Viable hybrids of varying fertility were made reciprocally between *Ph. vulgaris* and *Ph. coccineus*.

Smartt's data are matched by those from Kloz's serotaxonomic studies (Kloz et al., 1966). Antiserum against *Ph. vulgaris* seed protein fractions was obtained with *Ph. coccineus*, a less intense reaction with *Ph. acutifolius*, and a weak reaction with *Ph. lunatus.* The pattern of affinities indicated is exactly the same as that given by Smartt's, but not our, hybridization results.

Through his data, Smartt supported Lamprecht's conclusion regarding the influence of the cytoplasm on segregation in the interspecific hybrids. Cytoplasmic constitution affected fertility and viability in the F_2 and backcross progenies of *Ph. vulgaris* (pistillate)× *Ph. coccineus* (staminate) hybrids and also had an effect on viability in the *Ph. vulgaris* (pistillate)× *Ph. acutifolius* (staminate). In the slightly fertile *Ph. vulgaris* (pistillate) × *Ph. coccineus* (staminate) hybrids, the distorted F_2 segregation was due, in Smartt's view, to the abortion of pollen and zygotes with inviable genotypes and genome/plasmon interactions. The differential loss of *Ph. coccineus* genes in sporogenesis and embryogenesis occurring in *Ph. vulgaris* plasmon were considered to be the main reasons for the distorted F_2 segregation (Smartt, 1970).

EMBRYO AXIS CULTURE

The factors that determine the modifications of the processes of development normally committed to the setting of the ovule that lead to embryo abortion are discussed by Raghavan (Raghavan, 1977). These factors include:

1. Embryo axis development. There is an early embryonic stage when developmental anomalies start to set in. Survival is generally erratic, and ovules harbor an undeveloped hybrid that has no more than a clump of sometimes highly vacuolated necrotic cells, resulting in a shrunken aborted seed.

2. Endosperm development. The embryo axis has nutritional dependence on the endosperm. In unsuccessful crosses, the endosperm starts to disintegrate soon after fertilization, and the embryo axis is deprived of the immediate source of food supply. The causes of endosperm malfunction are several and will not be discussed here (see Raghavan 1977).

These conditions can be overcome by culturing the embryo axis *in vitro*. In the genus Phaseolus, Honma (1955) was one of the first to use this technique. Honma cultured "nonviable" embryo axes obtained from developing seeds of a *Ph. vulgaris* × *Ph. acutifolius* cross; he succeeded in obtaining four intact plants that completed their biological cycle.

With the objective to transfer tolerance to low temperature from *Ph. vulgaris*, a hybridization program between both species was developed in the Netherlands (Braak and Kooistra, 1975). Hybrid plants were obtained only when *Ph. vulgaris* was used as the pistillate parent. The resultant seeds were harvested in the early developmental stages, followed by embryo axis culture. Several parameters were measured and compared with the behavior of *Ph. vulgaris* embryo axes in culture. One of the parameters was embryo length. The percentage of transplantable seedlings increased with the increase of embryo axis length, but this percentage remained appreciably lower in the hybrid embryos than in the *Ph. vulgaris* embryos. In the hybrids, the maximum attainable axis length is restricted by early seed abortion, which allows only a moderate percentage of transplantable seedlings to be expected. Actually about 10% of the seedlings could be transplanted, this figure being considered satisfactory (Braak and Kooistra, 1975), but only three of them developed into adult plants.

In using techniques *in vitro*, care should be exercised in the excision of the embryo of the axis itself and in the choice of nutrient medium.

Excision of embryo axis. Aseptic procedures are to be followed in the excision of embryo axes from either mature to immature seeds. The excised embryo axes should then be transferred to the sterile nutrient medium. Hard-coated seeds should be surface-sterilized and soaked in water for a few hours to facilitate removal of the embryo axes. Sometimes after soaking, the seeds may need to go through a second sterilization. In all cases, with mature or immature seeds (removed from the developing pod) the embryo axes should be excised intact without damage; with small embryo axes, operations should be done under a dissecting microscope.

Nutrient medium. Selection of the medium to sustain continued growth of embryo axes *in vitro* is of utmost importance. The nutrient solution varies with plant species, and the younger the embryo axis is, the more complex it should be. Embryo axes from mature seeds can be grown in inorganic salt medium supplemented with sucrose. However, embryo axes from the early stages of seed development in the pods require also vitamins, growth regulators, amino acids, and sometimes natural endosperm extracts such as coconut milk. Mannitol as an osmoticum is also recommended to be added to the medium to simulate the osmotic pressure of the ovular sap in which the embryo axis was embedded.

Immature embryos from several *Phaseolus* interspecific crosses were cultured *in vitro* in a medium containing Murashige & Skoog (M&S) mineral salt, sucrose, and vitamins. The effects of glutamine and gibberellin also were examined (Mok et al., 1978). Glutamine was effective in increasing the survival rate, but gibberellin had no apparent effect. Plants derived from *Ph. vulgaris* × *Ph. lunatus*, *Ph. vulgaris* ×*Ph. acutifolius*, and *Ph. acutifolius*× *Ph. vulgaris* crosses were obtained. An influence of the genotypes of both parents on the growth rate and final size of these hybrid embryos was observed.

In our laboratory, defined media were developed to culture embryo axes excised from mature or immature common bean (*Ph. vulgaris*) seeds (Crocomo et al., 1979). Mature seeds are soaked in sterile distilled water and the embryo axis excised. These are surface sterilized using sodium hypochlorite (10%), followed by three washings in sterile deionized water, and germinated on solid agar (0.8%) medium. The basal medium contains M&S mineral salts and vitamins. Growth regulators are added in three different ratios, giving three different media: E, E_2, and E_4, with the following constitutions. (1) Medium E = basal medium plus kinetin 1mg/l; indole acetic acid (IAA) 5mg/l; (2) medium E_2 = kinetin 0.25mg/l; IAA 1.3mg/l; (3) medium E_4 = kinetin 0.25mg/l; naphthaleneacetic acid (NAA) 0.1mg/l; 2, 4-dichlorophenoxyacetic acid (2, 4-D) 1, 5mg/l; IAA 1mg/l.

Medium E promotes only geotropic root differentiation. Only callus proliferation occurs in embryo axes cultured on medium E_4, which has four growth regulators. Embryo axes cultured on medium E_2 develop into complete plantlets with shoots and roots. These can be subsequently transplanted to an aqueous nutrient solution, maintained in a growth chamber (under controlled conditions), and finally transferred to soil.

By these results, we show that development of embryo *Ph. vulgaris* axes in culture can be controlled by the manipulation of the kinds and concentrations of growth regulators, attaining either callus proliferation, root morphogenesis, or intact and fertile plants.

Embryo axes cultured *in vitro* can be used to study the host-parasite relationship, as was done by Padmanabhan (1967), who used embryos from *Ph. vulgaris* seeds. The embryos were cultured in Nitsch's medium enriched with vitamins. The incorporation of the fungal toxin, fusaric acid, from *Fusarium*, into the culture medium was demonstrated to interfere with water entry into germinated embryos of *Ph. vulgaris* and induced characteristic wilting of embryonic leaves. The sensitivity of the embryos to the toxin permits better comprehension of the biochemical action of the toxin in the adult plant.

Symbiosis of *Ph. vulgaris* seedlings, obtained from excised embryo axes cultured *in vitro* (medium E_2) inoculated with *Rhizobium phaseoli* has been reported (Gallo et al, 1978). The nodules formed in these seedlings *in vitro* were characterized by a high nitrogenase activity as measured by the acetylene reduction technique.

BIOCHEMICAL STUDIES ON *PHASEOLUS* EMBRYO AXES *IN VITRO*

Isoenzyme patterns have been used to characterize bean cultivars and varieties (Cury, 1980). Isoenzymes of esterase, leucil aminopeptidase (LAP), malic dehydrogenase (MDH), peroxidase, and acid phosphate are among those analyzed at our laboratory.

Recently we have obtained biochemical data pertaining to the degradation of storage protein and isoenzyme activities during *Ph. vulgaris* embryo axis development in culture (Crocomo et al., 1981). The following parameters were examined: growth curve, soluble proteins, storage proteins, and the isoenzymes esterase, leucil aminopeptidase, malic dehydrogenase, and peroxidase. After a lag phase of three days in the three media, media E and E_2, referred to above, had a stationary phase at 20 days while the roots (medium E_4) continued to grow exponentially. The soluble proteins decreased in cells and tissue of embryos cultured in the three media.

To follow storage protein degradation during the development of the embryo axes in culture in the several media, polyacrylamide gel electrophoresis (PAGE) technique was used. The main band (Rm 0.12) of the storage proteins was totally utilized in 24 hours in the three media. Other bands were totally or partially utilized depending on the type of tissue developed according to the medium; other bands were not utilized at all, and others appeared after 48 hours.

As far as isoenzymes are concerned, use of the PAGE pattern showed that some isoenzymes (e.g., esterases, Rms 0.44 and 0.48) remained active during development of the embryo axis in culture, whereas others were synthesized and others degraded. Esterase Rm 0.64 remained active for 48 hours in embryos developed in medium E_4. Activity of MDH Rm 0.15 was detected only after 5 days in medium E_4 and 5 days later was inactive, while MDH Rm 0.58 was active during the whole period of the experiment, have only quantitative differences among cell and tissue in the three media. The LAP Rm 0.92 activity remained the same during embryo development in either medium. LAP Rms 0.57 and 0.70 had qualitative and quantitative differences in each medium.

The excised embryo did not have any activity of peroxidase isoenzymes, which was detected only after 72 hours in culture, the more active being band Rm 0.32, which had a quantitative difference depending on the medium. Bands Rm 0.41 and Rm 0.44 were detected in tissue growing in medium E_4. Tissue in medium E could be characterized by band Rm 0.59.

REFERENCES

Al-Yasiri, S., and D. P. Coyne. 1964. Effect of growth regulators in delaying pod abscission and embryo abortion within the interspecific cross *Phaseolus vulgaris* × *Phaseolus acutifolius*. Crop Sci. 4:422–35.

Al-Yasiri, S., and D. P. Coyne. 1966. Interspecific hybridization in the genus *Phaseolus*. Crop Sci. 6:59–60.

Anderson, K., K. T. Shanmueam., S. T. Lim., L. N. Csonka, R. Tait, H. Hennecke, D. B. Scott., S. S. M. Hom, J. F. Haury, A. Valentine, and R. C. Valentine. 1980. Genetic engineering in agriculture with emphasis on nitrogen fixation. Trends in Biochem. Sci. 5:35–39.

Bemis, W. P., and N. Kedar. 1961. Inheritance of morphological abnormalities in seedlings of two species of *Phaseolus*. J. Hered. 52:171–78.

Braak, J. P., and E. Kooistra. 1975. A successful cross between *Phaseolus vulgaris* and *Phaseolus ritensis* Jones with aid of embryo culture. Euphytica 24:669–79.

Burkart, A. 1952. Las leguminosas argentinas silvestres y cultivadas. B. Aires Acme Agency. 2d ed. 569 pp.

Coyne, D. P. 1964. Species hybridization in *Phaseolus*. J. Hered. 55:5–6.

Crocomo, O. J., L. A. Gallo, G. S. Tonin, and N. Sacchi. 1979. Developmental control of *Phaseolus vulgaris* using embryo axis culture. Energ. Nucl. Agic. Piraciciaba-SP. 1:55–58.

Crocomo, O. J., J. A. Cury, and N. Sacchi. 1981. Degradation of storage protein and synthesis of isoenzymes during *Phaseolus* embryo development in culture. Abhdlg. Akad. Wiss. DDR, Abt. Math., Naturwiss, Tech. n. 5n., pp 247–48.

Cury, J. A. 1980. Contribuicao ao estudo de isoenzimas em feijao (*Phaseolus vulgaris* L.), Ph.D. Thesis, Piracicaba-SP, Brazil.

Flowers, T. J., P. F. Troke, and A. R. Yeo. 1977. The mechanism of salt tolerance in halophytes. Ann. Rev. Plant. Physiol., 28:89–121.

Gallo, L. A., A. P. Ruschel, and O. J. Crocomo. 1978. Nitrogenase activity of nodulated plants obtained from embryos axis culture. Rev. Bras. Bot. 1:131–32.

Hubbeling, N. 1957. New aspects of breeding for disease resistance in beans (*Phaseolus vulgaris* L.). Euphytica 6:111–41.

Honma, S., and O. Heeckt. 1959. Interspecific hybrid between *Phaseolus vulgaris* and *Ph. lunatus*. J. Hered. 50:233–37.

Honma, S. 1955. A technique for artificial culturing of bean embryos. Proc. Am. Soc. Hort. Sci. 65:405–6.

Ibrahim, A. M., and D. P. Coyne. 1975. Genetics of stigma shape, cotyledon position, and flower color in reciprocal crosses between *Phaseolus vulgaris* L. and *Phaseolus coccineus* (Lam.) and implications in breeding. J. Amer. Soc. Hort. Sci. 100:622–26.

Kedar, N., and W. P. Bemis. 1960. Hybridization between two species of *Phaseolus* separated by physiological and morphological blocks. Proc. Amer. Soc. Hort. Sci. 76:397–402.

Kloz, J., E. Klozova, and V. Turkova. 1966. Shemotaxonomy and genesis of protein characters with special reference to the genus *Phaseolus*. *Preslia* (Praha) 38:229–36.

Lamprecht, H. 1935. Ron och synpunkter vid foradling av koksvaseter. Beretin. Nordisk. Jord. br. Forsker Foren 5th kongr. Kobenhavn. July 1935. 4-7 Hefte: 538–46.

Lamprecht, H. 1941. The limit between *Phaseolus vulgaris* and *Ph. multiflorus* from the genetical point of view. 7th Int. Genet. Congr. (Edinburgh, 1939), pp. 179–80.

Lamprecht, H. 1948a. The genetic basis of evolution. Agric. Hort. Genet. Landskrona 6:83–86.

Lamprecht. H. 1948b. Zurlosung des artsproblems neue und bisher bekannte ergebnisse der kreuzung *Phaseolus vulgaris* × *Ph. coccineus* L. und reziprok. Agri. Hort. Genet. Landskrona 6:87–145.

Lorz, A. P. 1952. An interspecific cross involving the lima bean *Phaseolus lunatus* L. Science 115:702–3.

Mafra, R. C. 1981. Agriculture de sequeiro no tropico semirido. Secret. Agr. do estado de pernambucana de pesq. agrop. - IPA, Recife-PE, Brazil 56 pp.

Magalhaes, A. A., and A. A. Millar. 1978. Efeito do deficit de agua no perido repordutivo sobre a prodacao do feijao. Pesq. Agrop. Brasileira 13:55–60.

Measures, J. C. 1975. Role of amino acids in osmoregulation of non-halophilic bacteria. Nature 257:398–400.

Mendel, G. 1866. Versuche uber pflanzenhybriden. Verhandlungen des natur-forschenden vereins in Brunn. 4:3–47. (English translation. *In* Experiments in plant hybridization, Oliver and Boyd, Edinburgh, 1965. 95 pp.)

Miranda, P., P. R. F. De Britto, and J. B. Cabral. 1981. Deficiencia hidrica na cultura do feijoeiro comum (*Phaseolus vulgaris* L.). *In* Empresa Pernambucana De Pesquisa Agropecuaria-IPA Projeto Feijao; relatorio anual de Pesquisa. Recife-PE, Brazil, 94 pp.

Miranda, P., P. R. F. De Britto, J. B. Cabral, and M. L. Pimmentel. 1981. Deficiencia hidrica na cultura do feijoeiro macassar (*Gigna uniquiculata* L. Walp.). *In* Empresa Pernambucana de Pesquisa Agropecuaria-IPA. Projeto Feijao; relatorio anual de Pesquisa, Recife-PE, Brazil. 94pp.

Mohan, S. T. 1982. Estudos preliminares de cruzamentos interspecificos em *Phaseolus*. *In* I Renuiao Nacional de Pesquisa de Feijao (Embrapa). 361pp. Goiancia-Go-Brazil.

Mok, D. W. S., M. C. Mok, and A. Rabakoarihanta. 1978. Interspecific hybridization of *Phaseolus vulgaris* with *Ph. lunatus* and *Ph. acutifolius*. Theor. Appl. Genet. 523:209–15.

Padmanabhan, D. 1967. Effect of fusaric acid on *in vitro* culture of embryos of *Phaseolus vulgaris* L. Current Sci. 36:214–15.

Raghavan, V. 1977. Applied aspects of embryo culture. *In* J. Reinert and Y. P. S. Bajaj (eds.), Plant cell, tissue, and organ culture, pp. 375–97. Springer-Verlag, Berlin.

Smartt, J. 1970. Interspecific hybridization between cultivated American species of the genus *Phaseolus*. Euphytica 19:480–89.

Thomas, C. V., R. M. Manshardt, and J. G. Waines. 1983. Tepaines as source of useful traits for improving common bean. Desert Plants 5:43.

Thomas, C. V. and J. G. Waines. 1984. Fertile backcross and allotetraploid plants from crosses between tepany beans and common beans J. Hered. 75:93–98.

Tschermark-Seysenegg, E. 1942. Uber bastarde zwischen Fisole (*Phaseolus vulgaris* L.) und Feuerbone (*Phaseolus multiflorus* Lam.) und ihre eventuelle praktische verwertbarkeit. Zuchter 14:153–64.

Vieira, C. 1967. O feijoeiro-comm: cultura, doencas e melhormento. Vicosa-MG, Brazil. 220pp.

JULES JANICK

Embryogenics: The Technology of Obtaining Useful Products from the Culture of Asexual Embryos

8

INTRODUCTION

Traditional agriculture is based upon field cultivation of selected plant species for the production of specific organs, such as fruits, seeds, or tubers, that accumulate useful metabolites. This technology might be circumvented *in vitro* if these metabolites could be produced economically. The requirements for such a system involve a combination of two processes: (1) proliferation and growth of a suitable tissue or organ, and (2) efficient synthesis of an economically desirable product. The product should be of high economic value because growth of plant tissue *in vitro* requires a carbon source for energy and specialized growth systems are essential (Goldstein et al., 1980).

Cell cultures have been suggested as a means to synthesize desirable metabolites, but in most cases they either do so only at much reduced rates or not at all (Nickell, 1980). The regulatory mechanisms responsible for this repression are not understood, and as a result the use of cell cultures for the production of plant products *in vitro* has been severely limited (Jones, 1974). The culture of organs is an alternative system for the synthesis *in vitro* of desirable metabolites. In particular, the culture of asexual embryos offers the possibility for production *in vitro* of valuable products normally produced in cotyledons.

Asexual embryogenesis, the process of embryo initiation and development from cells that are not the direct products of gametic fusion, is a natural phenomenon in many species (Raghavan, 1976; Tisserat et al., 1979). Leeuwenhoek described multiple embryos in orange seed in 1719, and the nucellar origin of polyembryony was documented by Strasburger in 1878. Natural asexual embryony *in vivo* is now known to be a widespread phenomenon associated with cells and/or tissues of the nucellus and integuments, megagametophyte, embryo, and endosperm. Although a number of researchers had evidence for asexual embryogenesis in cell and callus cultures of carrot and datura in the 1950s, confirmation of this phenomenon was made independently by Reinert (1958) and Stewart, Mapes, and Smith (1958). Subsequent studies with many species suggest that asexual embryogenesis in cultured cells and tissues may be a universal trait whose occurrence depends on the interaction of an appropriate tissue with an appropriate induction stimulus.

The intriguing nature of asexual embryogenesis has engendered a growing research interest. Although the regeneration potential of this phenomenon was immediately recognized, new exploitable benefits were predicted as literature accumulated. The applied potential of asexual embryogenesis is summarized in table 1.

In this paper the uses of asexual embryos for the synthesis of metabolites will be emphasized. The production *in vitro* of lipids from *Theobroma cacao* L. will be considered as a test system.

TABLE 1

AGRICULTURAL USES OF ASEXUAL EMBRYOGENESIS *IN VITRO*

1. Regeneration
   a. development of seedlings via precocious germination
   b. synthesis of artificial seed via encapsulation
2. Germplasm Preservation via Cryogenic storage
   a. embryogenic cells or tissues
   b. asexual embryos
3. Direct Synthesis of Metabolites

SEED METABOLITES OF *THEOBROMA CACAO*

Mature zygotic seeds of *T. cacao* represent the raw material for the production of two important products: the expressible lipids referred to as cocoa butter, and the remaining component collectively referred to as cocoa or chocolate solids (Zoumas and Finnegan, 1979). The seed coat (shell) has little value and is considered a contaminant. Although the entire zygotic embryo is used, the cotyledon and not the embryonic axis represents the product of value.

Cocoa butter although a natural product is well defined chemically (Chacko and Perkins, 1964; Coleman, 1961; Meara, 1949; Scholfield and Dutton, 1959; Vander Wal, 1960; Youngs, 1961). It represents storage lipids, predominantly triglycerides that exist primarily as the triacylglycerols of palmitic, stearic, and oleic acids that normally accumulate at the later stages of embryo maturation. The melting characteristics of cocoa butter make it the preferred seed fat for the manufacture of chocolate. At room temperature (20°C) cocoa butter is a solid that starts to soften at 30°C, melting completely at slightly below body temperature. The melting characteristics of cocoa butter are due to its unique glyceride composition, consisting of 2.7 to 12.5% trisaturated, 67.5 to 81.3% monounsaturated, 15.3 to 27.0% diunsaturated, and 0.7 to 2.1% triunsaturated triglycerides. The glyceride fatty acid composition of cocoa butter consists of 24.4 to 28.6% palmitic, 34.2 to 36.2% stearic, 33.4 to 38.1% oleic, and 1.8 to 3.6% linoleic acid. Cocoa butter is more valuable than cocoa solids because it has multiple uses, and more butter than solids is used in the manufacture of chocolate, leading to a surplus of cocoa solids. Because cocoa lipids represent the more chemically defined as well as more valuable product, it was monitored as the yield product in the culture of asexual embryos.

Attempts to culture cotyledonary cells of *T. cacao in vitro* for the production of cocoa butter was attempted by Tsai and Kinsella (1981 a,b). A patent (Cocoa bean cell culture, 4, 306,022) was obtained for this system. Although cultured cells produced fatty acids associated with cocoa butter their amount and proportion did not resemble those found in the storage lipids of mature seed (table 2).

Asexual embryogenesis in cacao was first reported by E. B. Esan (1977) at the Vth International Cocoa Research Conference at Ibadan, Nigeria in September 1975 but the proceedings were not published until 1977. The phenomenon was reported independently by Pence, Hasegawa, and Janick in 1979. The concept that asexual embryos could be used in a "nonagricultural" method to produce cacao cotyledons for useful products was filed in a U.S. patent on 13 October 1978 by Jules Janick and Valerie C. Pence and awarded on 27 May 1980 (4, 204, 366). Subsequent patents were awarded on the tissue itself (4, 301, 619) and on the proliferation and development of embryos (4, 291, 498). An additional patent was issued on 8 October 1985 for the production of asexual embryos by embryogenic callus (4,545,147).

There are two independent processes involved in the production *in vitro* of cocoa lipids (Janick et al., 1982):

1. the initiation and proliferation of asexual embryos; and
2. the development and maturation of asexual embryos *in vitro*.

TABLE 2

FATTY ACID COMPOSITION OF ORGANS AND TISSUES OF CACAO

SOURCE	FATTY ACID DISTRIBUTION (MOLE %)				REFERENCE
	Palmitic 16:0	Stearic 18:0	Oleic 18:1	Linoleic 18:2	
Cocoa butter	27.7	31.3	36.8	3.8	Janick et al., 1982
Immature embryo (pink/white)	32.5	8.7	25.4	31.0	Janick et al., 1982
Cotyledonary callus (7 wk)	29.3	5.2	22.3	34.0	Tsai & Kinsella, 1981b
Cotyledonary cell suspension (14 days)	28.3	2.4	11.5	48.6	Tsai & Kinsella, 1981b
Asexual embryo (46 days)*	28.8	28.6	32.2	9.4	Kononowicz & Janick, 1984b

* High sucrose protocol.

INITIATION AND PROLIFERATION OF
ASEXUAL EMBRYOS OF *THEOBROMA CACAO*

Asexual embryos of cacao occurs from tissues of immature zygotic embryos ranging in size from 2.5 to 18 mm (white to white/pink), but maximum frequency occurs from white transluscent embryos about 6.5–7.0 mm (Pence et al. 1979, 1980). This is equivalent to 100–120 days after pollination (Cheeseman, 1927). Asexual embryogenesis occurs in a low frequency in a basal medium containing Murashige and Skoog (1962) salts and the following (in mg/liter): casein hydrolysate, 1000; i-inositol, 100; glycine, 2; pyridoxine·HCl, 0.5; nicotinic acid, 0.5; and thiamine·HCl, 0.1) without growth regulators, but frequency increases with auxin (NAA, IAA, or 2,4-D) and coconut water (10%). There is evidence that coconut water and auxin may have a synergistic effect increasing embryogenesis (table 3).

TABLE 3

INTERACTION OF AUXIN AND COCONUT WATER ON THE
INDUCTION OF ASEXUAL EMBRYOGENESIS IN CACAO
FROM SPADE-SHAPED ZYGOTIC EMBRYOS

Auxin	CW	Embryogenic Frequency (%)
0	0	4
0	+†	12
+*	0	8
+	+	63

SOURCE: Pence et al., 1980.
* + = 8 μM NAA, 2, 4D and 80 μM 1AA.
† + = 10% CW.

Asexual embryos arise from zygotic embryos through 3 distinct processes:

1. A "non-budding" process whereby structures resembling embryos or parts of embryos differentiated from internal meristematic cotyledonary tissue (Pence et al., 1980).

2. A budding process in which asexual embryos pass through the normal stages of embryogenesis beginning with the initiation of a globular structure on a suspension-like stalk that arises from epidermal tissue on cotyledonary veins or the embryonic axis (Pence et al., 1980) as shown in figure 1.

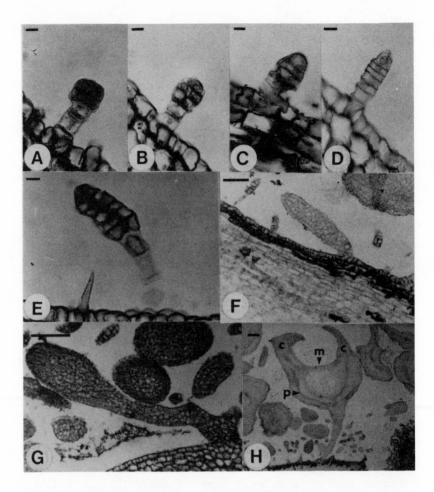

Fig.1. Photomicrographs illustrating the initiation of "budding" type asexual embryos. (A) and (B) Hair-like structure with a 4- to 10-celled globular body on a 3- or 4-celled stalk (bar = 0.01 mm). (C) and (D) Hair-like structures formed by transverse divisions only (bar = 0.01 mm). (E), (F), and (G) Developing asexual embryos still attached to the stalk with the upper portion becoming multicellular (E, bar = 0.01 mm; F and G, bar = 0.1 mm). (H) Developing a characteristic embryo morphology. Cotyledons (c) and apical meristem (m) are clearly discernible along with the procambial strands (p) (bar = 0.1 mm). (SOURCE: Pence et al., 1980)

3. A process whereby embryos develop from embryogenic callus (Kononowicz et al., 1984) as shown in figure 2.

Development of Budding Centers

Asexual embryos of some clones continue to proliferate in the absence of growth regulators by budding and forming continuously propagating cultures referred to as "budding centers" (BC). Of 50 clones originally established, 2 clones (BC 21 and 46) continued to proliferate after 5 years through hypocotylary budding. In addition, some budding centers spontaneously produce embryogenic callus which initiates asexual embryos in low frequency (BC 5, 36, and 46).

Embryogenic Callus

Embryogenic callus of BC 5, 36, and 46 is pale yellow and friable. An associated white callus which turns brown is nonembryogenic (Kononowicz et al., 1984).

The initiation of embryos in embryogenically-competent callus occurs at low frequency in a hormone-free basal medium. High concentrations of 2,4-D plus coconut water stimulate callus production and supress embryo initiation with maximum frequency (percentage of cultures with embryos) and maximum intensity (number of embryos per embryogenic culture) occurring at 10^{-3} to 10^{-2} mg/liter 2,4-D. Embryos originate from meristematic tissue at the periphery of callus clumps (fig. 2A) that are associated with nodular growth (fig. 2B) arising from small cells with dense cytoplasm, centrally located nucleus, and prominent nucleoli. Asexual embryos during development either remain embedded in the callus (fig. 2C,E) or are connected through suspensor-like structures of varying morphology (fig. 2D,F).

The response of embryogenic callus to gibberellic acid (GA) is complex and clone specific (Kononowicz and Janick, 1984c). For example, GA stimulated embryogenesis in BC 5 but not 36. The GA antagonists butanedioic mono-(2,2-dimethylhydrazide) (daminozide) and (2-chloroethyl) trimethylammonium chloride (CCC) depressed embryogenesis in both clones but 2'-isopropyl-4'-dimethylamino-5'-methylphenyl-1'-peperdine carboxylate methyl chloride (AMO 1618) stimulated embryogenesis with BC 5 at very low concentrations (0.1 mg liter $^{-1}$). Under certain circumstances embryogenic callus completely reverts to embryos, but the definitive stimulus of this "high frequency" embryogenesis is unresolved. Embryogenesis from callus is stimulated when sucrose is replaced by glucose or fructose (Elhag, Whipkey, and Janick, in press).

Embryogenic callus retains embryogenic competence in liquid culture.

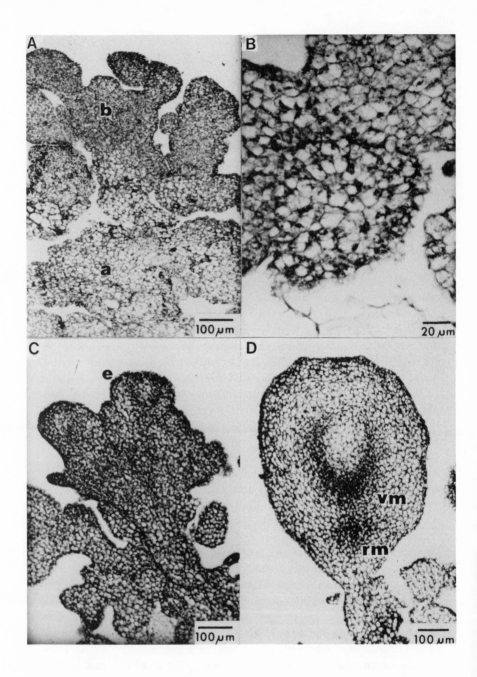

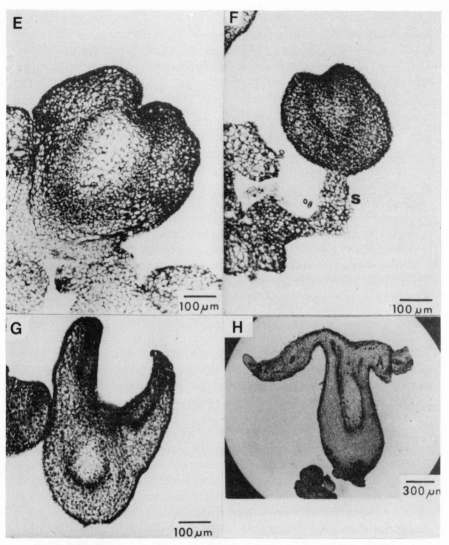

Fig. 2. Development of asexual embryos of cacao via callus. (A) Cross section of embryogenic-competent callus derived from budding centers. The older portion is in the center (a) with new nodular growth (b) on the periphery (b). (B) A close-up of small compact spherical nodules in the new callus. These compact nodules are loosely attached to one another and give the callus a friable texture. These nodular bodies resemble proembryos. (C) An older nodule showing an early stage of embryo (e) formation. (D) Globular embryo showing the formation of root meristem (rm) and vascular meristem (vm). (E) Heart-shaped embryo loosely connected to the callus mass. (F) Heart-shaped embryo with a suspensor-like connection to the callus mass. (G) Torpedo stage. (H) "Walking stick" embryo stage showing cotyledonary development. (SOURCE: Kononowicz et al., 1984)

Embryogenic callus, unlike nonembryogenic callus, remains in clumps and does not form cell suspensions. Embryogenic callus subcultured in liquid medium for one year spontaneously reverted to embryos in one instance.

Clonal Variation

There is evidence of clonal variation in budding centers and callus. Normal asexual embryos are transluscent and pale even in low light. Some developing embryos turn dark green and opaque under low light. Dark green embryos do no develop normally and do not accumulate lipids when presented with the appropriate signal (high sucrose). In some embryos the "dark green" effect is mosaic with a sector of cotyledons in a dark green nondevelopmental phase. It may be that the dark green phenotype is associated with a transition of proplastids from a metabolic phase involved in lipid synthesis to a photosynthetic phase associated with germination.

Morphological variants (abnormal thickened cotyledons) have been observed in BC 36. Another variant in budding centers is a shift to high embryogenic capacity linked to a developmental block at the globular stage.

Synchrony

The successful commercialization of the system requires a combination of rapid proliferation of asexual embryos, preferably in a liquid system, and synchronous development of embryo initiation and development. Synchronization is particularly important because embryo receptivity, the ability of embryos to develop normally, is a sensitive phenomenon. There appears to be a narrow window in the developmental pathway in which embryos have maximum responsivity. Embryos transferred in the globular stage do not develop cotyledons and cannot accumulate lipids, whereas embryos transferred too late adapt poorly to the accumulation phase of development. Embryo stage has a critical effect on subsequent development. Thus, translucent, but not opaque, cotyledonary embryos respond to the sucrose stimulus. Further, embryo development can be suppressed by stress such as transfer to fresh solid media before transfer to liquid systems.

EMBRYO DEVELOPMENT

In Vivo

Embryo development *in vivo* is characterized by a series of morphological and physiological changes. The earliest stage is marked by quiescence of the zygote with rapid endosperm development. The first embryo development occurs 40–50 days after fertilization (Cheeseman, 1927). The second 50–70 days involves rapid growth and development of the embryo as it passes from a globular body to cotyledonary formation. The final 50–60 days is marked by cotyledonary growth characterized by increase in size, change in color, accumulation of storage lipids, and increase in alkaloids. During this latter stage, there is an associated change from heterotrophic to autotrophic growth and a loss of embryo dormancy culminating in rapid germination at maturity.

Cacao embryo development based on dry weight accumulation can be divided into three phases: an initial lag phase (0–100 days), an accumulation phase (100–160 days), and a maturation phase (160–180 days). During the last two phases, the rate of lipid, alkaloid, protein, and anthocyanin accumulation is constant with respect to dry weight (Wright et al., 1982) as shown in figure 3. The seed maturation phase is characterized by a decrease in overall dry weight accumulation and not by a quantitative change in dry weight partitioning. This suggests that there is a single program in the accumulation of embryo constituents. Once the program begins, the rate of accumulation is constant in respect to increases in dry weight. This is true of each fatty acid. Different constant rates of accumulation of each fatty acid (fig. 4A) account for the observed changes in mole percentage ratios observed (fig. 4B).

In Vitro

Embryos cultured in vitro can be shifted to the accumulation phase by the addition of high concentrations of sucrose to the basal medium (Pence et al., 1981a,b). The most efficient protocol (Pence et al., 1981a) has been to place spade-shaped embryos (equivalent to 100–120 day old post-pollination) in liquid basal medium containing 3% sucrose with culture tubes placed in a Rollodrum apparatus rotating at 5 rpm for 10 days and then shift to basal medium with elevated sucrose, i.e., 2 days each with 9, 15, and 21% sucrose and 30 days with 27% sucrose, for a total of 46 days in liquid medium. Growth of embryos in liquid is maximal with 3% sucrose. The high sucrose protocol after day 10 induces the suppression of growth and the simultaneous accumulation of anthocyanins (Pence et al., 1981a), lipids (Pence et al., 1981b), and alkaloids (Paiva and Janick, 1983) and proteins mimicking normal development *in vivo*. Zygotic and asexual embryos respond similarly to this protocol (Kononowicz and Janick, 1984a). Fatty acid and triglyceride accumulation of asexual and zygotic embryos *in vitro*

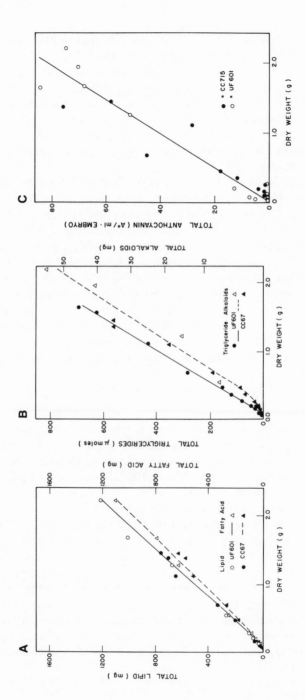

Fig. 3. Changes in metabolites as dry weight increases in zygotic embryos of cacao developing *in vivo*. (A) Lipids and fatty acids. (B) Triglycerides and alkaloids. (C) Anthocyanins. (SOURCE: Wright et al., 1982)

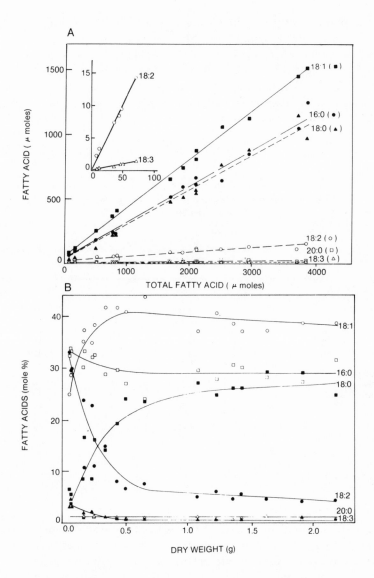

Fig. 4. Changes in fatty acids in zygotic embryos of cacao developing *in vivo*. (A) Relative rate of accumulation of fatty acids as total fatty acids increase in developing embryos. (B) Fatty acid content (mole %) vs. embryo dry weight. (SOURCE: Wright et al., 1982)

and zygotic embryos *in vivo* follow a similar pattern of development. This can be demonstrated by comparing embryos cultured *in vivo* and *in vitro* in respect to the mole percentage of each fatty acid (table 4) or triglyceride (table 5), or by the rate of accumulation of fatty acids as total fatty acids increase (fig. 5). Based on the accumulation pattern, the quality of lipids from asexual embryos should closely resemble commercial cocoa butter when the fatty acid content on a gram dry weight basis approaches 50%.

TABLE 4

EMBRYO GROWTH AND LIPID PRODUCTION OF ASEXUAL EMBRYOS
OF CACAO CULTURED IN LOW AND HIGH SUCROSE COMPARED WITH
ZYGOTIC EMBRYOS *IN VIVO*

VARIABLE	ASEXUAL EMBRYO SUCROSE PROTOCOL		ZYGOTIC EMBRYO
	3%	3→27%*	*IN VIVO*
Embryo growth			
Fresh wt (g)	2.5	1.0	1.5
Dry wt (g).....................	1.0	0.4	1.0
FW/DW (%)	8.0	40.0	68.0
Lipid production			
Lipid/DW (%)	4.1	26.0	48.0
Fatty acid/DW (%)	1.2	24.0	45.0
FA/L (%)	29.0	92.0	93.0
FA composition (mol %)			
Palmitic 16:0	34.0	26.0	29.0
Stearic 18:0	11.0	29.0	33.0
Oleic 18:1	5.0	34.0	34.0
P+S+O.........................	50.0	89.0	96.0
Linoleic 18:2	40.0	9.0	4.0
Linolenic 18:3	11.0	>1.0	>1.0

SOURCE: Pence et al., 1980; Janick et al., 1982; Kononowicz and Janick, 1984a.
*Filter-sterilized sucrose.

An analysis of lipid quality of embryos with a fatty acid content of about 30% by differential scanning calorimetry indicated that cocoa butter–like material was produced by culture *in vitro* of embryos, but some low melting components remained indicating that lipid quality was not identical to cocoa butter. Fatty acid quality equivalent to cocoa butter cannot be expected until fatty acid in asexual embryos reaches at least 50% on a dry weight basis.

TABLE 5

TRIGLYCERIDES OF ASEXUAL EMBRYOS OF CACOA CULTURED
IN HIGH SUCROSE PROTOCOL COMPARED TO COMMERCIAL
COCOA BUTTER

TRIGLYCERIDE	CONTENT (MOL %)	
	Asexual Embryos* (3→27% sucrose)	COCOA BUTTER
Monounsaturated		
DOS	33.3	36.8
POP	23.9	18.8
SOS.....................	17.9	22.6
Total	75.1	78.2
Diunsaturated		
SOO	6.5	8.9
POO	8.6	8.9
Total	15.1	17.8
Polyunsaturated		
PLP.....................	6.2	2.2
PLO	3.5	1.8
Total	9.7	4.0

* 25.3% fatty acids using filter-sterilized sucrose (Kononowicz and Janick, 1984b).

The sucrose effect on the accumulation phase of embryo development is a general "turn on" characterized by simultaneous increases in anthocyanin, alkaloid (caffeine + theobromine), and lipid content on a dry weight basis. This developmental signal is independent of embryo size. However, *stage* is critical. For example, embryos grown beyond 10 days in 3% sucrose lose their responsiveness to the high sucrose signal.

Although sucrose, glucose, or fructose all support proliferation and growth of asexual embryos (Janick et al., 1984), only sucrose stimulates the accumulation phase of embryos development. Glucose not only does not support lipid accumulation but small amounts of glucose in combination with sucrose inhibit accumulation (Kononowicz and Janick, 1984b).

The protocol for development of asexual embryos is characterized by large variability from experiment to experiment in embryo dry weight as well as fatty acid production on a dry weight basis. Much of this variation appears to be related to embryo state, but even under the best of conditions after 46 days in the high sucrose protocol embryos do not reach maturity—characterized by a fatty acid content on a dry weight basis of 50%. The best individual asexual embryos reached 39% fatty acid, but because of embryo variability the highest mean in any single experiment was 25% fatty acid

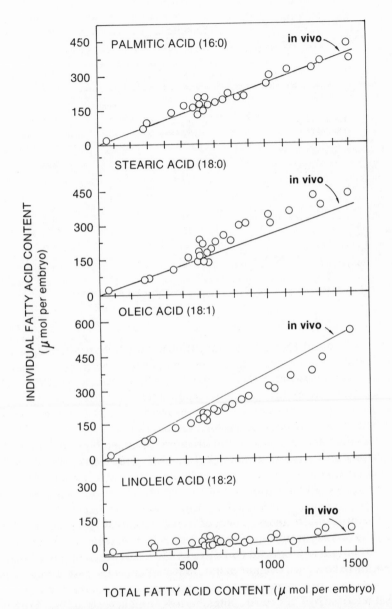

Fig. 5. Content of individual fatty acids as compared with total fatty acids of a population of 100-day-stage zygotic embryos (open circles) developing *in vitro* in high sucrose supplemented medium. The solid line is derived from different stages of zygotic embryos developing *in vivo*. (See fig. 2B). (SOURCE: Kononowicz and Janick, 1984C)

(table 6). This lack of complete development is not due to insufficient time in 27% sucrose. The development of asexual embryos in high sucrose reaches a peak at 30 days in a high sucrose. When embryos are left for more than 30 days, fatty acid content on a dry weight basis does not increase appreciably or decline. Further, medium replenishment does not improve fatty acid accumulation.

TABLE 6

PERFORMANCE OF ASEXUAL AND ZYGOTIC EMBRYOS
CULTURED IN THE SUCROSE PROTOCOL

Source of embryo	Fatty Acid Dry Wt. Basis (%)	P+S+O (Mol %)	Acid (Mol %)
Zygotic *in vivo*	52.0	94.0	3.0
	Best Individual Embryos		
Zygotic *in vitro*	43.4	93.9	4.6
Asexual *in vitro*	39.2	92.9	6.3
	Best Experimental Mean		
Zygotic *in vitro*	32.4	90.3	7.7
Asexual in vitro	25.3	89.6	9.4
	Mean (12 experiments)		
Asexual *in vitro*	16.9	85.3	12.0

Various factors affecting fatty acid content *in vitro* were tested on the assumption that some factor in the protocol apparently arrests or prevents embryo development. Nutrient modifications included total MS salts, inorganic and organic nitrogen, potassium and phosphorus as well as growth regulating substances including coconut water, benzylamino purine, naphthaleneacetic acid, gibberellic acid, and abscisic acid. Environmental factors included light and temperature.

None of the nutrient modifications including organic and inorganic salts or growth substances increased fatty acid accumulation over the basal medium plus sucrose from 3 to 27% in the standard protocol (Wright, Kononowicz, and Janick, 1984). Light or darkness was not critical. A series of experiments with temperatures ranging from 10° to 35°C indicated fatty acid quantity and quality was optimum at 26°C (Wright, Janick, and Hasegawa, 1984).

The fact that some embryos (zygotic or asexual) reach high levels of fatty acid content concommitant with high fatty acid quality indicates that the state of the embryo (embryo receptivity) is a critical factor. Clearly stage or state of the embryo before initiation of the high sucrose protocol is important. However, the absence of complete development in any embryo indicates that the protocol does not allow complete development. The factor(s) that limit the system are unknown.

GERMINATION OF ASEXUAL EMBRYOS

Immature, asexual embryos of cacao rarely germinate. However, soaking spade-shaped asexual embryos in distilled water or renewing a liquid medium induced radical development in up to 40% of embryos, suggesting that removal of one or more inhibitors promoted precocious germination (Wang and Janick, 1984). The leachate from asexual embryos inhibited seed germination of lettuce. The inhibitor in immature asexual embryos is unknown, but it is not abscisic acid (Wang and Janick, 1985).

CONCLUSIONS

Asexual embryos of *Theobroma cacao* can proliferate *in vitro* and be induced to develop toward maturity *in vitro*. However, full development has not been achieved. The limits to development appear to be due to embryo state before the sucrose signal, but the possibility exists that there is still a missing component in the sucrose protocol. Attempts to improve the sucrose protocol by alteration in the medium or the addition of growth regulation have been unsuccessful. The complete development of asexual embryos will depend on both improving receptivity of asexual embryos and modification of the culture protocol.

The fact that some individual embryos have developed as far as they have (tables 1, 6) suggests that complete development is within reach. Forecasts of economic feasibility must be deferred until complete development is obtained. Clearly, the production *in vitro* of cocoa butter will require a greater understanding of the metabolic regulation of lipid biosynthesis.

The commercial realization of asexual embryo culture *in vitro* will ultimately depend on the suitability of the metabolic product and the economics of the system. Improved economic feasibility may be obtained with the *in vitro* culture of asexual embryos of jojoba (*Simmondsia chinensis* Link). Cotyledonary storage lipids (liquid wax) of jojoba is more expensive than cocoa butter, and the liquid wax quality of jojoba does not change appreciably with zygotic or asexual embryo development (see Wang and Janick,

1986a,b,c). It is clear that the odds for success will be enhanced greatly if the value of the product is high and product quality is unaffected by *in vitro* culture.

ACKNOWLEDGMENT

I gratefully acknowledge colleagues, students, and technicians who have carried out various portions of these studies. These include: H. M. Elhag, P. M. Hasegawa, S. L. Kitto, A. K. Kononowicz, H. Kononowicz, N. R. Leopold, M. Paiva, V. C. Pence, Y. C. Wang, A. Whipkey, D. C. Wright, and M. Wyncott.

LITERATURE CITED

Chacko, G. K., and E. G. Perkins. 1964. Glyceride structure of cocoa butter. J. Am. Oil Chem. Soc. 41:843.

Cheeseman, E. E. 1927. Fertilization and embryogeny in *Theobroma cacao* L. Ann. Bot. 41:107–26.

Coleman, M. H. 1961. Further studies on the pancreatic hydrolysis of some natural fats. J. Am. Oil Chem. Soc. 38:685–88.

Elhag, H. M., A. Whipkey, and J. Janick. 1986. Induction of somatic embryogenesis from callus in *Theobroma cacao* in response to carbon source and concentration. Rev. Theobroma 16 (in press).

Esan, E. B. 1977. Tissue culture studies on cacao (*Theobroma cacao* L.). A supplementation of current research. *In* Proceedings V International Cacao Research Conference, pp. 116–25. Cacao Research Institute of Nigeria, Ibadan, Nigeria, 1977.

Goldstein, W. E., L. L. Lasure, and M. B. Ingle. 1980. Product cost analysis. *In* E. J. Staba (ed.), Plant tissue culture as a source of biochemicals, pp. 191–234. CRC Press, Boca Raton, Fla.

Janick, J., A. K. Kononowicz, and S. Kitto. 1984. Effect of carbon source on proliferation and development of asexual embryos of *Theobroma cacao*. HortScience 19:859 (abstr.).

Janick, Jules, D. C. Wright, and P. M. Hasegawa. 1982. *In vitro* production of cacao seed lipids. J. Am. Soc. Hort. Sci. 107:919–22.

Jones, L. H. 1974. Plant cell culture and biochemistry: studies for improved vegetable oil products. *In* B. Spencery (ed.), Industrial aspects of biochemistry, pp. 813–33. Federation of European Biochemical Societies, London.

Kononowicz, A. K., and J.Janick. 1984a. *In vitro* development of zygotic embryos of *Theobroma cacao*. J. Am. Soc. Hort. Sci. 109:266–69

Kononowicz, A. K., and J.Janick. 1984b. The influence of carbon source on the growth and development of asexual embryos of *Theobroma cacao* L. Physiol. Plant. 61:155–62.

Kononowicz, H., and J.Janick. 1984c. Response of embryogenic callus of *Theobroma cacao* L. to gibberellic acid and inhibitors of gibberellic acid synthesis. Z. Pflanzenphysiol. 113:359–66.

Kononowicz, H., A. K. Kononowicz, and J. Janick. 1984. Asexual embryogenesis via callus in *Theobroma cacao* L. Z. Pflanzenphysiol. 113:347–58.

Meara, M. L. 1949. The configuration of naturally occurring mixed glycerides. Part V. The configuration of the major component glycerides of cocoa butter. J. Chem. Soc. 1949:2154–57.

Murashige, T., and F. Skoog. 1962. A revised medium for rapid growth and bioassays with tobacco tissue cultures. Physiol. Plant. 15:473–97.

Nickell, L. G. 1980. Products. *In* E. J. Staba (ed.), Plant tissue culture as a source of biochemicals, pp. 235–69. CRC Press, Boca Raton, Fla.

Paiva, M., and J. Janick. 1983. *In vivo* and *in vitro* production of alkaloids in *Theobroma cacao* L. Acta Hort. 131:265–73.

Pence, V. C., P. M. Hasegawa, and J. Janick. 1979. Asexual embryogenesis in *Theobroma cacao* L. J. Am. Soc. Hort. Sci. 104:145–48.

Pence, V. C., P. M. Hasegawa, and J. Janick. 1980. Initiation and development of asexual embryos of *Theobroma cacao* L. *in vitro*. Z. Pflanzenphysiol. 98:1–14.

Pence, V. C., P. M. Hasegawa, and J. Janick. 1981a. *In vitro* cotyledonary development and anthocyanin synthesis in zygotic and asexual embryos of *Theobroma cacao*. J. Am. Soc. Hort. Sci. 106:381–85.

Pence, V. C., P. M. Hasegawa, and J. Janick. 1981b. Sucrose-mediated regulation of fatty acid biosynthesis in asexual embryos of *Theobroma cacao* L. Physiol. Plant. 53:378–84.

Raghavan, V. 1976. Experimental embryogenesis in vascular plants. Academic Press, London.

Reinert, J. 1958. Untersuchungen über die Morphogenese an Gewebekulturen. Ber. Dtsch. Bot. Ges. 71:15.

Scholfield, C. R., and H. J. Dutton. 1959. Glyceride structure of vegetable oils by countercurrent distribution. IV. Cocoa butter. J. Am. Oil Chem. Soc. 36:325–28.

Steward, F. C., M. O. Mapes, and J. Smith. 1958. Growth and organized development of cultured cells. I. Growth and division of freely suspended cells. Am. J. Bot. 45:693–703.

Tisserat, B., E. B. Esan, and T. Murashige. 1979. Somatic embryogenesis in angiosperms. Hort. Rev. 1:1–78.

Tsai, C. H., and J. E. Kinsella. 1981a. Initiation and growth of callus and cell suspension of *Theobroma cacao* L. Ann. Bot. 48:549–57.

Tsai, C. H., and J. E. Kinsella. 1981b. Tissue culture of cocoa beans (*Theobroma cacao* L.): changes in lipids during maturation of beans and growth of cells and calli in culture. Lipids 16:577–82.

Vander Wal, R. L. 1960. Calculation of the distribution of the saturated and unsaturated acyl groups in fats, from pancreatic lipase hydrolysis data. J. Am. Oil Chem. Soc. 37:18–20.

Wang, Y.-C., and J. Janick. 1984. Inducing precocious germination in asexual embryos of *Theobroma cacao*. HortScience 19:839–41.

Wang, Y.-C., and J. Janick. 1985. Characterizing the germination inhibitor from asexual embryo leechate of *Theobroma cacao*. J. Amer. Soc. Hort. Sci. 110:113–17.

Wang, Y.-C., and J. Janick. 1986a. Somatic embryogenesis in jojoba. J. Amer. Soc. Hort. Sci. 111:281–87.

Wang, Y.-C., and J. Janick. 1986b. Sucrose concentration and osmolarity as factors affecting *in virto* wax accumulation in jojoba embryos. HortScience 21 (in press).

Wang, Y.-C., and J. Janick. 1986c. *In vitro* production of jojoba liquid wax by zygotic and somatic embryos J. Amer. Soc. Hort. Sci. 111 (in press).

Wright, D. C., A. K. Kononowicz, and J. Janick. 1984. Factors affecting *in vitro* fatty acid content and composition in asexual embryos of *Theobroma cacao* L. J. Am. Soc. Hort. Sci. 109:77–81.

Wright, D. C., W. D. Park, N. R. Leopold, P. M. Hasegawa, and J. Janick. 1982. Accumulation of lipids, proteins, alkaloids, and anthocyanins during embryo development *in vivo* of *Theobroma cacao* L. J. Am. Oil Chem. Soc. 59:475–79.

Wright, D. C., J. Janick, and P. M. Hasegawa. 1984. Temperature effects on *in vitro* lipid accumulation in asexual embryos of *Theobroma cacao* L. Lipids 18:863–67.

Youngs, C. G. 1961. Determination of the glyceride structure of fats. J. Am. Oil Chem. Soc. 38:62–67.

Zoumas, B. L., and E. J. Finnegan. Chocolate and cocoa. 1979. *In* Encyclopedia of chemical technology, 3d ed., 6:1–19. John Wiley & Sons, New York.

Y. P. S. BAJAJ AND MANJEET S. GILL

Biotechnology of Cotton Improvement

9

INTRODUCTION

Biotechnology is currently being applied to several major crop plants. In this chapter we use cotton as an example of the progress that is possible for crop improvement through biotechnology. After an introduction to cotton as a crop, we offer a summary of recent progress in cotton biotechnology.

Cotton belongs to the genus *Gossypium* (family Malvaceae) and is the most important natural source of fiber currently used in the textile industry. It is also a valuable source of edible oil. The oil content of different cultivated species varies from 14.6–25.6% of dry weight (Pandey, 1976). Cottonseed meal contains a high percentage of protein and is rich in essential amino acids like lysine, methionine, tryptophan, and other amino acids (El-Nockrashy et al., 1969) and is primarily used as a fertilizer and cattle feed. The hull, which is the outer covering of the seed from which the cotton fibers arise, is used as roughage in cattle feed.

The genus *Gossypium* has 39 species, including 4 cultivated species (Mirza and Shaikh, 1980). These species have been categorized into 7 genome groups on the basis of the cytology of interspecific hybrids (Beasley, 1942; Phillips and Strickland, 1966; Edwards and Mirza, 1979). The A genome contains only two diploid cultivated species (*G. arboreum* L. and *G. herbaceum* L.). In the B, C, D, and E genomes there are 4, 9, 12, and 4 species, respectively. *G. longicalyx* has been included in the F genome and *G. bickii* in the G genome. There are six tetraploid species, including the cultivated types *G. hirsutum* L. and *G. barbadense* L., all of which have a mixture of the A and D genomes.

History and Distribution

Cotton has been cultivated in the warmer climates of the world since prehistoric times. India, where cotton has been an important product for more than 3,000 years, was one of the first countries to develop a cotton industry. Cotton was grown and used for clothing in Brazil, Peru, and Mexico long before the 1600s (table 1).

TABLE 1

IMPORTANT COTTON–GROWING COUNTRIES IN THE WORLD

	Total Area 1,000 hac	Yield kg/hac	Production 1,000 MT.
World	33,171	1,270	42,111
Egypt	523	2,675	1,400
Mexico	348	2,820	980
USA	5,250	1,213	6,377
Brazil	2,145	842	1,806
USSR	3,115	3,197	9,960
China	5,000	1,624	8,121
Pakistan	2,095	1,002	2,100
India	8,000	525	4,200
Turkey	676	1,840	1,248

SOURCE: FAO Production Year Book, 1981

The only cotton with spinnable lint that grows wild is *G. herbaceum* var. *africanum*, and this is probably the ancestor of all linted cottons in both the Old and New World. *Africanum* cotton still exists in the wild in parts of South Africa, and is a true woody perennial. Indus civilization provides the earliest archeological record we know of the existence of cotton cloth. It was in Asia that the great expansion of the Old World or Asiatic cottons, derived from *Africanum*, took place. With the selection pressure of man, cotton has evolved quickly. Presently there are two major species of cultivated cotton, *G. herbaceum* and *G. arboreum*, in the Old World.

The New World cottons (tetraploids) are natural amphidiploids that contain a set of chromosomes from one of the two A genome species and a chromosome set from a species of the D genome group (Beasley, 1940a; Harland, 1940; Skovsted 1934, 1937). Gerstel (1953) has demonstrated that the A genome of the natural amphiploid differs from *G. herbaceum* by two reciprocal translocations and from *G. arboreum* by three reciprocal translocations, thus indicating that *G. herbaceum* contributed the A genome to the New World cotton. Cytogenetic studies (Phillips. 1962, 1963) and seed electrophoresis studies (Cherry et al., 1970; Johnson and Thein, 1970) have indicated that *G. raimondii* is the most closely related of the known D species to the contributor of the New World D genome. This species is found on the Pacific foothills of the Andean Cordillera in Peru.

At present the cultivation of the Old World diploids is restricted to Asia and some parts of Africa. *G. barbadense* includes cultivars grown in the Nile River drainage area (Egyptian cotton), in Peru (Tanguis cotton), and in restricted areas of the southwestern United States (Pima cotton). About 90% of the commercial cotton is obtained from *G. hirsutum*, which is grown in almost all the cotton-growing areas of the world.

Diseases

Cotton suffers a great deal of loss by various insect pests and diseases. Sometimes because of insects such as jassids or bollworms or diseases such as wilt, root rot, or bacterial blight, the entire crop may be destroyed. Some of the important diseases and pests affecting cotton are summarized in table 2.

TABLE 2

SOME COMMON DISEASES AND PESTS OF COTTON

Disease/Pest	Causal Organism
1. Bacterial blight	*Xanthomonas malvacearum*
2. Ascochyt blight	*Ascochyta gossypii*
3. Verticillium wilt	*Verticillium alboatrum*
4. Root rot	*Phymatotrichum omnivorum*
5. Anthracnose	*Glomerella gossypii*
6. Rhizoctonia root rot	*Rhizoctonia solani*
7. Fusarium wilt	*Fusarium oxysporum* f. *vasinfectum*
8. Root knot nematode	*Meloidogyne incognita*
9. Boll weevil	*Anthonomas grandis*
10. Spotted bollworm	*Earias* spp.
11. American bollworm	*Heliothis* spp.
12. Pink bollworm	*Pectinophora gossypiella*
13. Leaf worm	*Alabama argillacea*
14. Aphid	*Aphis gossypii*
15. Thrips	*Thrips tabaci*
16. Flea hopper	*Psallus seriatus*
17. Tarnished and rapid plant bugs	*Lygus hesperus*
18. Jassid	*Amrasca biguttula biguttula*

Wild Species

In the genus *Gossypium*, there are 36 wild lintless or short-linted species. A number of wild species (table 3) possess characters including disease resistance, insect-pest resistance, drought resistance, and superior fiber qualities that are lacking in the cultivated types. If these traits could be incorporated in cultivated types, it would greatly enhance cotton production in the world.

TABLE 3

SOME WILD SPECIES OF COTTON POSSESSING USEFUL CHARACTERS

Species	Genome	Useful Characters
1. *G. anomalum* Wawrex Wawra & Peyr	B_1	Lint quality, resistance to black arm and jassid
2. *G. harkensii* Brendeg.	D_2^{-2}	Resistance to drought, source of cytoplasmic male sterility, resistance to spider mites
3. *G. raimondii* Ulbr.	D_5	Resistance to black arm and jassids, high density of seed hairs, and drought tolerance
4. *G. armourianum* Kearn.	D_2^{-1}	Resistance to black arm and boll worm, increased number of ovules per loculus
5. *G. thurberi* Tod.	D_1	Resistane to boll worms and potentiality for imparting high fiber strength
6. *G. aridum* (Rose & Standl.) Skov.	D_4	Drought tolerance
7. *G. somalense* (Gurke) J.B. Hutch.	E_1	Resistance to boll worm
8. *G. stocksii* Mast. in Hook.	E_2	Resistance to drought
9. *G. tomentosum* Nutt. ex Seem.	AD	Resistance to drought and jassids, lint quality, and strength
10. *G. bickii* Prokh.	$2G_1$	Resistance to insects, Gossypol-free seeds.

Objectives of Cotton Breeding

In cotton, genetic improvement is necessary both for increased yield and for upgrading the quality of lint. For cotton improvement, the following characters should be considered:

Yield. The yield of seed cotton per unit area is a complex character. It is dependent upon the number of plants per unit area, the number of bolls per plant, and the weight of seed cotton per boll. Development of short-statured, less leafy, compact sympodial types will be useful in increasing the number of plants per acre. Boll weight is also determined by the number of locules, the number of seeds per locule, and the weight of lint per seed.

Yield of lint is calculated from the data on yield of seed cotton and ginning percentage.

Earliness. The development of early maturing types is necessary because of the double cropping procedures used in many cotton-growing areas. In addition, early varieties are in the ground for a shorter period of time and can therefore escape much of the damage done by pink bollworms. Earliness can be measured by the position of the first fruiting (sympodial) node on the main stem, or the number of days taken from sowing to the date by which bolls burst in about 50% of the plants.

Fiber quality. The ultimate index of quality in cotton is determined by its spinning performance. Among those fiber properties that contribute most to the spinning value are fiber length, fineness, and strength. Fiber uniformity and maturity are also important. Depending upon fiber length, cottons can be classified into short (19 mm or below), medium (20–21.5 mm), superior medium (22–24 mm), long (24–26 mm), and superior long (27 mm or above). Fiber fineness is usually expressed in terms of weight per unit length of fiber.

Insect resistance. A number of insects attack cotton (table 2), and in some cases these can cause complete crop loss. The evaluation of breeding lines for insect resistance requires 3 steps: (1) the creation of maximum and uniform infestation under natural or artificial conditions, (2) development of a sampling plan for insect counts and for grading, and (3) classification of plants into grades. A number of characters like high gossypol content, presence of hair on stem and leaves, nectarless property, frego bract, okra leaf, earliness, and so on, are known to reduce the incidence of several insect pests. Breeding for resistance could involve the incorporation of these traits in the cultivated varieties.

Disease resistance. Various diseases cause reduction in yield by affecting germination, killing the plants, lowering their productivity, and also by affecting the quality of lint. The major diseases that attack cotton are wilt, root rot, bacterial blight, and anthracnose (table 2). The foliar diseases can cause losses up to 20%, and the more lethal diseases like wilt and root rots can cause more serious damage. For bacterial blight several strains of the pathogen have been identified, and a number of resistance genes have been found.

Conventional Methods of Cotton Breeding

Usually different types of breeding procedures have been established in self-pollinated versus cross-pollinated crops. In cotton though, because it cross-pollinates naturally, no standard breeding methods have been adopted. Conventional cotton breeding usually consists of introduction, selection, and hybridization as in self-pollinated crops, but different procedures are usually practiced. Plant introductions either have been grown as introduced or have been improved by natural or artificial selection. Selection is required to maintain varietal purity because of the partial cross-pollination as well as to aid in new variety development. Selection may be based on single plant selections or mass selection may be practiced by bulking seed from open-pollinated or selfed plants selected on the basis of appearance. Crosses between different varieties or strains can be used to produce segregating populations from which selections can be isolated.

All the breeding methods that have been used in cotton are based on selection in variable populations. Hybridization and mutation breeding have often been used to introduce variability into populations. Until now, mostly intraspecific hybridization has been practiced, and very little success has been achieved in utilization of the wide variability existing in wild germplasm. Mutation breeding also has not been very successful in cotton.

Need to Incorporate Unconventional Methods

Extensive efforts are being made to develop plant types that can cope better with various environmental hazards, insects, and diseases, and can give stable yield from year to year. However, conventional methods do not seem to be sufficient to bring about any breakthrough.

The prime requisite for the success of any crop improvement program is the presence of sufficient genetic variability for the traits that are to be improved. When the required variability is not present in a species, then it has to be induced by mutations or we have to look toward some related species for characters that are absent in the cultivated types. In the genus *Gossypium*, a number of wild species (table 3) have useful attributes that can be transferred from Old World diploids to the cultivated tetraploids. However, despite numerous attempts made, ther have been very few useful gene transfers in cultivated cottons. Success in this direction has been hampered because of various incompatibility barriers like embryo/endosperm abortion (Weaver, 1957, 1958; Govila, 1970; Pundir, 1972), very low boll setting, reduced viability and low germination of hybrid seeds (Sreerangasamy and Raman, 1963; Mehboob Ali et al., 1964), seedling lethality (Phillips, 1977; Phillips and Reid, 1975), or plant lethality (Phillips and Merritt, 1972). Even if interspecific hybrids are obtained, it is very difficult to secure any recombinants. Thus there is a very strong need to resort to some unconventional techniques that can ensure hybridization between different species so that hybrid plants can be produced in large numbers and utilized for growing a large number of selfed or backcrossed progenies. In this way the chances of recovering useful recombinants may then be increased. In addition, since New World cotton is tetraploid, the induction of mutations through culture of haploid cells and protoplasts (Bajaj, 1983a) would be ideal.

BIOTECHNICAL APPROACHES

During the last decade, tremendous progress has been made in the area of plant cell, tissue, and organ culture (Evans et al., 1983). The situation has been brought to a point where these *in vitro* methods form a very important component of biotechnology that has far-reaching implications in crop improvement programs. Following are some of the important opportunities that could be realized through *in vitro* culture:

1. Clonal propagation.

2. The culture of excised meristems has been used for virus elimination, and this has resulted in an increased yield in crops such as potato (Mellor and Stace-Smith, 1977).

3. Induction of mutations at the cellular level, especially those of haploid nature, would be useful for the induction of desirable mutants.

4. Wide hybridization through embryo and ovule culture both at the interspecific as well as intergeneric level has been achieved in a number of crops through the culture of the hybrid ovules and embryos (Bajaj et al., 1978, 1982; Gosal and Bajaj, 1983; Gill and Bajaj, 1984a,b,). *In vitro* fertilization techniques have also been employed to overcome incompatibility (Zenkteler and Melchers, 1978).

5. Production of haploids through anther culture, and their subsequent diploidization to homozygosity has enabled the early release of varieties in crops such as rice, wheat, and so on (Bajaj, 1983b).

6. Through the fusion of protoplasts, incompatibility barriers could be broken, and this would ensure the production of somatic hybrids and

cybrids. One could regenerate male sterile lines through cybrids, an area that is fruitful to pursue (Gleba and Evans, 1983).

7. Callus cultures obtained from somatic and sexual cells are a very rich source of genetic variability (D'Amato, 1977; Krikorian et al., 1983). Furthermore, hybrid callus can provide exchange of genetic material between the chromosomes belonging to different genomes. The genetically diverse plants regenerated from such somaclones could then be incorporated into breeding programs.

The above-mentioned approaches when applied to cotton would go a long way for its improvement. A brief review of the major accomplishments for *in vitro* culture of cotton is summarized in table 4. The main objective for the use of *in vitro* methods is to generate genetic variability, and eventually to incorporate this into cotton improvement programs. Some of these approaches are discussed below:

Wide Hybridization

Breeding new and improved varieties must proceed from basic material in which there is sufficient genetic variability. In all cases breeding work is dependent upon variability within a single species, and the conventional breeding techniques are ideal for the maintenance of existing varieties and for their improvement by cross-hybridization between varieties. However, new characteristics are not expected to appear, and one method to produce these is by hybridization with different related cultivated or wild species.

With embryo/ovule rescue and related cell culture techniques, interspecific hybrids have been obtained between diploid cultivated (*G. arboreum, G. herbaceum*) and wild species (*G. stocksii, G. anomalum*) of cotton. Normally, the flowers fall after cross-pollination; however, premature abscission could be prevented by application of growth regulators followed by the culture of immature hybrid embryos (Gill and Bajaj, 1984a,b). The F_1 hybrids obtained were transferred to the field and backcrossed with their respective cultivated parents. The backcrossed progeny showed considerable variation for several morphological characters (Gill and Bajaj, unpubl.).

The daily application of a mixture of growth regulators (GA_3 50 mg/1 + NAA 100 mg/1) to the flowers immediately after crossing considerably enhanced their retention and ensured further development in all crosses (table 5).

TABLE 4

PROTOPLAST, CELL, TISSUE, AND ORGAN CULTURE STUDIES OF COTTON

EMBRYO CULTURE

Species	Explant	Medium	Response	Reference
G hirsutum L.	Hybrid embryos	White's (1939)	Seedlings with rudimentary cotyledons	Beasley (1940b)
G. hirsutum L.	27–day-old embryos	White's (1943)	Normal seedlings	Lofland (1950)
G. hirsutum L.	12–14-day-old embryos	White's (5 times) + adenine sulphate (40 mg/l) + CH (150 ml/l) + NaCl (7 gm/l) + sucrose (20 gm/l)	Seedlings in some cases	Mauny (1961)
G. hirsutum L.	12–14-day-old embryos	Same as above (Mauny, 1961) with calcium malate or ammonium malate	Vigor of seedlings increased	Mauny et al. (1967)
G. arboreum X G. stocksii	15-day-old embryos	MS + CH (250 mg/l) + IAA (1.5 mg/l) + Kin (0.5 mg/l) + sucrose 3%	Hybrid plantlets that matured upon transfer to field	Gill & Bajaj (1984a, b)
G. arboreum X G. anomalum G. herbaceum X G. stocksii				
G. arboreum X G. anomalum	15-day-old embryos	MS + CH (250 mg/l) + Kin (0.5 mg/l) + Kin (0.5 mg/l) + sucrose 3%	Hybrid plants backcrossed to G. arboreum and vice versa; considerable variation seen in backcross progeny	Gill & Bajaj (1985)

OVULE CULTURE

Species	Inoculum	Medium	Response	Reference
G. hirsutum L.	6-day post anthesis ovules	White's 1943 (WM) + CH (1,000 mg/l) + GA (5 mg/l)	1.8 mm.-long embryos after 80 days	Joshi (1960)
G. hirsutum L.	2-day-old ovules	Not given	Seedlings	Beasley et al. (1971)

G. hirsutum L.	6-day-old ovules	WM + CH (1,000 mg/l) + GA (5 mg/l)	Ovules surpassed size of ovules developed in vivo; no germination	Joshi & Johri (1972)
G. hirsutum L.	5-day-old ovules	Murashige & Skoog's (1962)	Embryos with cotyledons 1/3 normal size	Eid et al. (1973)
G. hirsutum L.	2-day-old ovules	Beasley & Ting's (1973) (BT)	Good ovule and fiber growth	Beasley & Ting (1973)
G. hirsutum L.	2-day-old ovules	BT + NH_4NO_3 (15 mM) + IAA (5 μM) + GA (5 μM) + Kin (0.05 μM) + sucrose (40 gm/l)	Mature embryos and seedlings	Stewart & Hsu (1977a)
G. hirsutum L. X G. arboreum X G. hirsutum X G. herbaceum	2-day-old ovules	BT + NH_4Cl (15mM) + 4% sucrose	Mature viable embryos	Stewart & Hsu (1978)
Hybrid ovules from crosses involving tetraploid and diploid cultivated and wild species	3-day-old ovules	MS + CH (250 mg/l) + IAA	Hybrid plants that matured upon transfer to field	Gill & Bajaj (1948a)

ANTHER CULTURE

Species	Inoculum	Medium	Response	Reference
G. hirsutum L.	Young Anthers	DBM-II (Gresshoff and Doy, 1972) + NAA (5 mg/l) + BA (1 mg/l)	Haploid callus obtained	Barrow et al (1978)
G. barbadense L.		LS (Linsmaeir and Skoog, 1965) + NAA (5 mg/l) + BA (1 mg/l)	Callus]produced	Barrow et al. (1978)
G. hirsutum L.		Schenk and Hildebrandt (1972) + 2,4-D (5 mg/l)	Root and shoot-like tissues developed from callus	Hsi and Wu. (1981)
G. arboreum X G. hirsutum			Callus obtained; proliferating anthers freeze-preserved (−196°C) and later resumed growth	Bajaj (1982)
G. arboreum		MS + 2,4-D (2.0 mg/l) + Kin (0.2 mg/l) + CW 7% + Sucrose 3%	Pollen embryogenesis, and a wide range of genetic variability in callus	Bajaj and Gill (1984)

TABLE 4 (continued)

ESTABLISHMENT OF CALLUS AND SUSPENSION CULTURES

Species	Inoculum	Initial Medium	Growth Response	Reference
G. hirsutum L.	Mesocotyl explants	Basal medium + 2,4-D (5 mg/l) + Kin (0.1 mg/l) + p-CPA (2 mg/l) + myoinositol	Soft, friable, and rapid-growing callus	Schenk and Hildebrandt (1972)
G. hirsutum L.	Leaf tissue	LS (Linsmaeir and Skoog, 1965) + 2,4-D (0.1 mg/l) + myoinositol (5 mg/l) + ascorbic acid (5 mg/l) + 3% sucrose	Callus growth increased with rise in temperature (up to 30°C) and light intensity	Davis et al. (1974)
G. hirsutum L. and G. barbadense L.	Cambial surface of peeled stem segments	LS + myoinositol (100 mg/l) + thiamine, HCl (0.4 mg/l) + IAA (5 μM) + BA (0.5 μM)	Slow-growing callus was obtained in 5–10 days; browning of callus was reduced when sucrose was replaced by glucose	Sandstedt (1975)
G. hirsutum L. and G. arboreum L.	Hypocotyl segments	MS + NAA (1 mg/l) + Kin (1 mg/l) + Adenine sulphate (40 mg/l) + morphactin	Morphactin at low conc. promoted callus growth and enhanced chlorophyll content of tissue	Rani and Bhojwani (1976)
G. hirsutum L.	2-day-old ovules	BT + CPA medium + CEPA	Extremely friable callus from micropyler end of ovules	Hsu and Stewart (1976)
G. arboreum L.	Hypocotyl and stem segments	MS (inorg.) + thiamine. HCl (0.4 mg/l) + myoinositol (100 mg/l) + IAA (2.0 mg/l) + Kin (1.0 mg/l) + 3% glucose	Rapid proliferation of callus from hypocotyl explants	Smith et al. (1977)
Six species	Hypocotyl explants	Same as above. (Smith et al. [1977])	Callus could be initiated and maintained with minor modifications of the hormone constitution	Price et al. (1977)
G. hirsutum L.	2-day-old ovules	BT + other phytohormones (ABA, CEPA, IAA, Kin, GA)	ABA inhibited callus growth; At optimum CEPA conc. IAA inhibited growth; Kin stimulated micropylar callus growth; GA caused highest callus growth	Stewart and Hsu (1977b)
G. barbadense L.	Cotyledon	BT + NAA (17 mg/l) + DDT (dithiothreitol) 50 mg/l	Toxic brown coloration of callus inhibited by DDT	Katterman et al. (1977)
G. hirsutum L.	Stem segments	Basal medium + various levels of hormones and sugars	Changes in pH (5.3–5.9) had no effect on callus growth;	Williams (1978)

Species	Inoculum	Medium for Callus Induction	Medium for Regeneration	Response	Reference
G. hirsutum L.	Stem and leaf segments	Schenk and Hildebrandt's (1972)		light stimulated growth and greening; thiamine was essential for growth; ascorbic acid decreased necrosis; IAA or NAA necessary for growth; 2,4-D inhibited growth; Zeatin induced maximum growth among cytokinins	Ruyack et al. (1979)
G. arboreum X G. hirsutum	Ovules and anthers	Schenk and Hildebrandt's + 2,4-D (5 mg/l)		Light, friable callus was obtained; fresh weight doubling time of the callus was 4–9 days; in suspension cultures, 6 days.	Bajaj (1982)
G. arboreum	Ovule	MS (inorg.) + thiamine. HCl (0.4 mg/l) + myoinositol (100 mg/l) + IAA (2.0 mg/l) + Kin (1 mg/l) + glucose 3% (Smith et al. 1977)		Proliferating ovules were freeze-preserved (-196°C); these survived and later resumed growth.	Bajaj & Gill (1984)
	Anther	MS + NAA (2 mg/l) + BAP (1 mg/l) + sucrose 3%		Genetic variability in callus ranging from haploid, diploid, triploid to highly polyploids and aneuploids	
	Hypocotyl	MS (inorg.) + thiamine. HCl (0.4 mg/l) + myoinositol (100 mg/l) + IAA (2.0 mg/l) + glucose 3%			
G. arboreum X G. anomalum X G. hirsutum X G. arboreum	Hybrid embryos	MS + IAA (1.5 mg/l) + CH (250 mg/l) + Kin (0.5 mg/l) + sucrose 3%		Hybrid callus showed a wide range of genetic variability	Bajaj and Gill (1984)

REGENERATION OF PLANTS

Species	Inoculum	Medium for Callus Induction	Medium for Regeneration	Response	Reference
G. hirsutum L.	Stem segments	Basal medium + various combinations and concentrations of auxins, cytokinins, and sugars	Basal medium and Zeatin	Nodulation occurred	Williams (1978)

TABLE 4 (continued)

REGENERATION OF PLANTS

Species	Inoculum	Medium for Callus Induction	Medium for Regeneration	Response	Reference
G. klotzschianum	Hypocotyl segments	MS (inorg.) + thiamine HC 1 (0.4 mg/l) + myoinositol (100 mg/l) + IAA (2.0 mg/l) + Kin (1.0 mg/l) + Glucose 3%	Preculture in a liquid medium containing 2iP (10 mg/l); then cultured in presence of 10–15 mM glutamine	Somatic embryoids differentiated	Price and Smith (1979)

ISOLATION AND CULTURE OF PROTOPLASTS

Species	Origin of Protoplasts	Enzyme Mixture	Culture Medium	Growth Response	Reference
G. hirsutum L.	Hypocotyl callus	4% meicelase P + 0.25% Drise-lase, + 0.4% macerozyme + 11% mannitol (pH 5.6)	B_5 salts and vitamins + inositol (100 mg/l) + 0.5% glucose + 1% sucrose + 2,4-D (0.5 mg/l) + NAA (0.2 mg/l), BAP (0.2 mg/l)	Cell wall regeneration and multicellular colonies produced	Bhojwani et al. (1977)
G. hirsutum L	Cotyledonary leaves	Xylanase (1.5%) + cytase (1%) + Onozuka P 1500 + 0.5 M D-mannitol	MS agar medium + 3% galactose	Cell wall regeneration and colony formation from epidermal protoplasts	Khasanov and Butenko (1979)
G. klotzschianum	Callus	Cellulysin (1%) + 0.5% macerase + 0.7 M mannitol (4.7 pH)	Murashige and Skoog's salts + inositol (100 mg/l) + thiamine HC1 (0.4 mg/l) + 1.5% glucose + 2iP (1 mg/l) + NAA (10 mg/l) + mannitol 0.6 M + 3% agar (pH 5.7)	Multicellular colonies formed in two weeks	Finer and Smith (1982)
G. hirsutum	Cotyledonary leaves	Cellulose Onozuka (1%) + macerozyme (0.5%) + mannitol (13%) (pH 5.6)		Good yield of green protoplasts obtained after 4 hrs incubation	Gill (1983)
G. arboreum and G. herbaceum	Ovule-derived callus	Cellulose Onozuka (2%) + macerozyme (0.5%) + mannitol (15%) (pH 5.6)		Good yield of protoplasts obtained after 16 hrs incubation	

TABLE 5

EFFECT OF GROWTH REGULATORS (GA₃ 50 µM) + NAA (100 µM) on BOLL RETENTION
IN VARIOUS INTERSPECIFIC CROSSES IN *GOSSYPIUM*

CROSSES	CONTROL			TREATED		
	No. of Flowers Crossed	No. of Bolls Shed	% Retention	No. of Flowers Crossed	No. of Bolls Shed	% Retention
G. arboreum × *G. stocksii*	21	17	19.1	86	32	62.8
G. arboreum × *G. anomalum*	27	23	14.8	102	35	68.3
G. herbaceum × *G. stocksii*	26	23	11.5	68	30	55.8
G. arboreum × *G. hirsutum*	11	11	0	28	17	60.7
G. hirsutum × *G. arboreum*	36	30	16.7	71	20	71.8

SOURCE: Gill and Bajaj, 1984b.

In the controls (untreated), the boll retention at fifteen days after pollination (DAP) was only 11.5% in *G. herbaceum*× *G. stocksii*, 14.8% in *G. arboreum* × *G. anomalum*, and 19.1% in *G. arboreum* ×*G. stocksii*. Flower shedding started 4 DAP. Abscission occurred at a point where the pedicel is attached to the stalk. When growth regulators were applied, boll retention was increased to 55.8% in *G. herbaceum* × *G. stocksii*, 62.8% in *G. arboreum*× *G. stocksii*, and 68.3% in *G. arboreum*× *G. anomalum*. Hence, the best response was observed for *G. arboreum* × *G. anomalum*. Following hormone application, the base of the pedicel became thick and strong, and abscission was prevented.

Hybrid embryos of these three crosses, i.e., *G. arboreum*× *G. stocksii*, *G. arboreum* × *G. anomalum*, and *G. herbaceum* × *G. stocksii* were cultured on MS + indoleacetic acid (1.5 mg/1) + kinetin (0.5 mg/1) + casein hydrolysate (250 mg/1) and 87.6 mM sucrose, and incubated in the dark, and showed an overall increase in size during the first two weeks. The cotyledons enlarged, and the radicle underwent elongation and penetrated the medium in most cases. Following exposure to light, the cotyledons turned green, with the first leaf appearing in 30–35 days following which the embryos developed into plantlets (fig. 1). In cases where the embryos did not produce seedlings, either the radicle failed to penetrate the medium or the root stopped growing after a few days.

The response of cultured embryos varied in different crosses (table 6). The best growth response was observed in *G. arboreum* × *G. anomalum*, although we have recovered more plants in *G. arboreum*× *G. stocksii*. The *G. arboreum* × *G. anomalum* embryos produced seedlings in 30–35 days, whereas in *G. arboreum*× *G. stocksii*, it required a minimum of 35 days. In *G. herbaceum* × *G. stocksii*, only 48% of the embryos produced healthy seedlings.

TABLE 6

In vitro Growth Response of Hybrid Embryos Cultured 15 Days after Pollination on MS + IAA (1.5 mg/1) + CH (250 mg/1) + Kin (0.5 mg/1) + Sucrose 3%

Cross	No. of Embryos Cultured	No. of Plants Formed	% Plant
G. arboreum × *G. stocksii*	31	22	71
G. arboreum × *G. anomalum*	74	51	69
G. herbaceum × *G. stocksii*	39	19	48

SOURCE: Gill and Bajaj, 1984a.

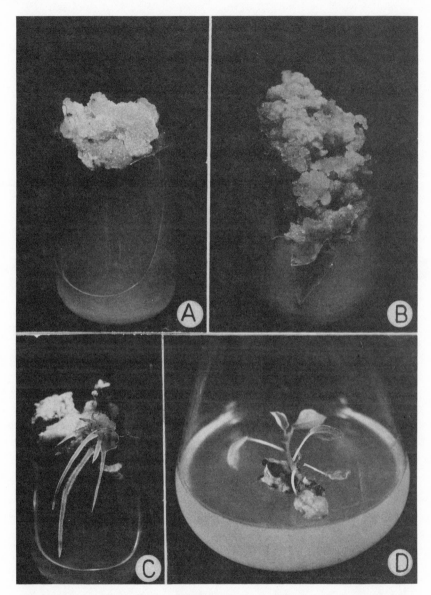

Fig. 1 *In vitro* studies on cotton. (A,B) Callus formation from immature embryos of *Gossypium arboreum* cultured on MS + NAA (1 mg/l) + BAP (0.5 mg/l), and MS + 2,4-D (0.5 mg/l) respectively. (C) 3-week-old culture of an excised embryo of *G. hirsutum* on MS + IAA (2 mg/l) + kin (1 mg/l); note the formation of well developed roots. (D) An interspecific hybrid plantlet (*G. arboreum* x *G. bickii*) obtained by the culture of immature embryo, 5-week-old culture.

TABLE 7

EFFECT OF LIGHT AND DARK TREATMENTS ON *IN VITRO* GROWTH RESPONSE OF COTTON
EMBRYOS CULTURED 15 DAYS AFTER POLLINATION ON MS + IAA (1.5 mg/l) +
CH (250 mg/l) + Kin (0.5 mg/l) + SUCROSE 3%

Species	Treatment	No. of Embryos Cultured	No. of Plants Formed	% Plant Formation
G. arboreum	Dark 15 days	36	27	75.0
	Light	32	14	43.7
G. herbaceum	Dark 15 days	33	20	60.6
	Light	34	13	38.2

SOURCE: Gill and Bajaj, 1984b.

The plants were transferred into pots at the 2–3 leaf stage with well-developed roots resulting in 80–90% survival. The hybrids grew normally, and the *G. arboreum × G. anomalum* hybrids in particular showed vigorous growth. The hybrids in all three crosses were perennial.

Hybrid ovules of these three crosses (table 8) were cultured 3 DAP on MS + IAA (1.5 mg/l) + kinetin (0.5 mg/l) + CH (250 mg/l) and doubled in size in 4–6 days of culture with callus formation observed within one week. There was considerable proliferation in some cases, and the resulting callus was soft and firable (fig. 2A–D). The callus formed from hybrid ovules with *G. arboreum* cv. G. 27 as the female parent was pinkish, whereas the ovules with *G. herbaceum* as the female parent formed a creamy-white callus. The ovules continued to grow along with the callus. The ovules germinated after about 50 days of culture (fig. 2D). In many cases

TABLE 8

GROWTH RESPONSE OF EXCISED OVULES OF VARIOUS INTERSPECIFIC
CROSSES OF $G^{\circ 4401}fl^6$= CULTURED 3 DAYS AFTER POLLINATION ON MS +
IAA (1.5 mg/l) + kinetin (0.5 mg/l) + CH (250 mg/l)

Hybrid	No. of Ovules Cultured	No. of Ovules Germinated	% Germination
G. arboreum × *G. stocksii*	145	41	28
G. herbaceum × *G. stocksii*	72	16	22.2
G. arboreum × *G. anomalum*	80	17	19.1
G. arboreum × *G. hirsutum*	68	23	34.7

SOURCE: Bajaj and Gill, 1984.

Fig.2 Growth response of hybrid ovules (*G. hirsutum* × *G. arboreum* and *G. arboreum* × *G. hirsutum*) on liquid and agar-solidified media. (A) 3 DAP ovules of *G. hirsutum* × *G. arboreum* after 14 days of culture on Stewart and Hsu's (1978) medium supplemented with IAA (0.5 mg/l). (B) Same, after 21 days of culture; note callus formation from some of the ovules. (C) Same, after 28 days of culture showing profuse callus formation. (D) *G. arboreum* × *G. hirsutum* ovule (3 DAP) after 55 days of culture on MS + IAA (1.5 mg/l) + Kin (0.5 mg/l) + CH (250 mg/l) showing callus formation as well as germination. (E) 12-week-old seedling derived from 3 DAP ovule of *G. hirsutum* × *G. arboreum* cultured on MS + IAA (1.0 mg/l) + Kin (0.2 mg/l) + CH (250 mg/l) after transfer to half strength MS + sucrose 1%.

cotyledons were either malformed or did not develop. A large number of seedlings were produced in *G. arboreum* × *G. hirsutum*, but with low survival (fig. 2E). On the other hand, the seedlings obtained from *G. arboreum* × *G. anomalum* ovules were vigorous, and showed the highest survival rate. Seedlings continued to develop on half-strength MS + 1% sucrose and could be raised to maturity. Root tips of *G. arboreum* × *G. hirsutum* seedlings contained the triploid (3x = 3n = 39) chromosome number (fig. 7C).

The hybrid plants obtained in this fashion were transferred to the field (fig. 3B) and backcrossed to the parent lines. The backcrossed progeny had considerable variation for various morphological characters (fig. 3C). Characters of the parents, F_1 hybrids, and the backcross progeny are summarized in table 9. There were considerable differences with regard to plant size, leaf shape, flower color, and bract shape. In general, the F_1 hybrids had most of the characters of the wild parents. All the hybrids flowered at a late stage, were perennial, and two of the hybrids (*G. herbaceum* × *G. stocksii* and *G. arboreum* × *G. anomalum*) flowered throughout the year. However, all the hybrids were self-sterile. A few seeds could be produced when the *G. arboreum* × *G. anomalum* hybrid was backcrossed to *G. aboreum*, either as male and female parent (Gill and Bajaj, unpubl.). When the F_1 was used as the female parent, the seeds resembled those of *G. anomalum* in shape and had very fine, strong fiber.

Backcross progeny of (*G. arboreum* × *G. anomalum*) × *G. arboreum* and the reciprocal hybrid exhibited substantial variation (fig. 3C, table 9). Most of the plants had red pigment (a character of *G. arboreum* cv. G27), but had most other characters typical of *G. anomalum*.

Anther Culture

Haploids are valuable for the production of homozygous lines in one generation at any stage of a breeding program. These homozygous lines can be evaluated directly in the field. Haploids can also be used to increase the efficiency of existing breeding methods through improved selection methods. In the past the main source of cotton haploids has been polyembryony in which one or both members of twin seedlings may be haploid (Blank and Allison, 1963; Chaudhari, 1978). However, the use of cotton haploids is dependent upon their production at will in large numbers, which is not possible using these conventional methods.

Barrow and colleagues (1978) first cultured anthers in an attempt to efficiently produce haploids. The optimal production of callus occurred from anthers with microspores in meiosis. Callus grew profusely, but no organo-

Fig. 3 Backcross progeny derived by crossing *G. arboreum* × *G. anomalum* hybrid with *G. arboreum*; note the variability in the plant morphology. (Gill and Bajaj, unpublished).

genesis was detected. In our studies (Bajaj and Gill, unpublished), the excised anthers of *G. arboreum* cultured at the uninucleate stage of pollen on MS + 2,4-D (2.0 mg/ 1) + kinetin (0.2 mg/ 1) + CW 7% underwent prolifera-

TABLE 9

Description of Morphological Characters of the Two Parents, F_1 and Backcross Progeny Obtained by Embryo Culture

Character	G. arboreum	G. anomalum	F_1	Backcross Progeny
Stem				
Leaf	Erect, reddish brown, hairy 5-lobed, with pointed ends, reddish, hairy	Erect, green, extremely hairy 3–5 lobed, green, hairy	Erect, slightly reddish, hairy 5–7 lobed with pointed ends, red, hairy	Erect, green or reddish, hairy 3–7 lobed, color green or reddish, hairy
Flower	Red, small size; bract with shallow teeth, broad, do not flare back; calyx unlobed; pedicel 1–2 cm long; anthers deep red	Light pink, large size; bract deeply toothed, 3 teeth, very narrow, do not flare back; calyx 5 lobed; almost sessile flower (pedicel 2–3 mm); anthers very light pink	Red, large size; bract 3 toothed, breadth intermediate flare back; calyx 5-lobed; pedicel about 1 cm long; anthers orange colored	Pink or red, size variable; bract with shallow or deep teeth, width variable (very narrow to broad), in some the bracts flare back; calyx 5-lobed; sessile or 1–2 cm long pedicel; anthers red, orange, or pink
Boll	Boll 3–4 loculed, brown color	Bolls 3 loculed, green color	Bolls 3 loculed, light brown color	Bolls 3–4 loculed, green or brown color
Seed	Blackish seed with long fiber	Light brown seeds with small fibers	Light brown seeds with fibers	Blackish or brownish seeds

tion and showed pollen embryogenesis (fig. 4A−E). The best response was of MS + NAA (2.0 mg/l + BAP (1.0 mg/l). Subjecting flower buds (before the culture of excised anthers) to low temperature (4°−6°C) for 48 hrs had a negative influence on callus formation, but enhanced the frequency of pollen embryogenesis.

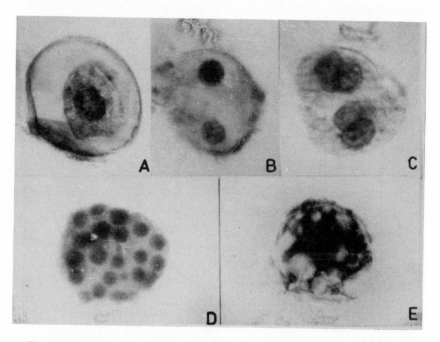

Fig. 4 Various stages of pollen embryogenesis in excised anthers of *Gossypium arboreum* cultured on MS + 2,4-D (2 mg/l) + kinetin (0.2 mg/l) + CW (7%) + sucrose (3%). (A-C) Uni-, bi-, and tetranucleate pollen from anthers cultured for one, two, and three weeks respectively. (D) A multinucleate pollen from 5-week-old culture. (E) A pollen-embryo from a 5-week-old cultured anther. (From Bajaj and Gill, unpublished.)

Establishment of Callus Cultures

Callus cultures have been initiated from embryos, ovules, anthers, and various explants of young *in vitro* grown seedlings of various species. Best callus growth was obtained from the ovules cultured on MS + IAA (1.5 mg/l) + kinetin (0.5 mg/l) + CH (250 mg/l) + sucrose 87.6 mM followed by embryos and anthers. Of the explants used (hypocotyl, cotyledons, roots), the hypocotyl was the best for callus growth. In general, callus induction was genotype-dependent. Of the three species used, callus formation was best for *G. herbaceum,* followed by *G. arboreum* and *G. hirsutum,* in that order. The diploid species invariably grew much faster than the tetraploids.

Callus obtained using this method was subcultured every 3–4 weeks. Cell suspension cultures could be established on media containing NAA (2.0 mg/1) + BAP (0.5 mg/1), or containing various concentrations of 2,4-D (0.2, 0.5, and 1.0 mg/1). The growth response of the three species is summarized in figure 5. Growth response was typically sigmoid. Increased concentrations (0.2–100 mg/1) of 2,4-D had an adverse effect on cell multiplication (fig. 6). At low concentrations of 2,4-D (0.2 mg/1), growth in *G. herbaceum* was faster than in *G. arboreum.* However, at higher concentrations (0.5 and 1.0 mg/1) the response was reversed. The comparison between hypocotyl and ovule-derived cell suspensions showed that ovular callus grew rapidly for all three species (fig. 5).

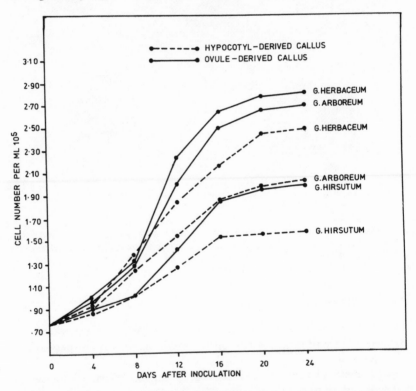

Fig. 5. Comparative growth of the ovule- and hypocotyl-derived callus cell suspensions of *G. arboreum, G. herbaceum,* and *G. hirsutum* grown for 24 days. Note the faster growth rate of ovule-derived callus in all the three species.

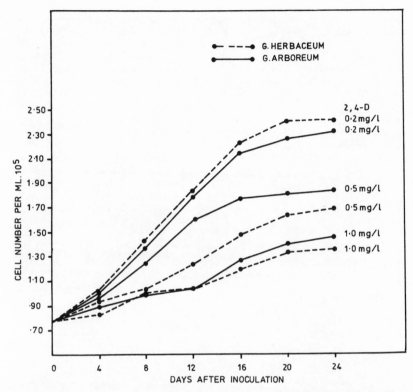

Fig. 6. Effect of various concentrations of 2,4-D (0.2–1.0 mg/l) on growth of callus cell suspensions of *G. herbaceum* and *G. arboreum* grown for 24 days. Note that with increase in the concentration of 2,4-D the growth in suspension.

The callus cultures raised from embryos, ovules, anthers, and hypocotyl segments in *G. arboreum* and *G. herbaceum* were fast growing. The ovule-derived callus grew faster than that from anthers and the hypocotyl in both species. *G. herbaceum* callus was white, and *G. arboreum* callus was red (the color of the donor variety G27). The extent of genetic variability for chromosome number is shown in figure 7A–F and 8A–C. The original chromosome number in both species was 2n = 26; however, in callus the number ranged from haploid (n = 13) to highly polyploid (>104).

The embryos of a cross *G. hirsutum* × *G. arboreum* were cultured on the medium of Smith and colleagues (1977). The resultant callus was subcultured periodically on MS + NAA (2 mg/l) + BAP (0.5 mg/l). The callus was creamy-white, soft, and friable. Cytological examination of one-year-old callus showed a wide range of chromosome numbers varying from less

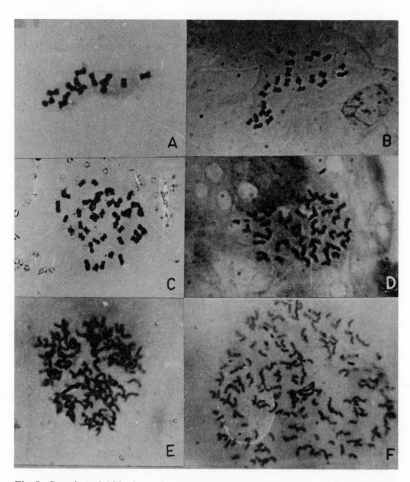

Fig. 7. Genetic variability in ovule- and embryo-derived callus of various diploid species (2n = 26), showing chromsome numbers varying from haploid to highly polyploid. (A) Haploid (n = 13); (B) diploid (2n = 26); (C) triploid (3n = 39); (D) tetraploid (4n = 52); (E) hexaploid (6n = 78); (F) highly polyploid. (From Bajaj and Gill, 1984.)

than 26 to more than 45 (table 12, fig. 8). However, the majority of the cells contained chromosome numbers between 2n = 26 and 41. The cells with less than 26, and more than 45 chromosomes were rare, i.e., 1.4 and 0.84% respectively. No polyploids were observed.

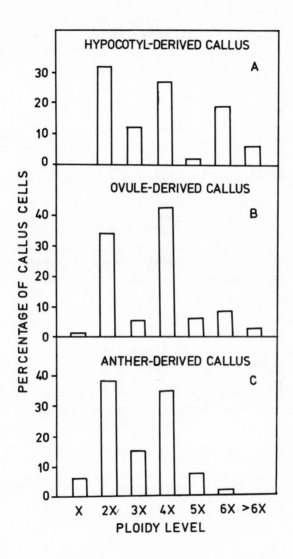

Fig. 8. Histograms showing the extent of genetic variability in callus cells obtained from excised hypocotyl, ovules and anthers of *Gossypium arboreum* (2n = 26). (From Bajaj and Gill, 1984.)

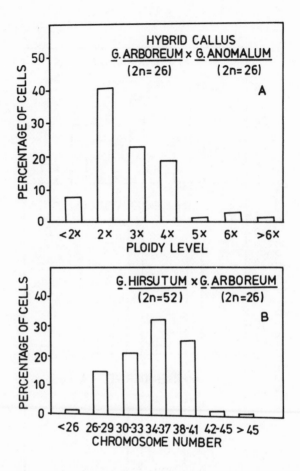

Fig. 9. Histograms showing the extent of genetic variability in the hybrid embryo callus cultures of *G. arboreum* (2n = 26) x *G. anomalum* (2n = 26) (A) and *G. hirsutum* (2n = 52) x *G. arboreum* (2n = 26) (B). The cells with aneuploid numbers are combined with the nearest euploid category in A. (After Bajaj and Gill, 1984.)

Somaclonal Variation

The success of any crop improvement program depends on the usable genetic variability in the base population. In this regard cell and tissue culture is known to be a rich source of variability (see Krikorian et al.,

1983). Callus tissues may undergo endomitosis, chromosome loss, translocation, polyploidy, aneuploidy, gene amplification, cryptic chromosome rearrangement, somatic gene rearrangement, mutations, and other genetic changes (Larkin and Scowcroft, 1981). Though most of these changes may not be of much significance, there may be some that can be selected and isolated. Plants can be regenerated from variant cell lines and evaluated under field conditions. This phenomenon has been exploited in sugar cane (Liu and Chen, 1976), potato (Shepard *et al.*, 1980), Brassica (Horak *et al.*, 1975), and maize (Green et al., 1977) for the improvement of these crops. In cotton (Bajaj and Gill, 1985) two approaches *in vitro*, i.e., interspecific hybridization and establishment of callus cultures, have resulted in a wide range of cultures ranging from haploid to highly polyploid, including many aneuploids.

Wild species of *Gossypium* possess a number of useful traits such as insect resistance, disease resistance, drought resistance, and superior fiber qualities. However, because of various incompatibility barriers, it has not been possible to introgress desirable genes from wild species into commercial crop cultivars. Even if hybridization between different species is effective, pairing and subsequent exchange of genetic material between chromosomes of two species occurs at a low frequency (Stephens, 1950). If a small segment of alien genome carrying desirable genes is transferred to the chromosomes of the cultivated species, then that trait can be stabilized in commercial crop cultivars. It has been speculated (Larkin and Scowcroft, 1981) that tissue culture may generate an environment for enchancing chromosome breakage and reunion events, and thus a tissue culture cycle of hybrid material may provide the means for obtaining the genetic exchange needed between two genomes in an interspecific hybrid. In addition, it is hoped that somatic hybridization can result in hybrids with genetic exchange. Such hybrids would be produced in cell culture in an environment known to produce chromosome rearrangements. In addition, somatic hybridization can be used to produce polyploid plants resulting in fertile F_1 hybrids and thus avoid the use of tedious chromosome doubling procedures.

CONCLUSIONS AND PROSPECTS

Cotton is the most important fiber crop being used in the textile industry. The production of lint per unit area is very low, and one of the important reasons for low yield is the susceptibility of the crop to insect diseases and various environmental stresses including drought and cold. Conventional methods of breeding have not been adequate to cope with these problems because of the lack of sufficient genetic variability among cultivated types. Desirable genes for insect and disease resistance, drought tolerance, and superior fiber qualities that are present in some wild *Gossypium* species cannot be transferred into cultivated types because of various incompatibility barriers. Therefore, there is a strong need to resort to unconventional biotechnology approaches that can be helpful to increase genetic variability.

In vitro techniques form an important component of biotechnology, and have the potential not only to improve present cultivars, but also for the synthesis of novel plants and early release of high-yielding plants resistant to various diseases, pests, stresses, temperature, and such.

It has been demonstrated that cell and tissue culture methods can be manipulated for the induction of genetic variability in cotton. Embryo/ ovule culture methods have been used to produce hybrids between different species. Irradiation of seeds and pollen in segregating generations may increase chances of transferring a few genes of the wild species into the cultivated types. *In vitro* environment may also be suitable for enhancing chromosome breakage and reunion events in the cultured cells. A tissue culture cycle of the hybrid material can enhance the frequency of genetic exchange between chromosomes from different species. The hybrid callus will then be a rich source of genetic variability. Once the problem of regeneration is solved, somatic hybrid cells and cellular genetic variants can be evaluated among regenerated plants, and may be included in breeding programs.

More concrete work needs to be done to induce the regeneration of complete plants from callus, pollen, and protoplasts. Once this is achieved, this system can then be utilized for more fruitful genetic engineering, somatic hybridization, and mutation studies.

REFERENCES

Afzal, M., S. M. Sikka, and A. Rahman. 1945. Cytogenetic investigations in some aboreum-anomalum crosses. Indian J. Genet., 5:82–91.
Bajaj, Y. P. S. 1979a. Technology and prospects of cryopreservation of germplasm. Euphytica 28:267–85.
Bajaj, Y. P. S. 1979b. Establishment of germplasm banks through freeze-storage of plant

tissue culture and their implications in agriculture. *In* W. R. Sharp et al. (eds.), Plant cell and tissue culture: principles and applications, pp. 745–74. Ohio State University Press, Columbus.

Bajaj, Y. P. S. 1982. Survival of anther and ovule-derived cotton callus frozen in liquid nitrogen. Curr. Sci. 51:139–40.

Bajaj, Y. P. S. 1983a. Haploid protoplasts. *In* K. L. Giles (ed.), Plant protoplasts, pp. 113–41. Int. Rev. Cytol. Suppl. 16, Academic Press, New York.

Bajaj, Y. P. S. 1983b. *In vitro* production of haploids. *In* D. A. Evans et al. (eds.), Handbook of plant cell culture. I. Techniques for propagation and breeding, pp. 228–87. Macmillan, New York.

Bajaj, Y. P. S. 1983c. Cryopreservation and international exchange of germplasm. *In* S. K. Sen and K. L. Giles (eds.), Plant cell culture in crop improvement, pp. 19–41. Plenum Press, New York.

Bajaj, Y. P. S. 1985. Cryopreservation of embryos. *In* K. Kartha (ed.), Cryopreservation of plant cells, tissues, and organs. CRC Press, Boca Raton, Fla. (In press.)

Bajaj, Y. P. S., and M. Bidani. 1980. Differentiation of genetically variable plants from embryo-derived callus of rice. Phytomorphology 30:290–94.

Bajaj, Y. P. S., K. S. Gill, and G. S. Sandha. 1978. Some factors enhancing the *in vitro* production and hexaploid triticale (*Triticum durum* × *Secale cereale*). Crop Improv. 5:62–72.

Bajaj, Y. P. S., and M. S. Gill. 1985. *In vitro* induction of genetic variability in cotton (*Gossypium* spp.). Theoret. Appl. Genet. 70:363–68.

Bajaj, Y. P. S., P. Kumar, M. M. Singh, and K. S. Labana. 1982. Interspecific hybridization in the genus *Arachis* through embryo culture. Euphytica 31:365–70.

Bajaj, Y. P. S., and J. Reinhert. 1977. Cryobiology of plant cell cultures and establishment of gene-banks. *In* J. Reinert and Y. P. S. Bajaj (eds.), Applied and fundamental aspects of plant cell, tissue, and organ culture, pp. 757–77. Springer-Verlag, Berlin.

Barrow, J., F. Katterman, and D. William. 1978. Haploid and diploid callus from cotton anthers. Crop Sci. 18:619–22.

Beasley, C. A., and I. P. Ting. 1973. The effects of plant growth substances on *in vitro* fiber development from fertilized cotton ovules. Amer. J. Bot. 60:130–39.

Beasley, C. A., I. P. Ting, and L. A. Feigen. 1971. Improvements in fiber yield may come from test tube cotton. Calif. Agri. 2:6–8.

Beasley, J. O. 1940a. The origin of American tetraploid *Gossypium* species. Amer. Nat. 74:285–86.

Beasley, J. O. 1940b. Hybridization of American 26-chromosome and Asiatic 13-chromosome species of *Gossypium*. Jour. Agric. Res. 60:175–87.

Beasley, J. O. 1942. Meiotic chromosome behaviour in species, species hybrids, haploids, and induced polyploids of *Gossypium*. Genetics 27:25–54.

Bhojwani, S. S., J. B. Power, and E. C. Cocking. 1977. Isolation, culture, and division of cotton callus protoplasts. Plant Sci. Lett. 8:85–89.

Blank, L. M., and D. C. Allison. 1963. Frequency of polyembryony in certain strains of *Gossypium hirsutum* L. Crop. Scie. 3:97–98.

Chaudhari, H. K. 1978. The use of semigamy in the production of cotton haploids. Bull. Torrey Bot. Club 105:98–103.

Cherry, J. P., F. R. H. Katterman, and J. E. Endrizzi. 1970. Comparative studies of seed proteins of species of *Gossypium* by gel-electrophoresis. Evolution 24:431–77.

Crocomo, O. J., and N. Ochoa-Alejo. 1983. Herbicide tolerance in regenerated plants. *In* D. A. Evans et al. (eds.), Handbook of plant cell culture. I. Techniques for propagation and breeding, pp. 770–81. Macmillan, New York.

D'Amato, F. 1977. Cytogenetics of differentiation in tissue and cell cultures. *In* J. Reinert and Y. P. S. Bajaj (eds.), Applied and fundamental aspects of plant cell, tissue, and organ culture, pp. 343–56. Springer-Verlag, Berlin.

Davis, D. G., K. E. Dusbabek, and R. A. Hoerauf. 1974. *In vitro* culture of callus tissues and cell suspensions from okra (*Hibiscus esculentus* L.) and cotton *G. Hirsutum* L.). In Vitro 9:395–98.

Edwards, G. A., and M. A. Mirza. 1979. Genomes of the Australian wild species of cotton. II. The designation of a new G-genome for *Gossypium bickii.* Can. J. Genet. Cytol. 21:367–72.

Eid, A. A. H., E. Delange, and L. Waterkeyn. 1973. *In vitro* culture of fertilized cotton ovules. I. The growth of cotton embryos. Cellule 69:361–71.

El-Nockrashy, A. S., J. G. Simmons, and V. L. Frampton. 1969. A chemical survey of seeds of the genus *Gossypium* Phytochemistry 8:1949–58.

Evans, D. A., W. R. Sharp, P. V. Ammirato, and Y. Yamada (eds.). Handbook of plant cell culture. I. Techniques for propagation and breeding. Macmillan, New York.

FAO production yearbook 1981. 35:135–36.

Finer, J. J., and R. H. Smith. 1982. Isolation and culture of protoplasts from cotton (*Gossypium klotzschianum* Andress) callus cultures. Plant Sci. Lett. 26:147–51.

Gerstel, D. U. 1973. Chromosomal translocations in interspecific hybrids of the genus *Gossypium.* Evolution 7:234–44.

Giles, K. L. (ed.). 1983. Plant protoplasts. Int. Rev. Cytology Suppl. 16. Academic Press, New York.

Gill, M. S., and Y. P. S. Bajaj. 1984a. In vitro production of interspecific hybrids in cotton. Curr. Sci. 53:102–4.

Gill, M. S., and Y. P. S. Bajaj. 1984b. Interspecific hybridization in the genus *Gossypium* through culture. Euphytica 33:305–11.

Gosal, S. S., and Y. P. S. Bajaj. 1983. Interspecific hybridization between *Vigna mungo* and *Vigna radiata* through embryo culture. Euphytica 32:129–37.

Govila, O. P. 1970. Fertilization and seed development in crosses between *Gossypium arboreum* and *G. raimondii.* Indian J. Genet. 30:152–56.

Green, C. E., R. L. Phillips, and A. S. Wang. 1977. Cytological analysis of plants regenerated from maize tissue cultures. Maize Genet. Newslett. 51:53–54.

Gresshoff, P. M., and C. H. Doy. 1972. Development and differentiation of haploid *Lycopersicon esculentum* (tomato). Planta 107:161–70.

Harland, S. C. 1940. New polyploids in cotton by the use of colchicine. Trop. Agric. (Trin.) 17:53–54.

Horak, J., J. Lustinec, J. Mericek, M. Kaminek, and D. Polackova. 1975. Regeneration of diploid and polyploid plants from the stem pith explants of diploid narrow stem kale (*Brassica oleracea* L.). Ann. Bot. 3:571–77.

Hsi, Y. L., and K. I. Wu. 1981. Callus formation induced from the anther culture of upland cotton (*Gossypium hirsutum* L.) (Abstract). *In* Proc. Symp. Plant Tissue Culture, pp. 239–40. Pitman Publishing, London.

Hsu, C. L., and J. M. Stewart. 1976. Callus induction by (2-chloroethyl) phosphonic acid on cultured cotton ovules. Physiol. Plant. 36:150–53.

Iyengar, N. K., V. Santhanam, and M. Leela. 1958. Further studies in *arboreum-anomalum* crosses. Indian Cott. Gr. Rev. 12:369–75.

Johnson, B. L., and M. M. Thein. 1970. Assessment of evolutionary afinities in *Gossypium* by protein electrophoresis. Amer. J. Bot. 57:1081–92.

Joshi, P. C. 1960. *In vitro* growth of cotton ovules. *In* Plant embryology: a symposium, pp. 199–204. Council of Scientific and Industrial Research, New Delhi.

Joshi, P. C., and B. M. Johri. 1972. *In vitro* growth of ovules of *Gossypium hirsutum*. Phytomorphology 22:195–207.

Kalyanaraman, S. M., and V. Santhanam. 1955. A note on the performance of some interspecific hybrids involving wild species of *Gossypium*. I. *Aboreum-anomalum* crosses. Indian Cott. Gr. Rev. 11:136–40.

Katterman, F., M. D. Williams, and W. F. Clay. 1977. The influence of a strong reducing agent upon the initiation of callus from the germinating seedlings of *Gossypium barbadense* L. Physiol. Plant. 40:198–200.

Khasanov, M. M., and R. G. Butenko. 1979. Isolated culture from cotyledons of cotton (*Gossypium hirsutum*). Fiziol. Rest (Mosc) 26:103–5 (in Russian).

Krikorian, A. D., S. A. O'Connor, and M. S. Fitter. 1983. Chromosome number variation and karyotype stability in cultures and culture-derived plants. *In* D. A. Evans et al. (eds.), Handbook of plant cell culture. I. Techniques for propagation and breeding, pp. 541–81. Macmillan, New York.

Larkin, P. J., and W. R. Scowcroft. 1981. Somaclonal variation: a novel source of variability from cell cultures for plant improvement. Theor. Appl. Genet. 60:197–214.

Lee, J. A. 1970. On the origin of haploid/diploid twinning in cotton. Crop Sci. 10:453–54.

Linsmaier, E. M., and F. Skoog. 1965. Organic growth factor requirements of tobacco tissue cultures. Physiol. Plant. 18:100–127.

Liu, M. C., and W. H. Chen. 1976. Tissue and cell culture as aids to sugarcane breeding. I. Creation of genetic variability through callus culture. Euphytica 25:393–403.

Lofland, H. B. 1950. *In vitro* culture of cotton embryos. Bot. Gaz. 3:301–7.

Mauney, J. R. 1961. The culture of immature cotton embryos. Bot. Gaz. 122:205–9.

Mauney, J. R., J. Chappel, and B. J. Ward. 1967. Effects of malic acid salts on growth of young cotton embryos *in vitro*. Bot. Gaz. 128:198–200.

Mehboob Ali, S., V. V. Krishnayya, and K. Satyanarayna Murthy. 1964. A note on an interspecific hybrid in cotton (*Gossypium arboreum* race *Indicum* × *G. stocksii*). Indian Cott. Gr. Rev. 18:107–8.

Mellor, F. C., and R. Stace-Smith. 1977. Virus-free potatoes by tissue culture. *In* J. Reinert and Y. P. S. Bajaj. (eds.), Applied and fundamental aspects of plant cell, tissue, and organ culture, pp. 616–35. Springer-Verlag, Berlin.

Mirza, M. A., and A. L. Shaikh. 1980. The currently recognised species of *Gossypium*: their relationship and description. Pakistan Cottons 24:283–97.

Murashige, T., and F. Skoog. 1962. A revised medium for rapid growth and bio-assays with tobacco tissues cultures. Physiol. Plant. 15:473–97.

Pandey, S. N. 1976. Utilization of cotton seed and its products. Cotton Devel. 6:4–9.

Phillips, L. L. 1962. Segregation in new allopolyploids of *Gossypium*. IV. Segregation in New World × Asiatic and New World × wild American hexaploids. Amer. J. Bot. 49:51–57.

Phillips, L. L. 1963. The cytogenetics of *Gossypium* and the origin of New World cottons. Evolution 17:460–69.

Phillips, L. L. 1977. Interspecific incompatibility in *Gossypium*. IV. Temperature conditional lethality in hybrids of *G. klotzschianum*. Amer. J. Bot. 64:914–15.

Phillips, L. L., and J. F. Merritt. 1972. Interspecific compatibility in *Gossypium*. I. Stem histogenesis of *G. hirsutum* × *G. Gossypxiodes*. Amer. J. Bot. 59:203–8.

Phillips, L. L., and M. A. Strickland. 1966. The cytology of a hybrid between *Gossypium hirsutum* and *G. longicalyx*. Can. J. Genet. Cytol. 8:91–95.

Phillips, L. L., and R. K. Reid. 1975. Interspecific incompatibility in *Gossypium*. II. Light and electron mircroscope studies of cell necrosis and tumorgenesis in hybrids of *G. klotzschianum*. Amer. J. Bot. 62:790–96.

Price, H. J., and R. H. Smith. 1979. Somatic embryogenesis in suspension cultures of *Gossypium klotzschianum* Andress. Planta 145:305–7.

Price, H. J., R. H. Smith, and R. W. Grumbles. 1977. Callus cultures of six species of cotton (*Gossypium* L.) on defined media. Plant Sci. Lett. 10:115–19.

Pundir. N. S. 1972. Experimental embryology of *Gossypium arboreum* and *G. hirsutum* L. and their reciprocal crosses. Bot. Gaz. 133:7–26.

Rani, A., and S. S. Bhojwani. 1976. Establishment of tissue cultures of cotton. Plant Sci. Lett. 7:163–69.

Reinert, J., and Y. P. S. Bajaj (eds.). 1977. Applied and fundamental aspects of plant cell, tissue, and organ culture. Springer-Verlag, Berlin.

Ruyack, J., M. R. Downing, J. S. Chang, and E. D. Mitchell, Jr. 1979. Growth of callus and suspension culture bells from cotton varieties (*Gossypium hirsutum* L.) resistant and susceptible to *Xanthomonas malvacearum* (E.F. gm) Poros. In Vitro 15:368–73.

Sandstedt, R. 1975. Habituated callus cultures from two cotton species. Beltwide Cotton Prod. Res. Conf. Proc. 26:368–73.

Schenk, R. U., and A. C. Hildebrandt. 1972. Medium and technique for induction and growth of monocotyledonous and dicotyledonous plant cultures. Can. J. Bot. 50:199–204.

Shepard, J. F., D. Bidney, and E. Shahin. 1980. Potato protoplasts in crop improvement. Science 28:17–24.

Skirvin, R. M. 1978. Natural and induced variation in tissue culture. Euphytica 27:241–66.

Skovsted, A. 1934. Cytological studies in cotton. II. Two interspecific hybrids between Asiatic and New World cottons. J. Genet. 28:407–24.

Skovsted, A. 1937. Cytological studies in cotton. IV. Chromosome conjugation in interspecific hybrids. J. Genet. 34:97–134.

Smith, R. H., H. J. Price, and J. B. Thaxton. 1977. Defined conditions for the initiation and growth of cotton callus *in vitro*. I. *Gossypium arboreum*. In Vitro 13:329–34.

Sreerangasamy, S. R. and V. S. Raman. 1963. Studies on the morphology and cytology of *Gossypium arboreum* X *G. anomalum* hybrids. Indian Cott. Gr. Rev. 17:353–59.

Stephens, S. G. 1950. The internal mechanism of speciation in *Gossypium*. Bot. Rev. 16:115–49.

Stewart, J. M., and C. L. Hsu. 1977a. In-ovulo embryo culture and seedling development of cotton (*Gossypium hirsutum* L.) Planta 137:113–18.

Stewart, J. M., and C. L. Hsu. 1977b. Influence of phytohormones on the response of cultured cotton ovules to (2-chloroethyl) phosphonic acid. Physiol. Plant. 39:79–85.

Stewart, J. M., and C. L. Hsu. 1978. Hybridization of diploid and tetraploid cottons through in-ovulo embryo culture. J. Heredity 69:404–8.

Turcotte, E. L., and C. V. Feaster. 1969. Semigametic production of haploids in pima cotton. Crop Sci. 9:653–55.

Weaver, J. B., 1957. Embryological studies following interspecific crosses in *Gossypium*. I. *G. hirsutum* X *G. arboreum*. Amer. J. Bot. 44:209–14.

Weaver, J. B., 1958. Embryological studies following interspecific crosses in *Gossypium*. II. *G. arboreum* X *G. hirsutum*. Amer. J. Bot. 45:10–16.

White, P. R. 1939. Potentially unlimited growth of excised plant callus in an artificial nutrient. Amer. J. Bot. 26:59–64.

White, P. R. 1943. A handbook of plant tissue culture. Jaques Cattell, Lancaster.

White, P. R. 1963. The cultivation of animal and plant cells. Ronald Press, New York.

Williams, M. D. 1978. Conditions for optimal growth and differentiation of *Gossypium hirsutum* L. tissue culture. Dissertation Abstract International B39: 2041B.

Zenkteler, M., and G. Melchers. 1978. In vitro hybridization by sexual methods and by fusion of somatic protoplasts. Theor. Appl. Genet. 52:61–90.

OTTO J. CROCOMO, NEFTALI OCHOA-ALEJO,
CHRISTINA H. R. P. GONCALVES, AND OSNY O. S. BACCHI

Development of Improved Sugar Cane Varieties through Genetic Engineering Techniques

10

ABSTRACT

In Brazil, sugar cane is at the heart of the renewable energy sources program since it is an economically important source of both sucrose and ethanol. Every effort toward its improvement in yield, sucrose content, pesticide tolerance, and disease resistance is desirable. The techniques of biotechnology can be applied at the cellular level from the protocol developed in our laboratories for regeneration of plants from cultures of sugar cane *in vitro*. Sugar cane plant regeneration, protoplast isolation and fusion, herbicide tolerance, and disease resistance are discussed in the light of data from our laboratories.

INTRODUCTION

Green plants manufacture food, fiber, oil, and produce energy using the building blocks made from CO_2, N_2, O_2, minerals, and water present in the environment. Plants are specialized in the production of one or more of these products, and among crop plants, sugar cane is one of the most important sources of food (sucrose) and energy (ethanol), especially now that we are facing a shortage of fuel due to the "petrol crisis." The world production of sugar from sugar cane in 1979 was 35×10^6 MT with Brazil occupying second place. In the agricultural year 1980/81, the total area cultivated with sugar cane in Brazil was 2.67 million ha, the South Central region being the main area where this crop is cultivated (Brugnaro & Lavorenti, 1981).

To attain the goal of 10.7×10^9 liters alcohol (ethanol), the area to be cultivated with sugar cane should increase to 5 million ha. To minimize this requirement for new acreage and to meet the goals for ethanol production, new varieties need to be released with increased yield, sugar content, disease resistance, and tolerance to pesticides and environmental stress.

The sugar cane plant is a "source-to-sink" system (fig. 1). Sucrose is manufactured in the leaf-blades (the "source") as an end product of the C4-photosynthetic cycle, through biochemical reactions catalyzed by sucrose synthetase and sucrose phosphate synthetase (Suzuki and Crocomo, unpublished). From the leaf blades, sucrose passes on to the storage parenchyma in the stalk (the "sink") through the leaf sheaths. The level of activity of invertase present in the leaf sheaths is a control mechanism that regulates the accumulation of sucrose in the stalks (Suzuki and Crocomo, unpublished; Sampietro et al., 1980).

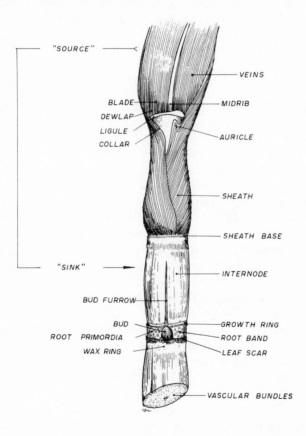

Fig. 1. Diagram of stalk and leaf of sugar cane showing the "source-to-sink" system.

Being an economically important crop, sugar cane has been the subject of breeding programs to obtain better varieties. Breeding and mutation methods have been the principle channels through which this induced variability is achieved. The manipulation of genetically variable populations leads to the identification of variants that can be selected for the desired characteristics. Very often a conventional breeding program takes more than 8–10 years to release a new sugar cane variety with a set of stable genes expressing the desired characteristics. The new approaches using *in vitro* techniques, such as recombinant-DNA and protoplast technology (Crocomo and Ochoa-Alejo, 1983) will lead to new genotypes that can be selected from cell cultures. The key step in the development of a new crop variety through this new technique is the ability to regenerate plants with high frequency from genetically modified cells using tissue culture (Heinz et al., 1977; Crocomo et al., 1979; Liu, 1981).

PLANT CELL CULTURE

Plant cell culture is one of the more significant techniques for plant improvement. Cell culture can be used for cloning or for somaclonal variation. For clonal propagation, meristematic explants are generally used because such cultures have a limited degree of genetic variability. Nonmeristematic explants or even single cells (from suspension cultures) can be used for production of genetic variability.

It has long been recognized that one of the greatest potentials for crop improvement is the production of new hybrids using protoplast fusion (Nickell and Torrey, 1969; Wittner, 1974). The development of techniques of somatic hybridization will provide alternative means to achieve genetic recombination and to increase genetic diversity (Gamborg et al., 1974). A more selective procedure that may be possible in the future is the transfer of isolated DNA or specified genes (Scowcroft, 1977) from one source into a recipient plant. Utilizing such novel approaches for genetic manipulation *in vitro* will depend upon the methods of culturing tissue, cells, embryos, anthers, and isolated protoplasts (and their fusion products) and of course the regeneration of plants from these cultures. Since somatic hybridization offers short-term potential to "create" hybrids with disease resistance, tolerance to stress conditions, and improved product quality and growth characteristics, this article will concentrate on prospects for somatic hybridization. The basic procedures involved in protoplast fusion and somatic hybridization in plants—protoplast isolation; protoplast fusion and growth of the hybrid cells; selection of the hybrid cells followed by their cloning; and regeneration of hybrid plants (Gamborg et al., 1979)—are described below.

Protoplasts

Protoplasts are wall-less plant cells. Absence of the wall facilities fusion with other protoplasts, providing the opportunity for somatic hybridization. Because of their single-cell nature, plant protoplasts have simplistically been likened to microbes as objects for the techniques of genetic engineering. Theoretically, what is needed is induction of a desired genetic change (e.g., through introduction of foreign DNA or fusion) into a population of protoplasts, followed by appropriate selection pressure first to recover a cell and, ultimately, a plant having desired characteristics (Larkin and Scowcroft, 1981a).

Protoplast Isolation

For protoplast isolation, plant tissue is treated with a mixture of enzymes that degrade the cell wall components and preserve the structure and viability of the resulting protoplasts. A wide range of plant tissues can be used to isolate protoplasts, the most convenient material being leaf mesophyll and cells from liquid cell suspension cultures. Donor plants for protoplast isolation must be grown under precisely controlled conditions of nutrition, temperature, light intensity, and photoperiod. Only when isolated from properly grown plants will protoplasts undergo divisions when placed in a culture medium, in which growth and proliferation are induced.

In our laboratories, protoplasts from shoot apices have been isolated from shoots of *Saccharum* spp. regenerated from callus induced from meristematic tissue maintained *in vitro* (Evans et al., 1980). Viable protoplasts were released from this callus when treated with 3% cellulase, 4% pectinase, and 1.5% hemicellulase. When these protoplasts were placed in culture medium, cell-wall regeneration was observed in 7–10 days. The callus was produced in 21 days.

Protoplast Fusion

In higher plants, sexual reproduction takes place by fusion between two gametes, which are highly specialized. This is followed by the fusion of gametic nuclei, resulting in the formation of the sexual hybrid. Unlike sexual hybridization, somatic hybridization combines not only the nuclei forming a hybrid nucleus but also the cytoplasms (Cocking, 1979a,b). The cytoplasm with organelles that contain DNA are important to the control of a cell's metabolic activities (Hsam and Larter, 1974). This leads to important changes that can be detected in regenerated somatic hybrid plants.

The importance of cytoplasmic mixing obtained from protoplast fusion for plant breeding is great. Liu (1981), in Taiwan, has been working on sugar cane protoplasts to modify the cell contents in an attempt to obtain somatic hybridization between sugar cane and soybean (*Glycine max*), a legume crop. He has succeeded in isolating protoplasts readily from sugar cane meristematic leaf tissue. It was noticed that about 7% of the freshly isolated protoplasts contained more than one nucleus, indicating that spontaneous fusion had occurred during protoplast isolation. Protoplast fusion can be induced by treatment with polyethylene glycol (PEG). Liu (1981) also isolated protoplasts from soybean root nodules. The intention was to transfer nitrogen-fixing genes from the legume to sugar cane through protoplast fusion.

The opportunity to increase cytoplasmic variability through protoplast fusion was only recently recognized. A cytoplasmic hybrid (cybrid) will be obtained when two protoplasts are fused. The stability of the hybrid nuclei is greatly influenced by the species fused. If for some reason there is a loss of the chromosomes of one of the species, or if one of the nuclei dies, no somatic hybrids are produced, but a range of novel cybrids can theoretically be obtained.

Enucleate subprotoplasts or nucleate microplasts (Bilkey and Cocking, 1980) have been fractionated from protoplasts and may provide suitable enucleated units to produce cybrids (Cocking, 1981). Using this technique to produce enucleate units, the transfer of cytoplasmically based herbicide resistance between sexually compatible or incompatible species may be facilitated, as has been demonstrated with respect to atrazine resistance (Gressel et al., 1983).

Regeneration of Plants

Although callus formation has been obtained from protoplasts derived from sugar cane cells (Evans et al., 1980; Maretzki and Nickell, 1973), protoplasts have been more difficult to culture and to regenerate plants from than cultured cells. Callus induction followed by plant regeneration of sugar cane has been obtained by several groups (Heinz et al., 1977; Crocomo et al., 1979; Liu, 1977). The protocol developed in our laboratories (Crocomo et al., 1979) permits callus induction from inner leaves of sugar cane shoots obtained from nodal cuttings, followed by shoot induction and rooting. Disinfected internode explants of ca. 0.5 cm are aseptically placed in glass bottles containing conditioning medium for the initiation of callus proliferation. Once formed, callus tissue is subcultured onto a shoot induction medium for the development of shoots. This latter process requires several weeks depending on the variety tested at a frequency of 10–40 shoots surgically separated and subcultured on a rooting medium. Depending on the sugar cane variety, rooting occurs at a high frequency. Once rooted, sugar cane plantlets are transferred to pots containing a mixture of sand and vermiculite in a high-humidity environment, then to soil in pots in a greenhouse before field trials.

The protocol has three obvious applications: (1) production of a large number of plantlets within a short time interval under controlled environment; (2) high-frequency regeneration of plantlets throughout the year without seasonal restrictions; (3) production of plantlets from nonflowering clones. This technique can be used to propagate mutants derived from mutation breeding programs or plants regenerated from genetically altered protoplasts, and shortens the duration of time required for field tests and the eventual establishment in plantations of new varieties.

Somaclonal Variation

Tissue culture *per se* appears to be a novel source of genetic variability. Plants derived *in vitro* from cell culture of somatic tissue are called somaclones (Larkin and Scowcroft, 1981a). Somaclonal variation has been attributed to variation already present in the plant material itself and/or to the changes that occur during the tissue culture cycle (Nickell and Heinz, 1973). The latter involves the establishment of a more or less dedifferentiated cell mass (e.g., callus), proliferation of a number of cell generations, followed by regeneration of plants (Shepard et al., 1980).

Examples of variations that have been observed include differences in the morphogenetic pattern of cell populations (Chaturvedi and Mitra, 1975) and cell growth behavior (Blakely and Steward, 1964; Davey et al., 1971) or karyotype changes (Murashige and Nakano, 1967; Bayliss, 1980). On the other hand, genetic variability has also been detected by alternations in the capacity to synthesize and accumulate secondary products (Zenk et al., 1977; Tabata et al., 1978; Constabel et al., 1981).

Cell suspensions prepared from single-cell cultures derived from 5 varieties of sugar cane were shown to have variation in chromosome number and karyotype after 6 years in culture (Heinz et al., 1969). Cells in liquid cultures derived from variety F1164 (2n = 108) had 2n chromosome numbers between 92 and 191 (Liu, 1981).

Following passage through a tissue culture cycle, the genetically unstable sugar cane variety H50-7209 gave rise to somaclones with a high frequency of genetic variability, whereas callus-derived plantlets of the genetically stable var. H37-1933 remained stable with 2n = ca. 106 (Heinz and Mee, 1971). Karyotype variation has been observed previously in other sugar cane somaclones (Liu and Chen, 1976). Callus-derived plants from var. F156 (2n = 114) ranged from 86 to 108 and var. F164 from 88 to 108. On the other hand, the genetically stable var. F146 gave rise to somaclones with a less degree of genetic variability than unstable clones.

Among 13 somaclones derived from varieties Fl156 and F164, sugar cane production, sugar production, and stalk number of the somaclone 70-6132 was higher in 32%, 34%, and 6% respectively than the donor (var. F164), and 20%, 16%, and 8% respectively higher than the variety that is considered to be the best cultivated in Taiwan (Liu and Chen, 1978).

Isoenzyme Patterns

Isoenzymatic profiles obtained through polyacrylamide disc gel electrophoresis have been used in sugar cane mainly for identification of varieties and clones (Thom and Maretzki, 1970). The enzymes that have been studied are esterase, peroxidase, and amylase in leaf extracts of plants of the same age (Hammil and Heinz, 1970). As with variation in chromosome number, variations in the electrophoretic pattern of isoenzymes of sugar cane tissues have been detected (Heinz et al., 1969). However, isoenzyme patterns can be related to disease resistance (Shannon, 1968), drought resistance (Mali and Mehta, 1977), and other characteristics not detected morphologically.

Recently the electrophoretic pattern of the isoenzymes of esterase and peroxidase of several Brazilian varieties of sugar cane has been analyzed. In preliminary work (Goncalves et al., 1981), mature buds, green leaves, and roots were used. The different tissues presented qualitative differences among the varieties. Buds were the best tissue to compare isoenzyme profiles among eleven varieties (fig. 2). The roots and leaves (but not buds) had variation in the isoenzyme patterns depending on different physiological age and environmental conditions. To detect changes due to the influence of physiological age, leaves of two morphologically similar varieties, CB41-14 and CB41-76, were examined before and after true root formation. No qualitative differences were detected between the two varieties; the isoenzyme patterns differed only quantitatively. Some bands were more intensely stained in the extracts of plants without true roots than in those with true roots.

SUGARCANE
VARIETIES

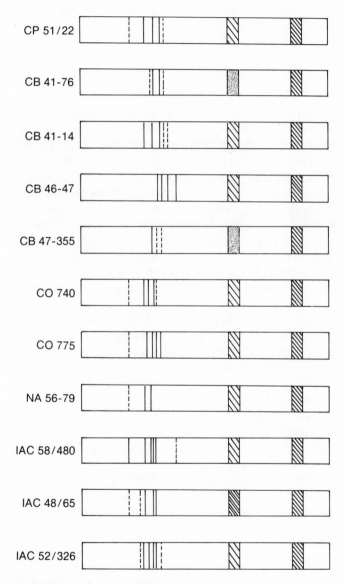

Fig. 2. Diagram of esterase isoenzymes profile of 12-month-old sugar cane buds.

The genetic constitution of regenerated plants from callus has also been studied using the isoenzyme pattern. The esterase zymogram of the callus-derived line 70-6132 had two bands less than its donor, cv. Fl164 (28 bands). The callus-derived line 70-6136 was similar to its donor (cv. Fl164) in number of bands, but different in the pattern of bands (Liu and Chen, 1976).

Herbicide Tolerance

Sugar cane tissue cultures in the plant regeneration stage are suitable to study the physiological and biochemical effects of herbicide and to rescue genetic variants having tolerance to toxic compounds such as herbicides.

Using the protocol developed at our laboratories that permits selection for herbicide tolerance using callus-derived plantlets (Crocomo and Ochoa-Alejo, 1983), the effects of several levels of the herbicides Ametryn (atrazine), Dalapon (aliphytic), and the sugar cane ripener Mefluidide on cells of three different sugar cane varieties (NA56-79, IAC48-65, CB41-76) maintained *in vitro*, were compared (Crocomo et al., 1981). Cells (callus) of variety CB41-76 had a higher sensitivity to Dalapon, Ametryn, and ripener than those of varieties IAC48-65 and NA56-79 (fig. 3). Growth of callus of CB41-76 was lower in presence of 40 ppm Ametryn, 20 ppm Dalapon, or Mefluidide. In the presence of Dalapon and Mefluidide, the callus mass was dark brown in color, followed by necrosis. Callus growth of variety IAC48-65 was inhibited at 80 ppm of either Dalapon or Mefluidide. These chemicals were less effective in IAC48-65 than in callus of variety CB41-76. On the other hand, 80 ppm Ametryn induced a discrete callus growth inhibition in variety NA56-79, with Dalapon or Mefluidide having no detectable effect on callus growth.

As can be seen in figure 4, a decrease in protein content in the cells of the three varieties was observed in the presence of Ametryn. Variety CB41-76 was more affected, followed by varieties NA56-79 and IAC48-65. Ametryn caused drastic increase in reducing sugar content in cells of variety CB41-76 (fig. 5). On the other hand, Dalapon present in the culture medium increased the reducing sugar level of variety CB41-76 but to a lesser extent than the Ametryn-induced increase. In variety IAC48-65, the effect of the two herbicides is reversed.

Disease Resistance

Tissue culture is also an excellent tool to study host-parasite relationship and to select for disease-resistant plants. This approach is based on some

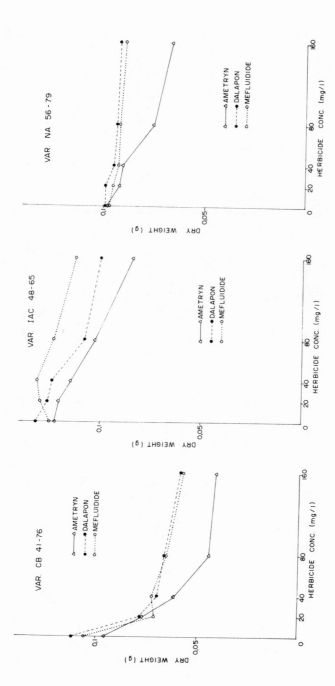

Fig. 3. Dry weight of the callus of 3 sugar cane varieties in the presence of Ametryn, Dalapon, and the ripener Mefluidide.

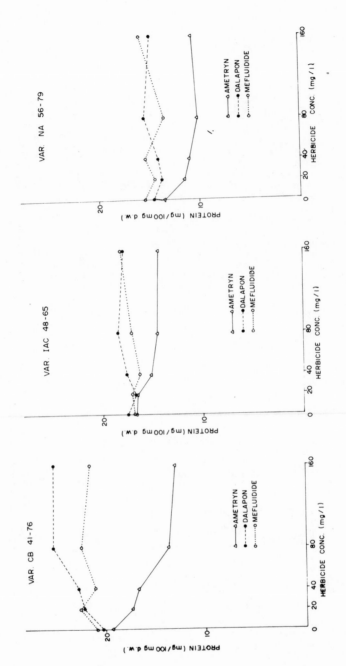

Fig. 4. Effect of Ametry, Dalapon, and the ripener Mefluidide on the protein content of sugar cane varieties.

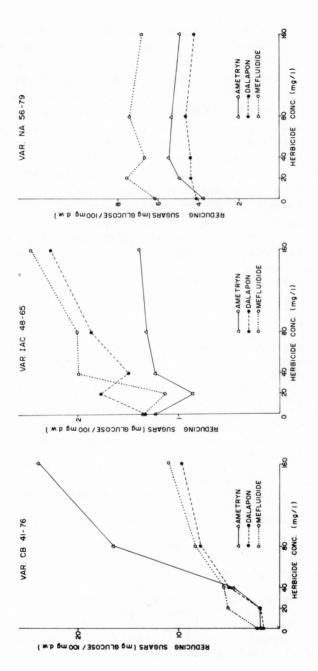

Fig. 5. Effects of Ametry, Dalapon, and the ripener Mefluidide on the reducing of sugar content on sugar cane varieties.

observations that pathogens do not grow on callus obtained from explants derived from resistant plant species, but grow on callus from sensitive plants (Chen et al., 1979). The regenerated plants from resistant callus would be assayed for disease resistance. This approach is of great importance for sugar cane improvement, since it may permit the development of disease-resistant subclones. Eyespot-resistant somaclonal variant subclones from susceptible parent clones have already been developed (Heinz, 1973) as well as resistant subclones from Fiji-susceptible and downy mildew-susceptible parent clones (Krishnamurthi and Tlaskal, 1974; CSIRO, 1979; 1980). Some of these subclones also had enhanced sucrose yield when compared with the donor plant (Goncalves et al., 1981).

Larkin and Scowcroft (1981b) observed a high range of somaclonal variation in susceptibility to eyespot disease in sugar cane plants obtained through tissue culture techniques. A high number of tissue culture-derived plants had high resistance to the disease that was maintained through several asexual generations.

REFERENCES

Bilkey, P. C., and E. C. Cocking. 1980. Isolation and properties of plant microplasts: newly identified subcellular units capable of wall synthesis and division into separate microcells. Eur. J. Cell Biol. 22:502.
Blakely, L. M., and F. C. Steward. 1964. Growth and organized development of cultured cells. 7. Cellular variation. Am. J. Bot. 51:809–20.
Bayliss, M. W. 1980. Chromosomal variation in plant tissues in culture. Intern. Rev. Cytol. Suppl. 11B:113–44.
Brugnaro, C., and N. A. Lavorenti. 1981. Evolucao do setor sucro-alcooleiro. Coord. Plan. Araras Planalsucar. Piracicaba, SP, Brasil. 25pp.
Chaturvedi, H. C., and G. C. Mitra. 1975. A shift in morphogenetic pattern in *Citrus* callus tissue during prolonged culture. Ann. Bot. 39:683–87.
Chen, W. H., M.C. Liu, C. Y. Chao. 1979. The growth of sugarcane downy mildew fungus in tissue culture. Can. J. Bot. 57:528–33.
Cocking, E. C. 1981. Opportunities from the use of protoplasts. Phil. Trans. R. Soc. Lond. B. 292:557–68.
Cocking, E. C. 1979a. Somatic hybridization by the fusion of isolated protoplasts: an alternative to sex. In W. R. Sharp, P. O. Larsen, E. F. Paddock, and V. Raghavan (eds.), Plant cell and tissue culture: principles and applications, pp. 353–69. Ohio State University Press, Columbus, Ohio.
Cocking, E. C. 1979b. Prospects for the future: cell culture. In "Seed protein improvement in cereals and grain legumes, 2:395–402. Proc. of Symposium. I.A.E.A., Vienna, Austria.
Constabel, F., S. Rambold, K. B. Chatson, W. G. M. Kurz, and J. P. Kutney. 1981. Alkaloids production in *Catharanthus roseus* L. Dom. Part VI. Variation in alkaloids spectra of cell lines derived from one single leaf. Plant Cell Rep. 1:1–3.

Crocomo, O. J., and N. Ochoa-Alejo. 1983. Herbicide tolerance in regenerated plants. *In* D. A. Evans, W. R. Sharp. P. V. Ammirato, and Y. Yamada (eds.), Handbook of plant cell culture, 1:770–81. Macmillan, New York.

Crocomo, O. J., and N. Ochoa-Alejo. 1983. The potential contribution of plant cell and tissue culture to crop improvement. *In* L. W. Shemitt (ed.), Chemistry and world food supplies: *The New Frontiers*—CHEMRAWN II. Pergamon, New York.

Crocomo, O. J., W. R. Sharp, and M. T. V. Carvalho. 1979. Controle da morfogenese e desenvolvimento de plantas em cultura de tecido de cana-de-acucar: resultados experimentais. An. 1 Congr. Nac. Soc. Tec. Ac. (STAB) 1:241–43.

Crocomo, O. J., N. Ochoa-Alejo, C. H. R. P. Goncalves, and O. O. S. Bacchi. 1981. Tolerancia de variedades de cana-de-acucar a herbicidas utilizando a tecnica de cultura de tecidos. An. 2 Congr. Nac. Soc. Tec. Ac. (STAB), 2/4:21–40.

CSIRO, Annual Report (1979/1980). Division of Plant Industry. Pp. 66–74.

Davey, M. R., M. W. Fowler, and H. E. Street. 1971. Cell clones contrasted in growth, morphology, and pigmentation isolated from callus culture of *Atropa belladona* var. Lutea. Phytochemistry 10:2559–75.

Evans, D. A., O. J. Crocomo, and M. T. V. Carvalho. 1980. Protoplast isolation and subsequent callus regeneration in sugarcane. Z. Pflanzenphysiol 98:355–58.

Gamborg, O. L., F. Constabel, K. N. Kao, and K. Ohyama. 1974. Plant protoplasts in genetic modifications and production of intergeneric hybrids. *In* R. Harkham, D. R. Davies, D. A. Hopwood, and R. Home (eds.), Modifications of the information content of plant cells, pp. 181–96. Holland Press, Amsterdam.

Gamborg, O. L., K. Ohyama, L. E. Pelcher, L. C. Fowke, K. Kartha, F. Constabel, and K. Kao. 1979. Genetic modification of plants. *In* W. R. Sharp, P. O. Larsen, E. F. Paddock, and V. Raghavan (eds.), Plant cell and tissue culture: principles and applications, pp. 371–98. Ohio State University Press, Columbus.

Gamborg, O. L., F. Constabel, L. C. Fowke, K. N. Kao, K. Ohyama, K. K. Kartha, and L. E. Pelcher. 1974. Protoplast and cell culture methods in somatic hybridization in higher plants. Canad. J. Genet. Cytol. 16:737–50.

Goncalves, C. H. R. P., J. A. Cury, and O. J. Crocomo. 1981. Isoenzimas na identificacao e selecao de variedades de cana-de-acucar (*Saccharum* spp). Anais do 2 Congr. Nac. Soc. Tec. Ac. (STAB), 2/4:328–40.

Gressel, J., Y. Regen, S. Malkin, and Y. Kleifeld. 1983. Characterization of an S. triazine-resistant biotype of *Brachypodium clistachyon*. Weed Sci. 31:450–56.

Hammil, E., and D. J. Heinz. 1970. Studies of sugarcane isoenzymes. Ann. Rep. Exp. Sta. HSPA, pp. 13–14.

Heinz, D. J., G. W. P. Mee, and L. G. Nickell. 1969. Chromosome numbers of some *Saccharum* species hybrids and their cell suspension cultures. Am. J. Bot. 56:450–56.

Heinz, D. J., and G. W. P. Mee. 1971. Morphologic, cytogenetic, and enzymatic variation in *Saccharum* species hybrid clones derived from callus tissue. Amer. J. Bot. 58:257–62.

Heinz, D. J. 1973. Sugarcane improvement through induced mutations using vegetatively propagated plants. *In* Induced mutation in vegetatively propagated plants," pp. 53–59. I.A.E.A., Vienna, Austria.

Heinz, D. J., M. Krishnamurthi, L. G. Nickell, and A. Maretzki. 1977. Cell, tissue, and organ culture in sugarcane improvement. In J. Reinert and Y. P. S. Bajaj (eds.). Plant cell, tissue, and organ culture, pp. 3–17. Springer Verlag, Berlin.

Hsam, S. L. K., and E. N. Larter. 1974. Influence of source of wheat cytoplasm on the synthesis and plant characteristic of hexaploid triticale. Can. J. Genet. Cytol. 16:333–40.

Krishnamurthi, M., and J. Tlaskal. 1974. Fiji disease resistant *Saccharum officinarum* var. Pindar subclones from tissue culture. Proc. Intern. Soc. Sugarcane Technologists. 15:130−37.

Larkin, P. J., and W. R. Scowcroft. 1981a. Somaclonal variation a novel source of variability from cell cultures for plant improvement. Theor. Appl. Genet. 60:197−214.

Larkin, P. J. and W. R. Scowcroft. 1981b. Eyespot disease of sugarcane-host-specific toxin induction and its interaction with leaf cells. Plant Physiol. 67:408−14.

Liu, M. C. 1981. *In vitro* methods applied to sugarcane improvement. *In* T. A. Thorpe (ed.), Plant tissue culture: methods and applications in agriculture, pp. 299−323. Academic Press, New York.

Liu, M. C., and W. H. Chen. 1976. Tissue and cell culture as aids to sugarcane breeding. 1. Creation of genetic variation through callus culture. Euphytica 25:393−403.

Liu, M. C., and W. H. Chen. 1978. Tissue and cell culture as aids to sugarcane breeding. 2. Performance and yield potential of callus derived lines. Euphytica 27:273−81.

Maretzki, A., and L. G. Nickell. 1973. Formation of protoplasts from sugarcane cell suspension and the regeneration of cell cultures from protoplasts. *In* Protoplasts et fusion de cellules somatiques vegetables," 212:51−63. Colloq. Intern. C.N.R.S., Paris, France.

Mali, P. C., and S. L. Mehta. 1977. Effect of drought on proteins and isoenzymes in rice during germination. Phytochem. 16:643−46.

Murashige, T., and R. Nakano. 1967. Chromosome complement as a determinant of the morphogenetic potential of tobacco cells. Am. J. Bot. 54:963−70.

Nickell, L. G., and J. G. Torrey. 1969. Crop improvement through plant cell and tissue culture. Science 166:1068−70.

Nickell, L. G., and D. J. Heinz. 1973. Potential of cell and tissue culture techniques as aids in economic plant improvement. In A. M. Srb (ed.), Genes, enzymes and populations, pp. 109−28. Plenum, New York.

Sampiertro, A. R., M. A. Vattuone, and E. E. Prado. 1980. A regulatory invertase from sugarcane leaf sheaths. Phytochem. 19:1637−42.

Scowcroft, W. R. 1977. Somatic cell genetics and plant improvement. Adv. Agron. 29:38−81.

Shannon, L. M. 1968. Plant isoenzymes. Ann. Rev. Plant Physiol. 19:187−210.

Shepard, J. F., D. Bidney, and E. Shahin. 1980. Potato protoplasts in crop improvement. Science 28:17−24.

Tabata, M., T. Ogino, K. Yoshioka, N. Yoshikawa, and N. Hiraoka. 1978. Selection of cell lines with higher yield of secondary products. *In* T. A. Thorpe (ed.), Frontiers of Plant Tissue Culture, pp. 213−22. Int. Assoc. Plant Tissue Culture, Calgary.

Thom, M. and A. Maretzki. 1970. Peroxidase and esterase isozymes in Hawaiian sugarcane. Hawaiian Planter's Record 58:81−84.

Umiel, N., E. Brand, and R. Goldner. 1978. Streptomycin resistance in tobacco. I. Variation among calliclones in the phenotype expression of resistance. Z. Pflanzenzucht. 88:311−15.

Widholm. J. M. 1977. Selection and characterization of amino acid: analog resistant plant cell cultures. Crop Sci. 17:597−600.

Witwer, S. H. 1974. Maximum production capacity of food crops. Bioscience 24:16−24.

Zenk, M. H., H. El-Shagi, H. Arens, J. Stockigt, E. W. Weeler, and B. Deus. 1977. Formation on the indole alkaloids serpentine and amalicine in cell suspension culture of *Catharanthus roseus. In* W. Barz, E. Reinhard, and M. H. Zenk (eds.), Plant culture and its technical application, pp. 27−43. Springer Verlag, Berlin.

W. H.-T. LOH, S. C. HARTSEL, AND
L. W. ROBERTSON

Production of Secondary Metabolites in Plant Tissue Cultures: *In Vitro* Biosynthesis and Biotransformation of Cannabinoids

11

INTRODUCTION

Although the general public, and even most scientists, believe that all of our most important medicinal agents are produced by synthetic chemistry, there are still several valuable drugs that have to be obtained from higher plants (Farnsworth and Morris, 1976). This does not include the large number of antibiotics, enzymes, vitamins, and other natural products that are produced *de novo* by biosynthesis in microbial fermentations. Examples of prescription drugs that still must be extracted from higher plants include steroids (from diosgenin), codeine, atropine, reserpine, hyoscyamine, digoxin, digitoxin, scopolamine, pilocarpine, and quinidine. In addition, crude plant extracts commonly used in pharmaceutical preparations incude belladonna, ipecac, opium, rauwolfia, cascara, and digitalis. Production of drugs from both cultivated and wild plants is often costly and, in some cases, subject to interruption of supply. Specific examples of drugs from areas subject to political and economic upheavals incude opium, a source of codeine from *Papaver somniferum*, and quinidine, obtained from *Cinchona* bark (Schwartz, 1980).

One approach to more reliable production of highly valuable chemicals such as medicinal agents is the application of plant tissue cultures (Staba, 1980; Barz and Ellis, 1976). Potentially useful outcomes of large-scale plant tissue culture systems include: economical production of currently used plant drugs by *de novo* biosynthesis; production of novel drugs not normally produced in whole plants; production of useful drugs by biotransformation in cell cultures; and elucidation of biosynthesis pathways of secondary metabolites.

In this report, we shall review and contrast biotransformation by both microbial and plant tissue culture systems, with particular emphasis on the synthesis and transformation of cannabinoid compounds *in vitro*. The methodology of cannabinoid biotransformation by plant tissue culture, possible mechanisms of transformation, and elucidation of structures of novel metabolites also will be discussed.

BIOTRANSFORMATION WITH MICROBIAL SYSTEMS

The transformations of organic compounds (ethanol to acetic acid, for example) by microbial cultures have been known for hundreds of years (Sebek, 1982). It was the discovery of high-yielding biotransformations to useful steroidal intermediates by fungi and bacteria, however, which led to its commercial application in the production of useful drugs. Specifically, *Rhizopus arrhizus* and other fungi of the order Mucorales were found to convert progesterone to 11α-hydroxyprogesterone, a transformation that eliminated many difficult and costly chemical synthetic steps in the preparation of useful corticosteroids (Peterson and Murrary, 1952). Almost simultaneously, an unidentified actinomycete was reported to effect 16α-hydroxylation of progesterone (Perlman et al., 1952). The scientific and commercial success of microbiological transformations of steroids is evidenced by the volume of literature that has accumulated on the subject (Charney and Herzog, 1967; Iizuka and Naito, 1967 and 1981).

The successful extension of microbial transformation techniques into additional classes of bioactive compounds has been reviewed recently (Rosazza, 1982). Important classes of drugs that have been modified by microorganisms include antitumor agents, antibiotics, alkaloids, prostaglandins, and cannabinoids. Some of the very useful microbial transformation reactions, which are usually performed with stereospecificity and regiospecificity in mild culture conditions, include: hydroxylations; oxidations of alcohols to carbonyls, O-, N-, and S-dealkylations; reductions of double bonds; reductions of carbonyls to hydroxyls; hydrolyses of esters, amides, and glycosidic linkages; and acetylations and glycosylations (Rosazza, 1982; Kieslich, 1976). In addition to the obvious synthetic utility of microbial transformation for preparation of potentially useful drugs, these techniques also have been successfully employed to study and model mammalian metabolism of drugs (Rosazza, 1982; Smith and Rosazza, 1975). Many microorganisms are known to possess cytochrome P-450 monooxygenase enzyme systems and are, therefore, capable of transformations identical to those of intact animal and liver microsomal systems (Smith and Rosazza, 1975).

BIOTRANSFORMATION WITH PLANT TISSUE CULTURE SYSTEMS

Compared with microbial systems, plant tissue culture biotransformation suffers from lack of an economy of scale. Plant cells grow very slowly, with typical generation times of 15 to 35 hours, whereas bacteria double every 18 to 25 minutes. In contract to microbial culture, plant tissue culture media contain highly refined, and therefore expensive, components. Even under these restrictions, however, large-scale batch culture of plant suspension cells is possible with the development of air lift-type fermenters. Unfortunately, the biosynthesis of secondary metabolites in plant cells usually occurs during the stationary growth phase. By contrast, biotransformation usually takes place during active cell growth, though at rates far slower than measured for microbial systems. In view of the low yields expected, commercial application of plant tissue culture biotransformation systems will be restricted to very specific reactions, i.e., those that generate valuable medicinal agents that cannot be produced by microbes or by chemical synthesis. Surprisingly, there is a variety of compounds and systems that fits all the criteria listed above.

The production of indole alkaloids, e.g., serpentine and ajmalicine, from *Catharanthus roseus* suspension lines has sparked a great deal of interest because this class of compounds may be useful in the treatment of circulatory disorders (Zenk et al. 1977). Other reports include the accumulation of trigonelline in suspension cultures of *Trigonella poenu-graecum* at concentrations higher than found in the differentiated plant (Radwan and Kokate, 1980), production of rosmarinic acid from cell suspensions of *Coleus blumei* in media supplemented with aromatic amino acids (Razzaque and Ellis, 1977), and a variety of anthraquinones from *Morinda citrifolia* (Zenk, et al., 1975). Also interesting is the biosynthesis in plant tissue cultures of novel secondary metabolites (mostly alkaloids) that are not present in the intact plant (Steck, et al., 1971; Ikuta et al., 1974; Akasu et al., 1976).

Many of the early studies on transformation of bioactive compounds in plant tissue cultures dealt with converting progesterone and other steroid hormones by *Digitalis* species, *Dioscoria deltoidea, Nicotiana tabacum, Vinca rosea*, and other systems (Reinhard, 1974; Iizuka and Naito, 1981). More recent investigations have led to success in biotransformation of a variety of bioactive compounds, including the medicinally important cardenolides, better known as cardiac glycosides, in plant tissue culture systems (Staba, 1980; Barz and Ellis, 1976). Specific examples of cardenolide biotransformations include the 5β-hydroxylation and gitoxigenin by suspension cultures of *Daucus carota* (Veliky et al., 1980), and the 5β-hydroxylation of digitoxigenin by immobilized cells of *Daucus carota* (Jones and Veliky, 1981). Additional studies demonstrated the 12β-hydroxylation of a variety of cardiac glycosides by immobilized *Digitalis lanata* cells (Brodelius et al., 1979).

In general, most oxidation reactions in biotransformation probably are carried out by relatively nonspecific monooxygenases, such as cytochrome P-450, commonly used in catabolic processes. For instance, many of the pentyl side chain cannabinoid metabolites are initial, terminal, and subterminal hydroxylations (Robertson, 1982). Reductive biotransformations also have been noted in plant tissue culture systems, especially involving steroids (Furuya et al., 1971; Stohs and El-Olemy, 1972). In addition, glycosylations (Furuya et al., 1973) and esterifications (Furuya et al., 1971) of fatty acids have occurred also on steroidal compounds in some plant tissue culture biotransformation systems.

BIOSYNTHESIS AND BIOTRANSFORMATION OF CANNABINOIDS

Because of their psychotropic effects, marihuana and hashish are among the most popular recreational drugs of the twentieth century. The cannabinoids are a group of C_{21} terpenophenolic secondary metabolites that are formed exclusively in marihuana, *Cannabis sativa* L. (Mechoulam, 1973, Nahas and Paton, 1979). Four cannabinoids that have been subjected to intensive chemical, metabolic, pharmacological, and clinical study are shown in figure 1. The single chemical agent responsible for the psychotropic (or "high-inducing") effects of marihuana is (-)Δ^9-6a, 10a-trans-tetrahydrocannabinol (Δ^9-THC), which is shown with the dibenzopyran numbering scheme most often used by workers in the United States (Mechoulam et al., 1976). Unless otherwise indicated, this numbering system will be used throughout this report.

The closely related isomer (-)-Δ^8-6a, 10a-trans-tetrahydrocannabinol (Δ^8-THC), which is present in very small amounts in the marihuana plant but is readily synthesized, shares the psychotropic activity and most other biological activities of Δ^9-THC (Nahas and Paton, 1979). Cannabidiol (CBD), which is numbered as a substituted monoterpene, has no psychotropic activity, but does possess anticonvulsant (antiepileptic) properties

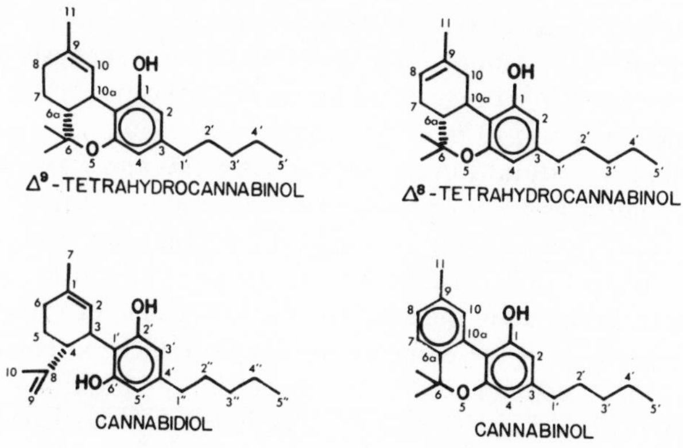

Fig. 1. Structure of four naturally occurring cannabinoids.

and other potentially useful actions (Archer, 1975; Nahas and Paton, 1979; Cohen and Stillman, 1976). Cannabinol (CBN) is probably an artifact of storage formed by air photooxidation of Δ^9-THC. However, it may modify the metabolism of certain of the other cannabinoids and is a potent inhibitor of prostaglandin biosynthesis (Cohen and Stillman, 1976).

Although the pathway of cannabinoid biosynthesis has not been established conclusively, Mechoulam (1970) has proposed that olivetol or olivetolic acid undergoes a condensation with geranyl pyrophosphate, a phosphorylated noncyclic monoterpene, forming cannabigerol or cannabigerolic acid respectively. Alternatively, Shoyama and colleagues (1975) have suggested that all the cannabinoids arise from the condensation of a 12-carbon polyketide (olivetolic acid precursor) and geraniol. It is postulated that cyclizations, oxidations, and allylic rearrangements, resulting from both enzymatic and non-enzymatic reactions, lead to the formation of all other cannabinoid compounds.

The analysis of fresh *Cannabis sativa* plant material indicates that most of the cannabinoids present are in the form of carboxylic acid derivatives, with the acid function substituted ortho or para to the phenolic groups. Therefore, most of the Δ^9-THC acid in the plant is present as Δ^9-THC acid A (2-carboxylic acid) or Δ^9-THC acid B (4 carboxylic acid), and CBD is present as CBD acid (3'-carboxylic acid). The presence of the carboxylic acid functionality strengthens the biosynthetic hypothesis of the 12-carbon polyketide intermediate, and most workers assume that cannabinoids are biosynthesized as carboxylic acids. The carboxylic acid groups are extremely labile in the cannabinoid acids, which spontaneously decarboxylate upon heating or storage, yielding the neutral cannabinoids shown in figure 1 (Mechoulam, 1970, 1973; Turner et al., 1980).

More than 60 naturally occurring cannabinoids and many hundreds of synthetic analogues are now recorded in the literature (Turner et al., 1980). Other natural cannabinoids generally differ from the four compounds shown in figure 1 in cyclization and oxidation state of the monoterpene ring and/or the length of the alkyl sid chains. The major reason for isolation and synthesis of these agents is to explore their potential as useful drugs and to understand better the metabolism and physiological activity of marihuana constituents in man. Documented biological activites of cannabinoids include the following: anticonvulsant (antiepileptic); antiemetic (counteract emetic effects of cancer chemotherapy); anti-glaucoma (decrease intraocular pressure); analgesic; anti-inflammatory; bronchodilatory; antimicrobial (A. A. Abdelaziz, C. I. Randles, and L. W. Robertson, submitted for publication 1986); and antihypertensive (Cohen and Stillman, 1976; Mechoulam and Lander, 1980).

Continuing investigations are aimed at maximizing the therapeutic effects of Δ^9-THC and other cannabinoids and minimizing the unwanted side effects (usually the psychotropic effects). Biotransformation of cannabinoids, using microbial and plant culture systems, offers an alternative and/or adjunct to solely chemical synthesis of new cannabinoids. In addition, it is hoped that biotransformation studies may also contribute additional insight into mechanisms and pathways involved in mammalian metabolism, as well as biosynthetic and transformation sequences in the plant. In this discussion, the term "microbial transformation" refers only to fungal and bacterial systems. Transformation by plant tissue culture systems will be discussed separately.

Microbial Cultures

Based on previous biotransformation reports, three classes of microorganisms were considered likely candidates for cannabinoid metabolism. Microorganisms that oxidized alkanes and alkyl benzenes will probably attack the alkyl side chain of cannabinoids. The nucleus of the cannabinoid molecule may be subject to transformation by microorganisms that convert phenols and other aromatic compounds. In addition, microorgansims that act on nonsteroidal isoprenoids may also attack the monoterpernoid moiety of cannabinoids (Robertson, 1982).

The microbial transformation products of cannabinoids described in the literature include 13 from Δ^9-THC, 14 from Δ^8-THC, 11 from CBD, and 5 from CBN (Robertson, 1982). Microbial transformations of Δ^8- THC and Δ^9-THC, summarized in figure 2, are presented here as an example of cannabinoid biotransformations. Most of the microorgansims that are found to transform the cannabiniods attack the alkyl side chains, with monohydroxylation at the 4'-carbon being most common, followed by the 3'- and 2'-positions. Terminal oxidations of Δ^9-THC by *Mycobacterium rhodochrous* has generated Δ^9-THC-5'-oic acid in yields of 30% to 40%, based on the amount of Δ^9-THC added. Subsequent beta-oxidation reactions produce 3'-hydroxy -Δ^9-THC-5'-oic acid, 4', 5'-bisnor-Δ^9-THC-3'oic acid, and a compound with a cinnamic acid-like side chain, 1', 2'-dehydro-4', 5'-bisnor-Δ^9-THC-3'-oic acid. Subterminal oxidation of Δ^9-THC by *Syncephalastrum racemosum* and *Aspergillus niger* leads to accumulation of 4'-hydroxy- and 4', 5'-bisnor-3'-hydroxy metabolites (Robertson et al., 1975, Robertson et al., 1978; Robertson, 1982). We also have reported the subterminal oxidation of olivetol by *Syncephalastrum racemosum* forming 4'-hydroxyolivetol as the major metabolite, as shown in figure 3 (Robertson, 1982). Further degradation of the olivetol side chain produces 4', 5'-bisnor-2'-hydroxyolivetol and 4', 5'-bisnor-olivetol-3'-oic acid (McClanahan et al., 1984). In addition, both a monomethyl olivetol and an olivetol dimer have been produced by transformation with *Fusarium roseum* (McClanahan and Robertson, 1985).

A second class of microbial metabolites results from allylic hydroxylations in the monoterpene ring at positions 8 or 11, and the corresponding dihydroxylated metabolites formed by combined attack on the monoterpene ring systems and alkyl side chains. Further oxidation leads to formation of ketone, aldehyde, or carboxylic acid functions from secondary or primary alcohols (Robertson, 1982). Thus far, only one fungus, *Fusarium roseum*, has been found to hydroxylate exclusively in the monoterpene portion of the molecule, forming the 8β-hydroxy-Δ^9-THC from Δ^9-THC in 12% yield (Robertson, 1982).

Although many of the microbial transformation studies briefly summarized here have been conducted in our laboratory, the following workers have also made significant contributions in the area. Binder (1976) found that cultures of *Cunninghamella blakesleeana* produced four monohydroxylated and two dihydroxylated metabolites when incubated with Δ^9-THC. Christie and colleagues (1978) identified two monohydroxylated and one dihydroxylated metabolites of Δ^9-THC formed by incubation with *Chaetomium globosum*. Biotransformation of Δ^8-THC by *Pellicularia fi-*

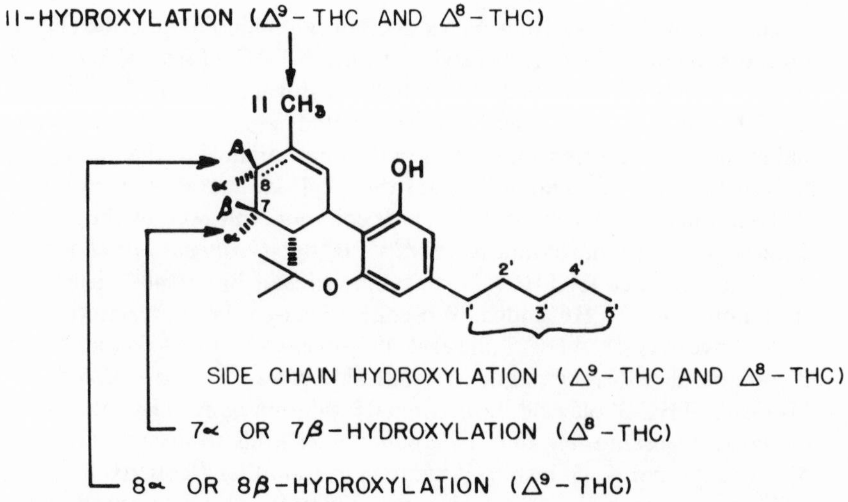

II-HYDROXYLATION (Δ^9-THC AND Δ^8-THC)

SIDE CHAIN HYDROXYLATION (Δ^9-THC AND Δ^8-THC)

7α OR 7β-HYDROXYLATION (Δ^8-THC)

8α OR 8β-HYDROXYLATION (Δ^9-THC)

OTHER OXIDATIVE BIOTRANSFORMATIONS:

1) PRIMARY ALCOHOL (II, 5') TO ALDEHYDE AND
   CARBOXYLIC ACID

2) SECONDARY ALCOHOL (7, 8, 1', 2', 3', 4') TO KETONE

3) SIDE CHAIN ALCOHOL TO CARBOXYLIC ACID
   (1', 2', 3', 4', 5')

4) EPOXIDATION (AT 9, 10 OR 8, 9)

NUMEROUS COMBINATIONS ARE POSSIBLE, INCLUDING
   CONJUGATION

Fig. 2. Microbial transformations of $\triangle^8$- and $\triangle^9$-tetrahydrocannabinol.

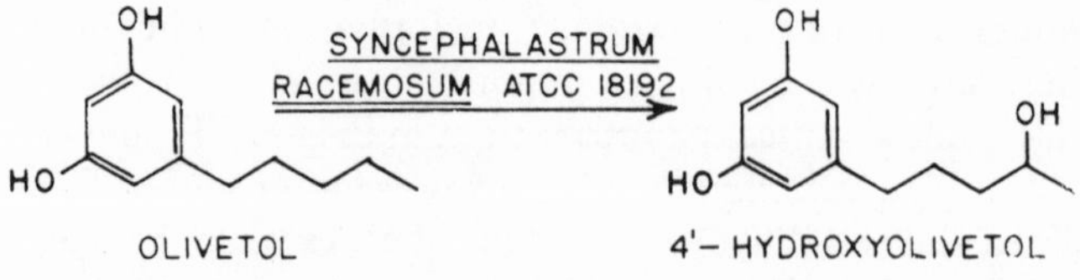

Fig. 3. Transformation of olivetol by *Syncephaslastrum racemosum.*

lamentosa f.s.p. *sasakii* and *Streptomyces lavendulae* was reported to form nine metabolites, consisting chiefly of dihydroxylated Δ^8-THC derivatives (Vidic et at., 1976). In addition, the transformation of the synthetic cannabinoids nabilone and $\Delta^{6a,10a}$-THC, by a variety of bacteria has been described (Archer et al., 1979; Fukuda et al., 1977).

In general, the results obtained from microbiological transformations are qualitatively similar to those reported in mammalian systems. This suggests that microbial cultures may be a reasonable model for the major mammalian routes of attack (Robertson, 1982; Smith and Rosazza, 1975). One important difference, however, is that side chain oxidations are favored quantitatively by microbial systems, whereas lower microsomal fractions preferentially attack the monoterpene portion of the molecule. A second important characteristic of the microbial systems is that they do not accumulate the conjugated metabolites, glucuronides and sulfates, which are commonly produced in mammalian metabolism of drugs (Robertson, 1982; Smith and Rosazza, 1975).

Two interesting types of non-enzymatically produced artifacts were discovered in the course of our microbial transformation studies. Cannabitriol, which has been isolated as a natural cannabinoid by several workers, was identified as a major air oxidation product of Δ^9-THC in highly oxygenated media (Robertson and Tsai, 1979a; Elsohly et al., 1977). In addition, a variety of nitrated cannabinoids, with nitro groups added in the 2-and 4-positions of the aromatic ring, were found in cultures of *Pseudomonas putida* grown on $NaNo_3$ or NH_4No_3. *Pseudomonas* reduces nitrate to nitrite, causing a non-enzymatic nitration during the acidic extraction process (Robertson and Tsai, 1979b).

Plant Tissue Culture

Veliky and Genest (1972) reported the first callus and suspension cultures derived from seedling roots of *Cannabis sativa*. Although phenolic compounds were detected by thin-layer chromatography (TLC) in extracts of the cell suspension culture, none of these were identified as cannabinoids by gas-liquid chromatography. These extracts, however, has significant antimicrobial activity against both gram-positive and gram-negative bacteria.

Since the flowering portions of *Cannabis* contain the highest cannabinoid levels, Hemphill and colleagues (1978) reasoned that this organspecific trait might continue to be expressed in tissue culture. Callus was induced from maturing bracts, and those leaves which subtend the floral organs, and vegetative leaves. The resulting callus cultures, however, produced no major terpenophenols. These workers also reported the presence of an unidentified phenolic compound in their cultures. Braut-Boucher and Petiard (1981) induced callus from a variety of plant tissues, using both drug-type and fibre-type *Cannabis* lines. They did not report the presence of any secondary metabolites in their cultures.

Since sucrose and glucose have been shown to repress the accumulation of secondary metabolites in microbial systems, Jones and Veliky (1980) tried to stimulate the biosynthetic pathway of plant tissue culture cells by alternating carbon sources. Callus and suspension cultures induced from *Cannabis sativa*, *Daucus carota*, and *Ipomoea* were gradually transferred to media containing glycerol or other carbon sources. No secondary metabolites were detected, but biotransformation of cardenolides by all three cell lines was unaffected by substitution of glycerol as the sole carbon source.

The only report of a secondary metabolite in tissue cultures of *Cannabis* came from Heitrich and Binder (1982). These authors suggested that $\triangle^8$-THC isolated from callus cultures was an extraction artifact of naturally occurring cannabinoids, $\triangle^9$-THC acid A and $\triangle^9$-THC acid B. They attributed the failure of previous attempts at biosynthesis to a poor choice of plant growth regulators and inadequate extraction procedures for isolation of carboxylic acids. Their callus extracts were dried and subjected to thermal decarboxylation to convert cannabinoid acids to the corresponding neutral cannabinoids prior to ethanol extraction. This contradicted our finding that cannabinoid acids were readily extractable from plant material with ethanol. Moreover, $\triangle^8$-THC was detected in significant amounts at only one point in the culture period, and rapidly disappeared thereafter.

Itokawa and colleagues (1977) reported limited biotransformation resulting from an alcohol oxidase activity in suspension cultures of *Cannabis sativa*. Although cannabinoids were present in the original explant, none could be detected in tissue culture. No cannabinoids were produced when the proposed precursor compounds, geraniol, nerol, olivetol, and ethyl olivetolate, were added to *Cannabis* suspension cultures. Extensive chemical analysis showed that primary and secondary allylic alcohols, such as geraniol, nerol, trans-cinnamyl alcohol, isophorol, and trans-verbenol, were oxidized forming the corresponding aldehydes.

Takeya and Itokawa (1977) subsequently repeated the biotransformations described above in a cell-free system. Callus cultures of *Cannabis* were homogenized in Tris buffer and incubated with one of four substrate alcohols for 4 to 15 hours. The corresponding aldehydes, resulting from cell-free biotransformation, were extracted with yields obtained ranging from 38% to 65%. Proton magentic resonance and circular dichroism analyses indicated that the enzyme system strongly favored one stereoisomeric form of the substrate over the other. For example, S-(-)-isophorol of the racemate was transformed to isophorone, and four isomers of (+)-trans-verbenol, specifically (+)- and (-)-trans-verbenol and (+)- and (-)-cis-verbenol, were preferentially biotransformed to (+)-verbenone.

We have also induced callus from both stems and leaves of young *Cannabis* plants (Loh et al., 1983). The optimal ratio of plant growth regulators differed according to the explant source, and more than one auxin/cytokinin combination was capable of callus induction and subsequent cell proliferation. Extensive TLC analysis indicated that the callus and derived suspension cultures contained no cannabinoids or cannabinoid precursors.

When *Cannabis* suspensions were fed with CBD, two major phenolics in addition to CBD were detected by TLC analysis after as little as two days of incubations (Hartsel et al., 1983). Gas chromatography/mass spectral (GC/MS) analysis of the extract indicated the presence of CBD as well as diastereoisomers of cannabielsoin (CBE). The structures of CBD and CBD acid and its metabolites, axial C_1-OH and equatorial C_1-OH isomers of CBE and CBE acid are shown in figure 4. Mass spectra of both CBE diastereoisomers are shown in figure 5. An identical transformation was carried out by a mature *Saccharum officinarum* L. suspension culture. Cannabidiol proved to be toxic to a suspension line of *Nicotiana tabacum* L., which browned and died shortly after the addition of the drug (unpublished).

Both CBD and CBE acids have been identified previously in *Cannabis sativa* plant material (Shani and Mechoulam, 1974; Kuppers et al., 1973; Grote and Spitellier, 1978; Bercht et al., 1973). Whether these compounds are natural products has been questioned due to the irregularity of their occurrence and difficulty with which they can be isolated (Turner et al., 1980). The status of CBE as a secondary metabolite, however, is strengthened by the ability of *Cannabis* suspension cultures to convert CBD to CBE.

Since *Saccharum officinarum* suspension cultures have been shown to perform the same transformation, one can infer that the mechanism involved is a general one rather than a specific one. A possible mechanism of the biotransformation of CBD and CBE would involve an epoxidation at the 1,2-double bond of CBD. If the epoxide becomes protonated, then a nucleophilic attack from the 2'-OH group could cause cyclization to CBE. An analogous Δ^9-THC epoxide has been proposed as a precursor of can-

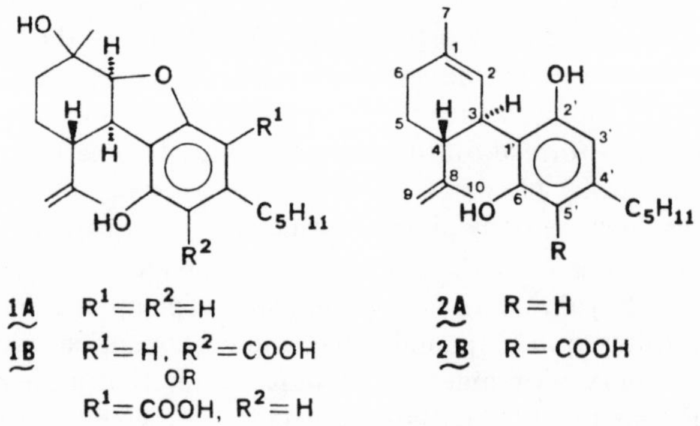

Fig. 4. Structures of cannabielsoin (1A) and cannabielsoic acid (1B) diastereoisomers and the precursor compounds, cannabidiol (2A) and cannabidiolic acid (2B).

nabitriol and 10-ethoxy-9-hydroxy-$\Delta^{6a,10a}$-THC, both of which are found in *Cannabis* plant extracts (Elsohly et al., 1977) Shani and Mechoulam (1974) have proposed a free radical intermediate for this transformation *in vivo*. The chemical synthesis of CBE involves a base-catalyzed attack of the phenoxy ion (Uliss et al., 1974). A cytochrome P-450 in *Catharanthus roseus* has both epoxidation and hydroxylation activity on geraniol, a presumptive cannabinoid precursor (Licht et al., 1979). Epoxidation of the 1,2 double bond of Δ^9-THC has been reported also in dog liver microsomal preparations (Widman et al., 1975).

The addition of olivetol, a presumptive cannabinoid precursor, to a *Cannabis* suspension culture led to the appearance of an additional major phenolic compound (Loh et al., 1983). Although the product has not yet been conclusively identified, mass spectral analysis indicated that this compound has a molecular weight of 210, which is below the range expected for a cannabinoid. The transformation product corresponds to the addition of 30 mass units to olivetol, which would be consistent with the addition of an oxygen and methylene group to the starting material. This may be explained by hydroxylation accompanied by methyl ether (methoxy) formation. Additional chemical and physical analyses are necessary in order to assign the structure completely. No similar transformation product has yet been identified from microbial transformation studies on olivetol, which revealed side chain hydroxylation products formation (figure 3; Robertson, 1982).

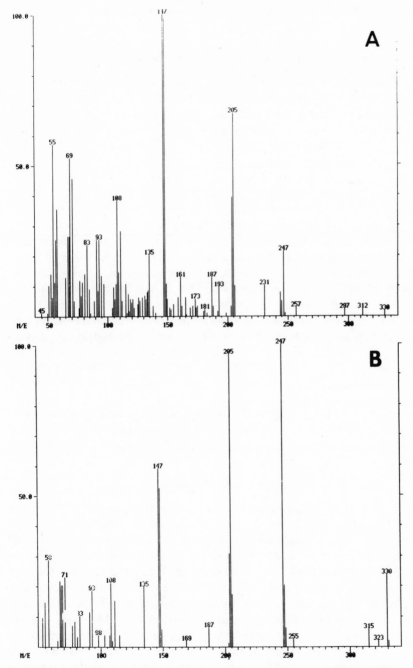

Fig. 5. Mass spectra of (A) C_1-OH axial and (B) C_1-OH equatorial isomers of CBE, showing the occurrence and relative concentration of various fragments.

BIOTRANSFORMATION IN CELL SUSPENSION CULTURES

Our review of the literature as well as our own experience suggests that the capability of biotransformation of a particular secondary metabolite is unrelated to whether the metabolite naturally occurs in that species. *Solanum tuberosum* and *Atropa belladonna* will act on a variety of steroidal hormones, *Cannabis sativa*, *Daucus carota*, and *Ipomoea* lines have been used in cardenolide transformation, and *Saccharum officinarum* cultures have been shown to transform cannabinoids and cannabinoid precursors. The methodology of cannabinoid transformation can serve as a model for plant tissue culture transformations in general.

Reproducible results require the development of a stable cell line maintained by regular subculture. The growth kinetics of such cultures are sufficiently stable to ensure that replicate cultures will reach stationary phase at the same time for biotransformation. Because cannabinoids are only slightly water-soluble, cultures were shaken at 125 rpm in Erlenmeyer flasks to ensure constant mixing.

Representative cultures were analyzed for the presence of secondary metabolites prior to adding substrate compounds. Some cultures were autoclaved to serve as controls. A flask containing sterile media also was included in the experiment to rule out non-enzymatic (i.e., air oxidation, photoxidation) reactions. Samples of cultures were analyzed regularly since biotransformation may occur as soon as 48 hours after the substrate compound is added (Hartsel et al., 1983). The suspension cells and spent media may be extracted separately to ascertain whether the transformed products accumulate.

ANALYSIS OF BIOTRANSFORMATION COMPOUNDS

Secondary metabolites can be extracted from suspension cultures by a variety of solvents. *Cannabis* callus and suspension cultures were extracted by first drying or lyophilizing the tissues followed by grinding in absolute ethanol. Cannabinoids were extracted by partitioning into four volumes of chloroform and dried *in vacuo*. If particularly polar metabolites were expected (i.e., from olivetol), ethyl acetate was substituted for chloroform. The filtrate from the extract was then dried *in vacuo*. Cannabinoids were purified further by column chromatography (Silica gel, Merck PF-254, i.d. 2 cm, h. 14 cm) using the solvent system that yielded the best separation on TLC plates. If further purifications were necessary, preparative TLC was carried out on the pooled fractions. Successful purification of some metabolites may be achieved using preparative gas-liquid chromatography (GLC). Conditions for GLC analysis are identical to those for gas chromatographic/mass spectral analysis and will be discussed later.

Thin-layer Chromatrophic Analysis

Canninoids can be detected easily by separating culture extracts by TLC and spraying the plate with Fast Blue BB reagent. When sprayed with a 0.5% aqueous solution of Fast Blue BB salt, which binds to the ortho or para position of free phenols, cannabinoid compounds turn a bright color. These colors can be intensified further by holding the TLC plate over a bottle of concentrated NH_4OH. The resulting colors are unique for each cannabinoid and cannabinoid precursor: $\triangle^8$-THC and $\triangle^9$-THC (bright red), CBD (orange), CBN (violet), olivetol (purplish red), cannabichromene (reddish orange).

The simplicity of the TLC/Fast Blue BB procedure permits constant monitoring for the appearance and accumulation of transformation products on small TLC plates, and amounts as small as 150 nanograms can be detected. Tentative identification of a biotransformation product can be made by comparing the R_f values of reference standards with transformation products on three or four different solvent systems. We have utilized the following systems for the separation of cannabinoids and cannabinoid precursors on silica gel TLC: hexane-diethyl ether (80:20), toluene-absolute ethanol (80:20), chloroform-absolute ethanol (92:8), and chloroform-methanol (80:20). Representative R_f values for cannabinoids and precursor molecules in three solvent systems are shown in table 1. $\triangle^8$- and $\triangle^9$-THC are too closely related structurally to be separated by TLC. Reference standards and sample extracts should be chromatographed separately to measure the individual retention times and also spotted together for possible identity.

Gas-Chromatographic and Mass Spectral Analysis

Gas-liquid chromatography (GLC) is primarily a method of separation, although constituent compounds can be identified on the basis of their retention times. For most metabolites and especially crude extracts, an SE-54 capillary column (fused silica, WCOT, 30 m, split 1:7) gives very satisfactory and reproducible separation. Glass columns (6ft × 2mm), packed with either 3 percent OV-1 or 3 percent OV-17, have also been used successfully. One microliter samples, dissolved in chloroform or hexane, were chromatographed isothermally at 210°C (injector 230°C, He flow 30ml/min). As with TLC, samples should be run with and without reference standards to demonstrate identity.

Very polar cannabinoids and cannabinoid precursors, such as olivetol and olivetolic acid, should be derivatized. Underivatized cannabinoid acids will lose their carboxyl groups when volatilized and become indistinguishable from their neutral counterparts in GLC. Silyation with bis-(trimethylsil) trifluoracetamide (BSTFA, Pierce Chemical Company) will also significantly shorten elution times and allow separation and analysis of highly polar metabolites such as hydroxy and carboxylic acid derivatives.

Mass spectral analysis is required for identification of novel metabolites. Often GLC is combined with a mass spectrometer (GC/MS) for separation and identification of individual components in complex mixtures. The GLC conditions reported above are all appropriate for GC/MS analysis. Electron impact MS, in which samples are ionized by a 70 eV electron beam as they enter the mass spectrometer, is most appropriate for general usage. For very large or unstable molecules, the gentler chemical ionization mode may be more useful. Molecules ionized with CH_4, or other reagent gases, at 20 eV frequently yield high molecular weight fragments, usually including the molecular ion. Low molecular weight fragments are poorly resolved by chemical ionization.

The presence of key compounds in a crude mixture can be followed easily by recording the mass spectra in a real-time three-axis mode, as seen in figure 6. With this type of analysis, one can identify the MS pattern of a specific metabolite or class of metabolites with a particular GLC peak. A standard mass spectrum can then be obtained from each significant GLC peak, as shown in figure 5. Identification is usually based on comparisons with mass spectra of known standards.

SUMMARY

A review of the literature has shown that biosynthesis and biotransformation by plant cell cultures has led to advances in our understanding of cell regulation and biological chemistry. The plant kingdom has a bewildering array of potentially valuable and grossly underexploited secondary metabolites. Many of these are drugs and may be produced in plant cell culture systems. Plant cell cultures also have biotransformation reactions that are unique as compared with plant or microbiological systems *in vivo*. A more systematic study of the biochemical mechanisms that regulate the biosynthesis of secondary metabolites and biotransformation of precursor substrates is needed to fully exploit this technology.

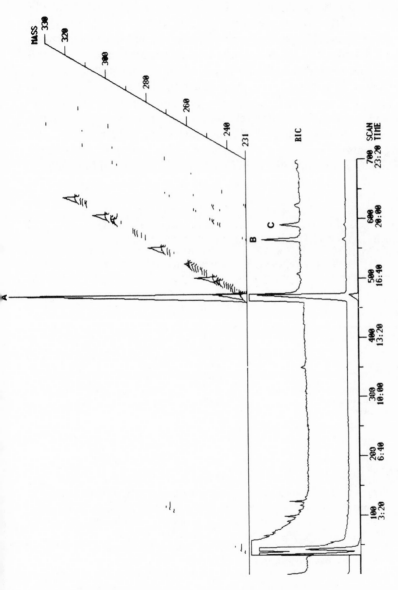

Fig. 6. Real-time gas chromatographic/mass spectral analysis of cannabidiol biotransformation in three-axis mode, showing the presence of (A) cannabidiol, (B) C_1-OH equatorial cannabielsoin, and (C) C_1-OH axial cannabielsoin.

In the area of cannabinoid chemistry, very interesting results have been reported in both biosynthesis and biotransformation of terpenophenolic compounds and their precursors. Biotransformation in both microbial and plant cell cultures has yielded a number of novel cannabinoid metabolites. When combined with mammalian and liver microsomal studies, these efforts may lead to an elucidation of important metabolic pathways in this class of compounds. In addition, the production of a variety of derivatives of Δ^9-THC, CBD, and other cannabinoids may contribute to the overall knowledge of structure-activity relationships of these biologically active molecules. Preliminary studies utilizing plant tissue cultures as biotransforming agents indicate that plant systems may produce useful transformation products not obtainable by microbial transformations or other means.

TABLE 1

R_f VALUES OF REFERENCE PHENOLIC STANDARDS (CANNABINOIDS, OLIVETOL, AND METABOLITES) USING TLC AND FAST BLUE BB SPRAY REAGENT IN THREE DIFFERENT SOLVENT SYSTEMS: A. CHLOROFORM-ABSOLUTE ETHANOL (92:8), B. HEXANE-DIETHYL ETHER (80:20), AND C. CHLOROFORM-METHANOL (80:20)

	SOLVENT SYSTEMS		
	A	B	C
Cannabidiol	0.74	0.61	0.80
Δ^9-Tetrahydrocannabinol	0.72	0.53	0.78
Cannabinol	0.71	0.49	0.79
Cannabigerol	0.70	0.44	0.78
Olivetol	0.25	0.08	0.62
Δ^9-Tetrahydrocannabinolic acid B	0.41	0.04	0.53
4'-Hydroxycannabidiol	0.29	0.03	0.62
4'-Hydroxyolivetol	0.13	0.00	0.29

REFERENCES

Archer, R. A. 1974. The cannabinoids: therapeutic potentials. *In* R. V. Heinzelman (ed.), Annual reports in medicinal chemistry, 9:253–59. Academic Press, New York.

Archer, R. A., D. S. Fukuda, A. D. Kossoy, and B. J. Abbott. 1979. Microbiological transformations of nabilone, a synthetic cannabinoid. Appl. Environ. Microbiol. 37:965–71.

Akasu, M., H. Itokawa, and M. Fujita. 1976. Biscoclaurine alkaloids in callus tissues of *Stephania cepharantha*. Phytochemistry 15:471–73.

Barz, W., and B. E. Ellis. 1976. Potential of plant cell cultures for pharmaceutical production. *In* J. L. Beal and E. Reinhard (eds.), Natural products as medicinal agents, pp. 471–507. Hippokrates, Stuttgart.

Bercht, C. A. L., R. J. Ch. Lousberg, F. J. E. M. Kuppers, C. A. Salemink, T. B. Vree, and

J. M. Van Rossum. 1973. *Cannabis.* VII. Identification of cannabinol methyl ester from hashish. J. Chromatogr. 81:163–66.

Binder, M. 1976. Microbial transformations of (-)-Δ'-3,4-trans-tetrahydrocannabinol by *Cunninghamella blakesleeana* Lender. Helv. Chim. Acta 59:1674–84.

Braut-Boucher, F., and V. Petiard. 1981. Sur la mise en culture *in vitro* de tissus de différents types chimique du *Cannabis sativa* L. Comptes Rendus 292:833–38.

Brodelius, P., B. Deus, K. Mosbach, and M. H. Zenk. 1979. Immobilized plant cells for the transformation of natural products. FEBS Lett. 103:93–97.

Burstein, S., E. Levin, and C. Varanelli. 1973. Prostaglandins and cannabis-II: Inhibition of biosynthesis by the naturally-occurring cannabinoids. Biochem. Pharmacol. 22:2905–10.

Charney, W., and H. L. Herzog. 1967. Microbial transformations of steroids: a handbook. Academic Press, New York. 728 pp.

Christie, R. M., R. W. Rickards, and W. P. Watson. 1978. Microbial transformation of cannabinoids. I. Metabolism of (-)-Δ^9-6a,10a-trans-tetrahydrocannabinol by *Chaetomium globosum.* Aust. J. Chem. 18:1799–1807.

Cohen, S., and R. C. Stillman (eds.). 1976. The therapeutic potential of marihuana. Plenum Press, New York. 515 pp.

Elsohly, M. A., F. S. El-Feraly, and C. E. Turner. 1977. Isolation and characterization of (+)-cannabitriol and (-)-10-ethoxy-9-hydroxy-$\Delta^{6a(10a)}$-tetrahydrocannabinol: two new cannabinoids from *Cannabis sativa* L. extract. Lloydia 40:275–80.

Farnsworth, N. R., and R. W. Morris. 1976. Higher plants—the sleeping giant of drug development. Amer. J. Pharmacy 14:46–52.

Fukuda, D., R. A. Archer, and B. J. Abbott. 1977. Microbiological transformations of $\Delta^{6a,10a}$-tetrahydrocannabinol. Appl. Microbiol. 33:1134–40.

Furuya, T., M. Hirotani, and K. Kawaguchi. 1971. Biotransformation of progesterone and prenenolone by plant suspension cultures. Phytochemistry 10:1013–17.

Furuya, T., K. Kawaguchi, and M. Hirotani. 1973. Biotransformation of progesterone by suspension cultures of *Digitalis purpurea* cultured cells. Phytochemistry 12:1621–26.

Grote, H., and G. Spitellier. 1978. Neue cannabinoide II. J. Chromatogr. 154:13–23.

Hartsel, S. C., W. H.-T. Loh, and L. W. Robertson. 1983. Biotransformation of cannabidiol to cannabielsoin by suspension cultures of *Cannabis sativa* L. and *Saccharum officinarum* L. Planta Medica 48:17–19.

Heitrich, A., and M. Binder. 1982. Identification of (3R,4R)-$\Delta^{1(6)}$-tetra-hydrocannabinol as an isolation artifact of cannabinoid acids formed by callus cultures. Experientia 38:898–99.

Hemphill, J. K., J. C. Turner, and P. G. Mahlberg. 1980. Cannabinoid content of individual plant organs from different geographical strains of *Cannabis sativa* L. J. Nat. Prod. 43:112–22.

Iizuka, H., and A. Naito. 1967. Microbial transformations of steroids and alkaloids. University Park Press, State College, Pennsylvania. 294 pp.

Iizuka, H., and A. Naito. 1981. Microbial conversions of steroids and alkaloids. Springer-Verlag, New York. 396 pp.

Ikuta, A., K. Syono, and T. Furuya. 1974. Alkaloids of callus tissues and redifferentiated plantlets in the Papaveraceae. Phytochemistry 13:2175–79.

Itokawa, H., K. Takeya, and S. Mihashi. 1977. Biotransformation of cannabinoid precursors and related alcohols by suspension cultures of callus induced from *Cannabis sativa* L. Chem. Pharm. Bull. 25:1941–46.

Jones, A., and I. A. Veliky. 1980. Growth of plant cell suspension cultures on glycerol as a sole source of carbon and energy. Can. J. Bot. 58:648–57.

Jones, A., and I. A. Veliky. 1981. Examination of parameters affecting the 5β-hydroxylation of digitoxigenin by immobilized cells of *Daucus carota*. European J. Appl. Microbiol. Biotechnol. 13:84–89.

Kieslich, K. 1976. Microbial transformations of non-steroid cyclic compounds. Georg Thieme Verlag, Stuttgart. 1,262 pp.

Kuppers, F. J. E. M., R. J. J. Ch. Lousberg, C. A. L. Bercht, C. A. Salemink, J. K. Terlouw, W. Heerma, and A. Laven. 1973. *Cannabis*-III. Pyrolysis of cannabidiol. Structure elucidation of the main pyrolytic product. Tetrahedron 29:2797–2802.

Licht, H. J., K. M. Madyastha, C. J. Coscia, and R. J. Krueger. 1979. Comparison of plant and hepatic cytochrome P-450 dependent monoterpene monooxygenases. *In* M. J. Coon, A. H. Conney, R. W. Estabrook, H. V. Gelboin, J. R. Gillette, and P. J. O'Brien (eds.), Microsomes, drug oxidations, and chemical carcinogenesis, pp. 211–15. Academic Press, New York.

Loh, W. H.-T., S. C. Hartsel, and L. W. Robertson. 1983. Tissue culture of *Cannabis sativa* L. and *in vitro* biotransformation of phenolics. Z. Pflanzenphysiol. 11:395–400.

McClanahan, R. H., and L. W. Robertson. 1984. Biotransformation of olivetol by *Syncephalastrum racemosum*. J. Nat. Prod. 47:828–34.

McClanahan, R. H., and L. W. Robertson. 1985. Microbial transformation of olivetol by *Fusarium roseum*. J. Nat. Prod. 48:660–63.

Mechoulam, R. 1970. Marihuana chemistry. Science 168:1159–66.

Mechoulam, R. (ed.). 1973. Marihuana: chemistry, pharmacology, metabolism, and clinical effects. Academic Press, New York. 409 pp.

Mechoulam, R., and N. Lander. 1980. Cannabis: a possible source of new drugs. Pharmacy International. 1980:19–21.

Mechoulam, R., N. K. McCallum, and S. Burstein. 1976. Recent advances in the chemistry and biochemistry of *Cannabis*. Chem. Rev. 76:75–112.

Nahas, G. G., and W. D. M. Paton (eds.). 1977. Marijuana: biological effects. Pergamon Press, New York. 777 pp.

Perlman, D., E. Titus, and J. Fried. 1952. Microbiological hydroxylation of progesterone. J. Am. Chem. Soc. 74:2126.

Peterson, D. H., and H. C. Murray. 1952. Microbiological oxygenation of steroids at carbon 11. J. AM. Chem. Soc. 74:1871–72.

Radwan, S. S., and C. K. Kokate. 1980. Production of higher levels of trigonelline by cell cultures of *Trigonella foenum-graecum* than by differentiated plants. Planta 147:340–44.

Razzaque, A., and B. E. Ellis. 1977. Rosmarinic acid production in *Coleus* cell cultures. Planta 137:287–91.

Reinhard, E. 1974. Biotransformation by plant tissue cultures. *In* H. E. Street (ed.), Tissue culture and plant science, pp. 329–62. Academic Press, London.

Robertson, L. W. 1982. Microbial transformation of cannabinoids. *In* J. P. Rosazza (ed.), Microbial transformation of bioactive compounds, 2:91–124. CRC Press, Boca Raton, Florida.

Robertson, L. W., S.-W Koh, S. F. Huff, R. K. Malhotra, and A. Ghosh. 1978. Microbiological oxidation of the pentyl side chain of cannabinoids. Experientia 34:1020–22.

Robertson, L. W., M. A. Lyle, and S. Billets. 1975. Biotransformation of cannabinoids by *Syncephalastrum racemosum*. Biomed. Mass Spectrom. 2:266–71.

Robertson, L. W., and M.-M Tsai. 1979a. Cannabitriol as an air oxidation product of ⁹-tetrahydrocannabinol. J. Nat. Prod. 42:694.

Robertson, L. W., and M.-M. Tsai. 1979b. Nitration of cannabinoids by cultures of soil bacterium. Presented at the 178th national meeting of the American Chemical Society, Washington, DC., 9–14 September 1979.

Rosazza, J. P. (ed.). 1982. Microbial transformation of bioactive compounds, vols. 1 and 2. CRC Press, Boca Raton, Florida. 133 pp. and 185 pp.

Schwartz, M. A. 1980. Prescription drugs in short supply: case histories. Marcel Dekker, New York. 130 pp.

Sebek, O. K. 1982. Notes on the historical development of microbial transformations. *In* J. P. Rosazza (ed.), Microbial transformations of bioactive compounds, 1:1–8. CRC Press, Boca Raton, Florida.

Shani, A., and R. Mechoulam. 1974. Cannabielsoic acids: isolation and synthesis by a novel oxidative cyclization. Tetrahedron 30:2437–46.

Shoyama, Y., M. Yagi, I. Nishioka, and T. Yamauchi. 1975. Biosynthesis of cannabinoid acids. Phytochemistry 14:2189–92.

Smith, R. V., and J. P. Rosazza. 1975. Microbial models of mammalian metabolism. J. Pharm. Sci. 64:1737–59.

Staba, E. J. 1980. Secondary metabolism and biotransformation. *In* E. J. Staba (ed.), Plant tissue culture as a source of biochemicals, pp. 59–97. CRC Press, Boca Raton, Florida.

Steck, W., B. Bailey, J. P. Shyluk, and O. L. Gamborg. 1971. Coumarins and alkaloids from cell cultures of *Ruta graveoleus*. Phytochemistry 10:191–94.

Stohs, S. J., and M. M. el-Olemy. 1972. Metabolism of progesterone by *Dioscorea deltoidea* suspension cultures. Phytochemistry 11:1397–1400.

Takeya, K., and H. Itokawa. 1977. Sterochemistry in oxidation of allylic alcohols by cell-free system of callus induced from *Cannabis sativa* L. Chem. Pharm. Bull. 25:1947–51.

Turner, C. E., M. A. Elsohly, and E. G. Boeren. 1980. Constituents of *Cannabis sativa* L. XVII. A review of the natural constituents. J. Nat. Prod. 43:169–234.

Uliss, D. B., R. K. Razdan, and H. C. Dalzell. 1974. Stereospecific intramolecular epoxide cleavage by phenolate anion. Synthesis of novel and biological active cannabinoids. J. Amer. Chem. Soc. 96:7372–74.

Veliky, I. A., and K. Genest. 1972. Growth and metabolites of *Cannabis sativa* cell suspension cultures. Lloydia 35:450–456.

Veliky, I. A., A. Jones, R. S. Ozubko, M. Przylaska, and F. P. Ahmed. 1980. 5β-hydroxygitoxigenin, a product of gitoxigenin produced by *Daucus carota* culture. Phytochemistry 19:2111–12.

Vidic, H. J., G. A. Hoyer, K. Kieslich, and D. Rosenberg. 1976. Microbiologische hydroxylierung von Δ^8-tetrahydrocannabinol. Chem. Ber. 108:3606–14.

Widman, M., M. Nordqvist, C. T. Dollery, and R. H. Briant. 1975. Metabolism of Δ^1-tetrahydrocannabinol by the isolated perfused dog lung: comparison with *in vitro* liver metabolism. J. Pharm. Pharmac. 27:842–48.

Zenk, M. H., H. El-Shagi, and V. Schulte. 1975. V. Anthraquinone production by cell suspension cultures of *Morinda citrifolium*. Planta Medica 1975 Suppl., pp. 79–101.

Zenk, M. H., H. El-Shagi, H. Arens, J. Stockigt, E. W. Weiler, and B. Deus. 1977. Formation of the indole alkaloids serpentine and ajmalicine in cell suspension cultures of *Catharanthus roseus*. *In* W. Barz, E. Reinhard, and M. H. Zenk (eds.), Plant tissue culture and its bio-technological application, pp. 27–43. Springer-Verlag, Berlin.

O. J. CROCOMO, O. G. BRASIL, AND
A. NATAL GONCALVES

Plant Cell Line Selection *In Vitro*

12

INTRODUCTION

The deliberate manipulation of genes of various organisms, through the techniques of genetic engineering, is aimed at introduction of permanent hereditary changes in the organism. Using *in vitro* techniques, based on recombinant-DNA technology, foreign genes from different sources can be bound with plasmid DNA and introduced within the host cells. On the other hand, plant cell culture can be used to introduce changes in genetic constitution brought about in cells by processes that resemble those occurring in nature. Both technologies can be used to engineer plants; however, recombinant DNA has only been successfully used in a few cases to express prokaryotic genes within plants (see Horsch et al., 1984).

Plant cell culture techniques may involve simple mutational alteration of the genetic material leading to an enhanced yield, or to improvement in the quality of useful secondary products. The manipulation of plant cells to achieve these goals involves the selection of plant cell lines, starting from cell cultures of chosen plant tissue or organ.

Production of secondary compounds by cell culture can be approached by inducing wild-type culture to produce a useful compound. However, a new approach has been proposed (Widholm, 1980) that implies the exploitation of the natural or induced variability found among plant cells, followed by selecting, cloning, and stabilizing those cell lines that feature desirable traits. Such cell lines may be selected at a low frequency from large populations. The cell line is indeed a cell culture bearing defined characteristics, although every cell in the "line" may not be identical.

It should be kept in mind that selection schemes are dependent upon which compound has to be produced. If the chemical is present in all plant species at all times, such as primary compounds like amino acids, the choice of the cell line to be used may not be crucial. On the other hand, if the desired compound is a secondary product and is present in just one or a few plant species, like pigments, alkaloids, phenolics, and some enzymes (e.g., papain), then the culture should be initiated from a plant species that normally produces the desired compound.

The knowledge of the physiological and biochemical mechanisms underlying the phenotypic expressions of favorable (cell lines) genotypes permits one to make a correlation between physiological responses and genetic potential. In higher organisms the phenotypic expression is regulated by hormones and in plants by the growth regulators including the phytohormones. The understanding of the hormone-genome interaction can lead to an understanding of the nature of such genetic switching devices.

CONTROL OF GENE EXPRESSION

Induction and Repression

The bacterial system. In the early 1900s, it was first demonstrated that certain enzymes present in yeast cells are formed only in the presence of their specific substrate. This effect was called "enzyme adaptation," and now after several studies carried out in bacterial systems mainly in *Escherichia coli* it is known as "enzyme induction." The now classical assay to detect the induction effect is to grow *E. coli* in the absence of β-galactoside; the bacteria cells will show only about 5 molecules per cell of the enzyme β-galactosidase. On the other hand, *E. coli* cells growing in the presence of the substrate may contain about 5,000 molecules of the enzyme. The increase of enzyme activity can be detected within 2–3 minutes after addition of the substrate.

The activation of an inactive presursor form of the enzyme already present in the cell or the *de novo* synthesis of new enzyme molecules might be the causes of the "induction effect." Both effects are supported by experimental data. Through immunological assays it has been demonstrated that the induced enzyme is antigenically different from any protein previously present in the cell. In turn, radioactive-labeling experiments confirmed that the induced enzyme is not derived from a precursor molecule synthesized before induction. Other studies have shown that if bacterial protein synthesis is inhibited with antibiotics, new enzyme molecules must be synthesized before induction occurs. Using RNA precursors, or using direct measurements of messengers, it has been shown that induction occurs by transcription of new RNA that is then translated into protein.

Specific inhibition of enzyme synthesis is known as "enzyme repression" and was discovered several years ago by Jacob and Monod (1961). In *E. coli* the synthesis of tryptophan synthetase is inhibited by its substrate tryptophan. The effects, as demonstrated by radioactive-labeling experiments, are repression of enzyme synthesis and not merely inhibition of enzyme activity. Experimental data have demonstrated that when the synthesis of an enzyme is repressed, synthesis of its mRNA no longer takes place and its remaining mRNA soon is degraded. As a consequence, no further protein molecules are synthesized. With subsequent cell division, those protein molecules already present in the cell are diluted out.

Usually a multienzyme system controls the functioning of a metabolic pathway, and in bacteria the genes governing the synthesis of the various enzymes of the system are often closely linked in a cluster on the genome. All the enzymes coded by the cluster of genes are usually induced or repressed together. The *lac* (lactose) genes of *E. coli* were the first system of this type to be described (Lewin, 1974) and since then used to exemplify the method by which induction is controlled in the cell.

The *lac* system is composed of three adjacent genes: *z*, *y*, and *a*. The *z* gene codes for the enzyme β-galactosidase, which catalyzes the hydrolysis of β-galactosides, such as lactose to glucose and galactose. The *y* gene specifies the protein β-galactoside permease. This is a membrane-bound protein that is part of the transport system responsible for the uptake of β-galactoside into the bacterial cells. The *a* gene codes for β-galactoside transacetylase, which catalyzes the transfer of an acetyl group from acetyl-CoA to galactoside.

Only the substrate of the enzyme, or in some cases other molecules closely related to it, can induce the *lac* system. The small molecules that cause enzyme induction are called inducer. In the repression system, the small molecules causing repression are termed *co-repressors*.

In an induction system as the *lac* operon, the same compounds are active or inactive as inducers of each enzyme, and the relative amounts of the enzymes synthesized in the presence of different inducers or various concentrations of one inducer are constant. This means that the coordinated synthesis of enzymes is governed by the inducer that interacts with some common controlling element. The fact that the specificity of induction does not depend upon the catalytic activity of the enzymes and that the rate of synthesis of different enzymes are under common control leads to the suggestion that the controlling element is not elaborated by the genes coding for the enzyme themselves.

These experimental observations made using *E. coli* led to the proposition by Jacob and Monod of the concept of *structural genes* and *regulator genes*, which are part of the *operon*. *Structural genes* are responsible for the synthesis of protein required by the cell, and the *regulator genes* control the synthetic activity of the *structural genes*.

In *E. coli* a mutation in a structural gene causes loss of activity of the enzyme coded by the structural gene. On the other hand, mutation in the other two regions, called *o* and *i*, can produce bacteria showing an abnormal ability to produce large quantities of *lac* enzymes in the absence of an inducer. These are constitutive mutants. Most of them synthesize more enzyme in the absence of the inducer than is produced by wild-type bacteria after induction; however, the relative proportions of the enzymes made are the same as in normal cells. This suggests that the *regulator loci* are the controls that respond to the inducer in wild-type bacteria.

The *i* gene governs the synthesis of a repressor protein, and its specificity is clear because it affects only the *lac* system. This specificity and the pleiotropy of the *i* gene, through which it controls the synthesis of all the three enzymes, implies that the repressor acts on some region of the system common to all structural genes. This is the *operator* region. If the *inducer* is absent, the repressor protein combines with the operator and as a result the complete *z y a* segment is switched off. When inducer is added, it inactives the repressor so that it can no longer combine with the operator; in that case the system resumes protein synthesis. This is *induction*.

The operator does not code for some product that can diffuse through the cytoplasm to govern the activity of the structural genes. According to Jacob and Monod's model, the operator controls the adjacent *z y a* segment by serving as a recognition element: the interaction of the operator sequence with repressor might prevent transcription of the adjacent genes. Both structural and regulator genes require transcription and translation to fulfill their functions, but the operator, which is a regulator site, functions by constituting a nucleotide sequence whose recognition by a protein controls the expression of adjacent genes.

The system as a whole, including both structural genes and regulator elements, is described by the term *operon*. Although the structural genes produce independent proteins, the operon behaves as a single unit in the transfer of genetic information from DNA to protein.

The interaction between repressor protein and operator can take place either (a) at the level of transcription of DNA into mRNA, or (b) at the level of translation of mRNA into proteins. If the first alternative takes place, the repressor must interact with the segment of DNA genetically defined as the operator. In the case of control at translation, the operator would have to be transcribed into messenger RNA and the repressor would act on a transcript representing all the structural genes.

In the model proposed by Jacob and Monod, all the systems of repression and induction appear to be controlled at the level of transcription, in such a way that recognition sites always comprise base pair sequences of DNA.

The plant system. The expressed genetic content and the large amount of DNA often found in the cells of eukaryotic organisms constitute a divergence called the "C" paradox. Some plant species like ferns and sugar cane show a very high number of chromosomes. There are fern species bearing 630 chromosomes (Brown and Bertke, 1974) and sugar cane varieties having about 200 chromosomes (Liu, 1981).

Probably all eukaryotes contain DNA base sequence in the their chromosomes that are represented many times. This feature of the genome has biological significance (Walbot, 1979). These sequences can be (Bendich, 1979): (a) repeated many thousands or millions of time per genome (highly repetitive sequences); (b) repeated a moderate number of times (repetitive sequences); (c) present once per haploid genome (unique copy DNA). In organisms with a low "C" value, the unique copy DNA specifies 1×10^4 to 5×10^4 genes.

The reiterated sequences are also called *satellite* or *simple-sequence* DNA and *intermediately repetitive DNA*. The satellite sequences are thought to be tandemly arranged, to be concentrated in centromeric heterochromatin, and to possibly serve a noncoding function in the folding, pairing, and movement of chromosomes (Walker, 1971). Intermediately repetitive DNA sequences, on the other hand, are less highly reiterated; these sequences are dispersively scattered and partially clustered throughout the genome.

The functions of those repeated sequences, which are interspersed with nonrepetitous DNA, are not yet understood. However, they are thought to serve in a gene regulatory capacity (Davidson and Britten, 1973). This model is an attempt to integrate the complexity of the DNA sequence organization and gene regulation in eukaryotes. The structural genes (unique copy DNA) would be contiguous to the repetitive sequence DNA; this would serve to regulate transcription of the adjacent gene. This model explains the coordinated regulation of structural genes dispersed throughout the genome. This feature makes Davidson and Britten's model very attractive since while in prokaryotes the genes of a particular biochemical pathway are often physically linked and coordinately regulated, in eukaryotes the genes of a particular biochemical pathway, such as a secondary product in plants, are located on different chromosomes.

Many plant species have a large component of intermediately repetitive DNA (Walbot, 1979). Species of higher plants with genomes of 5 to 98 pg of DNA per 2C nucleus contain about 80% repetitive sequences. On the other hand, plants with 1 to 4 pg of DNA per 2C nucleus have only 62% repetitive sequences.

There is experimental evidence that these repetitive sequences function as genes. It has been demonstrated that the structural RNAs, such as rRNA, and the mRNA for histone proteins, are transcribed from repetitive sequence DNA (Pardue, 1975). The transcription of mRNA is made by the repetitive sequences arranged as gene clusters in one or more chromosomal locations: the rRNA cistrons are found in the nucleolus region of plants such as *Zea, Happlopappus*, and *Phaseolus*. Transcription can be from repetitive or nonrepetitive DNA sequences. The rDNA is clustered, and the genes are separated by transcribed as well as nontranscribed spacer DNA. 5S rDNA sequences are found on the extreme ends of the long arms of most *Xenopus laevis* chromosomes arranged in clusters; there are about 24,000 copies of the 5S rDNA sequences per haploid chromosome complement. The tRNA genes in *Xenopus* are also arranged in clusters.

Genes that are present only once per haploid genome should be contained within the unique copy fraction of the genome. Nonrepetitive DNA sequences are responsible for the transcription of mRNAs of silk fibrin, ovalbumin, and globin (Bendich, 1979).

Probably mRNAs are derived from a larger initial transcript and the mRNA genes are separated by nontranscribed spacers. The higher molecular weight class of molecule than mRNA, called heterogeneous RNA (HnRNA), is the most likely mRNA precursor (Lewin, 1974). If so, HnRNA might contain some transcript from repetitive DNA sequences, which should be selectively lost during RNA processing.

To make clear what has been said and what will be discussed later, we should remember the main tenets of Watson and Crick's "central dogma" of molecular genetics, elucidated in the early 1950s (Watson and Crick, 1953).

1. All the information to build and operate a cell is encoded in the nucleotide sequence of the DNA that comprises the nuclear chromo-

somes and the cell organelles. Van't Hof and colleagues (1979) demonstrated that plant DNA is self-replicating. Work on ribosomal RNA genes has shown quantitative and qualitative relationships between specific gene products and DNA sequences (Kende and Tavares, 1968).

2. Expression of any information of the DNA requires its transcription into an intermediate RNA. Selected section of DNA at specific times permits selective use of some genetic information at that time; specific stimuli can influence the response of DNA sections. Bonner's classic experiments (Bonner et al., 1966) on pea chromatin demonstrated selective transcription of plant DNA.

3. The transcribed information is expressed (translated) as a specific protein; its amino acid sequence has been translated from the sequence of nucleotides of the messenger RNA that was used as template for protein synthesis.

The wheat germ system has been used for *in vitro* translation of several mRNAs including viral coat protein RNAs (Marcus, 1970); this fact elicits mRNA as an intermediate in genetic expression by plants.

Bearing this in mind and also that nucleus and cytoplasm interact, the flow of genetic information from the DNA sequences of nuclear genes to the cytoplasmic protein-synthesizing apparatus in eukaryotic cells involves the interaction of three cell compartments (Mans et al., 1979): nucleolus within the nucleus; nucleus within the cytoplasm; and cytoplasm.

In figure 1 the cell's genome is represented by a segment of DNA, and includes ribosomal RNA genes in the nucleolar compartment and the structural and regulatory genes in the nuclear compartment.

In eukaryotes there are three separate nuclear transcription systems (Mans et al., 1979). One of these systems is the *RNA polymerase I* (or *A*), which is associated with the nucleolar organizer region. This system is responsible for transcription of the ribosomal precursor RNA (pre rRNA) which is processed into ribosomal RNAs (rRNA). Ribosomal proteins associate with rRNAs, and the complex is transported into the cytoplasm, becoming functional ribosomes.

RNA polymerase III (or *C*) is another transcription system primarily concerned with the transcription of genes for precursor tRNA and 5S

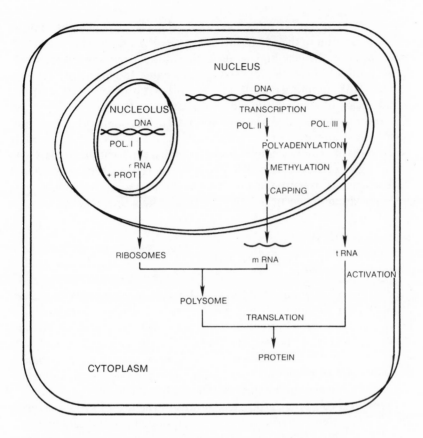

Fig. 1. Representation of genetic expression in eukaryotic cells.

RNA. The products of its action are transported to the cytoplasm to participate in the protein synthesizing machinery.

RNA polymerase II (or *B*) represents the second nucleoplasmic transcription system. Its products are heterogeneous in size and composition: they are the heterogeneous nuclear RNAs (HnRNA referred to above). The HnRNA represents transcripts of the structural genes of all the cell's proteins. RNA polymerase II and III are confined to the nucleas.

The metabolic processing of RNA polymerase II transcripts plays a significant role in the expression of genetic information in eukaryotic cells (Mans et al., 1979). The following processing reactions of HnRNA molecules can take place: (a) *polyadenylation*, which is the sequential addition of AMP moieties to their 3′-OH termini; (b) *methylation*, which is the 3′0-methylation of the penultimate nucleotide at the 5′ terminus of the newly formed mRNA molecule; (c) cleavage to smaller fragments or mononucleotides; (d) capping of the 5′ terminus of mRNA with 7 methylguanosine, as in the case of mammalian and viral messenger RNAs.

In this manner selected portions of some HnRNA molecules, once processed, are transported to the cytoplasm to serve as mRNA in the assembly polysomes; ribosomal subunits associate with amino acid-bearing tRNAs (resulting from the action of RNA polymerase III) and the mRNA under the influence of initiation factors. Functional polysomes are thus formed and translation of genetic information into protein then takes place.

HORMONE ACTION

Hormone Binding

In plants many specific developmental events can be initiated by the addition of plant hormones, which is best visualized when plant explants are grown *in vitro* under aseptic conditions. Indeed, experimental evidence has shown that plant hormones play a role in the regulation of gene activity during plant development. There are five groups of known plant hormones (growth regulators) that are involved in plant growth and development: cytokinins, gibberellins, auxins, abscisic acid, and ethylene.

Cytokinins can be incorporated into the structure of certain transfer RNAs. Exogenous cytokinins are not incorporated in significant amounts into RNAs, but the substitution of the various side chains on the N^6 position of the adenine portion is accomplished after the tRNA molecule is formed (Varner and Tuan, 1976). However, some active cytokinins, such as 9-methylbenzyladenine cannot be converted to the riboside triphosphate for incorporation into tRNA. Isopentenyl adenosine (IPA) was the first cytokinin identified as a constituent of tRNA. IPA is present only once in each of two tRNA^{Ser} species (Zachau et al., 1966). IPA was adjacent to the 3′ end of the anticodon. Tyrosine-tRNA (tRNA^{Tyr}) has been shown to contain thiomethylisopentenyladenosine (CH_3-IPA) adjacent to the 3′ end of the anticodon (Harada et al., 1968). The same kind of interaction between other natural or synthetic cytokinins and tRNA or other nucleotides can be inferred taking into account their metabolism, which is summarized below (Varner and Tuan, 1976):

Natural cytokinins
adenine in tRNA → IPA in tRNA
IPA riboside → adenosine → adenine
IPA → N⁶-(3-methyl-3-hydroxybutil) aminopurine
IPA riboside → zeatin riboside
zeatin → zeatin riboside → zeatin ribotide
zeatin riboside → adenosine; zeatin riboside → zeatin

Synthetic cytokinins
benzyladenine → benzyladenosine → benzyladenosine-5′-phosphate
benzyladenine → benzyladenine-7-glucoside
kinetin → kinetin riboside → kinetin riboside monophosphate

Among the *gibberellins*, GA_3 controls the formation and secretion of hydrolases in cereal grains. During the germination of barley grains, the mobilization of endosperm reserves is accomplished by the action of α-amylase, protease, and ribonuclease, which are synthesized in the aleurone tissue that surrounds the endosperm. The aleurone cells respond to the stimulus of GA that is formed in the embryo in the early stages of seed germination.

Amylase is the most predominant protein synthesized in the aleurone layer after treatment with GA_3 and is an important marker to study the mechanism of the hormone-controlled enzyme formation in this system (Varner and Tuan, 1976). The enhanced effect of GA_3 on α-amylase synthesis is blocked by actinomycin D, 6-methylpurine, and other RNA-synthesis inhibitors.

GA_3 affects membrane proliferation, which extensively occurs before the rapid increase of α-amylase activity. During the lag period, the endoplasmic reticulum proliferates concurrently with the formation of membrane-bound polyribosomes. The rough endoplasmic reticulum (RER) participates in the synthesis and export of secretory proteins, such as α-amylase in eukaryotic cells.

Two alternatives account for GA_3-enhanced synthesis of hydrolases (Varner and Tuan, 1976): (1) the synthesis is dependent upon the increased rate of formation (transcription and/or processing) of its specific mRNA; (2) the enhanced membrane proliferation mediated by GA_3 provides increased numbers of sites available for the synthesis of the enzymes, in which case there would be no need for the amount of α-amylase-specific mRNA to be increased in the presence of GA_3.

Gibberellic acid enhances RNA synthesis in isolated nuclei during the elongation of the pea epicotyl (Johri and Varner, 1968). There is an increase in the level of RNA polymerase associated with chromatin without a detectable increase in the amount of template DNA available, which was confirmed by DNA/RNA hybridization techniques: hybridizable RNA was not changed up to 50 hours after treatment of pea seedlings with GA (Thompson and Cleland, 1972). On the other hand, in cucumber hypocotyls the template activity of isolated chromatin can be increased by GA treatment (Johnson and Purves, 1970).

Gibberellic acid is also able to induce morphological reversions to the juvenile forms of growth when applied to the mature form of *Hedera helix* (Rogler and Dahmus, 1974). The juvenile forms produced are stable with time and show dramatic phenotypic differences from the mature form. To detect possible gene derepression during such reversion, the DNA/RNA hybridization competition techniques were used: no difference in the RNA population isolated from either juvenile or mature forms was observed. This implies that during the GA-mediated developmental process a post-transcriptional control mechanism would take place.

Auxins promote cell enlargement. Respiration, replacement of short-lived protein, RNA, and so on, are part of the machinery required to sustain cell enlargement.

An early effect of an auxin, indoleacetic acid (IAA), seems to be the promotion of secretion of cell wall-modifying enzymes and of cell wall precursors. If so, the sustained response of a growing tissue to IAA would require concomitant RNA synthesis and protein synthesis. The immediate response would be an increase in the rate of cell extension that probably would require neither RNA synthesis nor protein synthesis (Varner and Tuan, 1976). Gene activation might be required for sustained response, but not for the primary process.

The effect of *abscisic acid* (ABA) depends upon the continuous synthesis of RNA. This has been demonstrated when its inhibitory effect on the germination of cotton seeds was abolished in the presence of transcription inhibitors such as actinomycin D. The very early stages of cotton seed germination are not sensitive to actinomycin D, which suggests that the mRNA needed for germination preexists and that the effect of ABA could be via a transcriptional control point (Varner and Tuan, 1976).

Whereas gibberellic acid is a promoter, ABA is an inhibitor of wheat seed germination. No new mRNA or rRNA is associated with ribosomes, as has been shown by DNA-RNA hybridization experiments. This would indicate that the hormone regulates the use of mRNA already stored in the embryo. The hormone would play no role in the control of *de novo* synthesis of mRNA.

Hormone receptors

In animals the hormones in general are synthesized in defined organs and then translocated to other tissues and organs. There are two main classes of animal hormones, peptide hormones and steroid hormones. Both interact with target cells, called "receptor sites," that must have very high affinity and specificity for the hormone since animal hormones usually work at very low concentrations (10^{-10} to 10^{-12} M).

The peptide hormone binds to primary receptor sites located at the outer cell membrane (that cell is a target cell). On the inner side of the membrane, an enzyme, adenyl cyclase, is activated, forming cyclic AMP (cAMP), which acts as a "second messenger" inducing a number of biochemical events (see Jacobsen, 1983).

The steroid hormones enter their target cells and are recognized by soluble cytoplasmic proteins. The dimeric hormone-receptor complexes thus formed bind to acceptor sites in the chromatin, triggering the synthesis of new mRNA that leads to the production of new proteins (Jacobsen, 1983; Dodds and Hall, 1980).

These two models of animal hormone action cannot be applied in full to plant hormones. First of all, the peptide hormone model requires the action of an enzyme and of cAMP. Neither the corresponding enzyme nor cAMP has been found as yet in noteworthy amounts in plant cells (Amrhein, 1977). On the other hand, there is a similarity between the action of phytohormones and the steroids. However, the causality between hormone applications, hormone recognition, and hormone-induced gene activation has not been unequivocally demonstrated in plants. The membrane-bound sites of the latter have no cAMP-synthesizing activities (cf. Jacobsen, 1983).

As in animals, in plants the phytohormones interact with some component in the cell, the "receptor," which must have high specificity and affinity toward the hormone. However, there are some differences between plant hormones and animal hormones, the first being the fact that there are only five groups of known plant hormones. Besides that, the latter have simpler structures than those in animals. Also, in plants the site of synthesis of the hormone is not usually restricted to a specialized organ or tissue, the meristem being an exception: it appears to have greater synthetic capacity than other areas within the plant.

In plants almost every developmental process is controlled by hormones, and in many cases the same process may be controlled by different groups of hormones. As a result there is not usually one distinct target for a given plant hormone since at any one time many different plant tissues and organs are capable of responding to it. Recently, binding sites have been isolated that interact with all groups of plant growth regulators (Jacobsen, 1983). In animal systems a clear link between the interaction of the hormone (ligand) and the binding site (receptor) and the initiation of a sequence of distinct biochemical events exists resulting sometimes in dramatic changes such as morphogenesis or the regulation of the blood sugar levels. On the other hand, in plants, for the hormone binding systems known so far, no clear correlation seems to exist between the interaction of hormone and receptor sites and the onset of a biochemical or morphological change. Still, as in animal systems, the receptor must have not only a high affinity but also a high degree of specificity toward the molecule that will bind to it.

Auxin binding. Auxin binding sites have been isolated from maize, tobacco pith callus, cucurbits, soybean cotyledons, and bean seedlings (Dodds and Hall, 1980). Cell elongation is a response to some auxins, and it takes place rapidly, the lag time being a few minutes. This evidence favors the hypothesis that addition of auxin to a plant system (such as coleoptiles) stimulates the membrane-bound ATPase to pump protons, bringing about extension growth by affecting cell wall plasticity. In that case, protons would function as "second messenger" in the same manner as cAMP in animal systems.

The auxin binding sites appear to be associated with membranes. The auxin binding sites are proteins, as demonstrated by several experimental evidences. Auxin (naphthaleneacetic acid, NAA) binding sites are inactivated following treatment with mercuric ions or glutaraldehyde. Pronase, a proteolytic enzyme, only inhibits binding to some extent; however, pretreatment with phospholipase C followed by pronase eliminates all binding activity. The possible explanation would be that the binding site is buried in the membrane and is protected by membrane lipids (Dodds and Hall, 1980).

Ethylene binding. The ethylene binding site is membrane-bound. It is associated with subcellular particles. The most frequently studied ethylene binding site is that from *Phaseolus vulgaris* cotyledon tissue, which has been studied in both *in vitro* (cell-free system) and *in vivo* systems. As with auxin binding sites, the proteins responsible for the ethylene binding sites must be buried within the membrane system under the protection of lipids (Dodds and Hall, 1980).

Other phytohormones. Not much information can be gathered for the other classes of plant growth regulators, abscisic acid, gibberellins, and cytokinins. It seems that the binding site for abscisic acid may be located on the plasma membrane. In systems capable of extension, such as lettuce hypocotyls, interaction of gibberellins with cell wall material, which may represent a physiological binding site, has been demonstrated. On the other hand, tobacco cell fractions and the protonema of the moss *Funaria hygrometrica* have been used for isolation of cytokinin binding sites.

Callus tissue is a good plant system for studying the interaction between phytohormones and their sites (receptors). In most cases the phytohormones just "switch on" predetermined processes, but in other cases they appear to "determine" the developmental pattern as a whole. The most dramatic example is the well-known fact that the induction of root and/or shoot primordia in callus tissue can take place by merely manipulating the relative concentrations of auxins and cytokinins.

Secondary products can be formed in the cells of the whole plant by similar mechanisms. In the case of secondary products that are synthesized in special structures within the plants, these structures would be a consequence of the interaction between growth regulators and their receptors. Callus of papaya (*Carica papaya*) when grown in the presence of certain ratios of auxin/cytokinin produces laticifer vessels that later on will produce papain (Crocomo and Goncalves, unpublished)

PROPOSED ADAPTATIVE MECHANISM FOR PLANT CELL LINE SELECTION

It has already been stressed by other authors (Larkin and Scowcroft, 1981), that tissue culture can be *per se* "an unexpectedly rich and novel source of genetic variability" and not as it was visualized until just a few years ago a method of only cloning a particular genotype, thus enabling its rapid propagation. Very often phenotypic variants are observed among plants regenerated through cell culture. These variants have been termed somaclones (Larkin and Scowcroft, 1981), and are valuable adjuncts to plant improvement.

The variation among clones and subclones of a parental cell line can be spontaneous or induced. In the specific case of selection of cell lines to produce a particular secondary product, both situations can occur. The composition of the culture medium and environmental conditions can be manipulated to enhance its production. Stable cell lines can be derived from cell lines with altered genome. Even under the most favorable conditions, phenotypic variants can be detected and submitted to selection pressure *in vitro*. The culture media can be used to select cells able to divide and grow and not to regenerate plants. A variant cell having a reduced lag phase can have an increased capability to utilize nutrients from the medium, will divide and grow more rapidly, and will be a dominant cell in the culture (Sunderland, 1973; Bayliss, 1973, 1975).

The possibility of selecting cell lines resistant to antimetabolites is an indication that the cells in culture differ among themselves in the capability to produce metabolites, in membrane permeability, and in the sensitivity of their enzymes to inhibition (Binding, 1974; Widholm, 1974).

Phenotypic variability can result from a permanent alteration in base sequences of the DNA of the parent cell or from an alteration of the genetic expression ("epigenetic") that can or cannot be heritable. As it has been pointed out (Larkin and Scowcroft, 1981), somaclonal variation is found either in assexually propagated species or in seed-propagated, self-fertilizing species, in diploids as well as in nondiploids. In any case, we can visualize phenotypic variability in *in vitro*-regenerated plants as an expression of the genetic variability already preexisting in the somatic cells of the original explant, and which was induced by the chemical and physical conditions of the culture medium.

Regulatory genes can be altered in their action by the cell constituents since gene regulation is related to the synthesis of nuclear RNA to cytoplasm and protein to nucleus is established through the nuclear membrane (fig. 1). Cells that are adapted to a more intensive flux will respond more rapidly to a genetic stimulus, and will be stimulated to produce a certain metabolite in higher amounts. In this way the cell population will be continuously subjected to a selection pressure.

The possibility of isolating cell lines with mutations in genes opens up the alternative to obtain cell cultures selected for a desired goal, such as enhanced production of a secondary metabolite. This rationale is valid if the cytodifferentiation is a consequence of exchanges at the transcriptional and translational levels of gene expression.

The chemical constitution of each cell in a cell population could be responsible for the production of a secondary metabolite, since the chemical constituents can alter the action of regulatory genes. Gene regulation, being related to the synthesis of RNA in the nucleus and to the synthesis of protein in the cytoplasm in a flux regulated by the nuclear membrane, permits the appearance of cells that respond more rapidly to a genetic stimulus. A good example of genetic variability in eukaryotic cells in culture, and how they can be selected through phenotypic characteristics, is the selection of cell lines showing an increased production of anthocyanins (Yamamoto et al., 1982).

The effects of metabolites of aspartate metabolism on the growth of certain plants is another example. These effects are consistent with the presence of concerted and mutivalent inhibition of aspartokinase (Bryan, 1976). In rice tissue culture, the enzyme is regulated allosterically and synergistically by lysine plus threonine (Furahashi and Yatazawa, 1970) (fig. 2). On the other hand, traces of methionine release the inhibition. In corn cell cultures, the simultaneous addition of lysine and threonine, in equimolar concentrations, completely prevents callus growth (Green, 1982). Lysine and threonine would simultaneously inhibit aspartokinase and homoserine dehydrogenease, thus preventing the normal synthesis of lysine, threonine, and methionine. Hence, when lysine and threonine are added to the culture medium, growth inhibition results from the inability of the corn cells to synthesize methionine. The addition of methionine (or of homose-

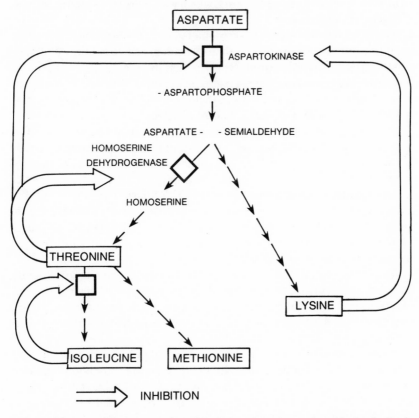

Fig. 2. Diagrammatic representation of the lysine, threonine, and methionine feedback control in higher plants.

rine) in the lysine-threonine culture medium reverts the inhibition permitting growth of the cultures.

This lysine-threonine inhibition has immediate application in the isolation of feedback-resistant mutants in tissue culture of corn, and it is a potential selection for mutants that could overproduce one or more of those three amino acids. These mutants would have altered feedback inhibition controls that are less sensitive to lysine-threonine inhibition (Green, 1982).

This type of variant could probably occur by a modification in the allosteric site of aspartokinase for both negative effectors, lysine and threonine, that characterize a mutation. On the other hand, the variant could result from cells in the culture that synthesize a higher amount (or higher specific activity) of aspartokinase to prevent the effect of certain exogenous concentrations of the negative effectors, which would probably characterize an "epigenetic" alteration.

The experimental evidence accumulated over the last few years on the regulation of the amino acid metabolism in plant cells in culture (Widholm, 1980) permits an explanation of the production of secondary metabolites, which are under the same type of regulatory control. Within a large population of cells, only those individuals that grow under certain culture conditions (chemical and physical) would be selected.

Several authors (Thimann 1972; Sharp et al., 1980) have shown that auxins and cytokinins can play an important role in the allosteric systems responsible for the biosynthesis of nucleic acids at the transcriptional and translational levels, which would allow for the synthesis of modified RNAs. The experimental results (Ingle, 1979; Sharp et al., 1980; Millikan and Ghosh, 1971; Domoney and Timmis, 1980) show differences in the biosynthesis of DNA and RNA that occur in different stages of differentiation. It is well known that the biosynthesis of ribonucleotides and deoxyribonucleotides are under allosteric control.

The genetic variability of somatic cells could be due to alterations in the allosteric mechanisms that regulate these biosynthetic pathways, or to alterations in the processing and translation of the genetic information governing cell development. Auxins and cytokinins would work on these mechanisms, modifying them.

Phytohormones alone or synergistically could provoke alterations in the mechanisms of biosynthesis and also in the ratios of purine and pyrimidine nucleotides and in the flux of genetic information, allowing for the appearance of different phenotypes that could then be selected for the desired trait. Depending upon the origin and the physiological stage of the cells and the quantity and quality of the stimulus given by the phytohormones, some cells in a population of cultured cells have different phenotypic capabilities.

The somatic cell genetics of plants would be a manifestation of allosteric regulation mechanisms in the biosynthesis and in the shifts of the ratio between and within purine and pyrimidine nucleotides, and in the transcription, processing, and translation of genetic information during cellular differentiation and development.

REFERENCES

Amrhein, N. 1977. The current status of cyclic AMP in higher plants. Annu. Rev. Plant Physiol. 28:123–32.

Bayliss, M. W. 1973. Origin of chromosome number variation in cultured plant cell. Nature 246:529–30.

Bayliss, M. W. 1975. The effects of growth *in vitro* on the chromosome complement of *Daucus carota* suspension cultures. Chromosoma 51:401–11.

Bendich, A. J. 1979. The nature of families of repeated DNA sequences in plants. *In* I. Rubenstein, R. L. Phillips, C. E. Green, and B. G. Gengenbach (eds.), Molecular biology of plants, pp. 1–30. Academic Press, New York.

Binding, H. 1974. Mutation in haploid cell culture. *In* K. J. Kasha (ed.), Haploids in higher plants: advances and potential, pp. 323–38. University of Guelph, Guelph, Ontario.

Bonner, J., M. E. Dahmus, D. Frawmbrough, R. C. Huang, K. Murashige, and D. H. H. Tuan. 1966. The biology of isolated chromatin. Science 156:47–56.

Brown, W. V., and E. M. Bertke. 1974. Textbook of cytology. 2d ed. C. V. Mosby, St. Louis. P. 361.

Bryan, J. K. 1976. Amino acid biosynthesis and its regulation. *In* J. Bonner and J. E. Varner (eds.), Plant biochemistry, pp. 525–60. Academic Press, New York.

Davidson, E. H., and R. J. Britten. 1973. Organization, transcription, and regulation in the animal genome. Quart. Rev. Biol. 48:565–613.

Doods, J. H., and M. A. Hall. 1980. Plant hormone receptors. Sci. Prog. Oxf. 66:513–35.

Domoney, C., and J. N. Timmis. 1980. Ribosomal RNA gene redundancy in juvenile and mature ivy (*Hedera helix*). J. Expt. Bot. 31:1193–2110.

Furuhashi, K., and M. Yatazawa. 1970. Methionine lysine-threonine-isoleucine interrelationships in the amino acid nutrition of rice callus tissue. Plant Cell Physiol. 11:569–78.

Green, C. E. 1982. Application of tissue and cell culture techniques to corn breeding. *In* H. D. Loden and D. Wilkinson (eds.), Report of 37th Annual Corn and Sorghum Res. Conf, pp. 52–66. ASTA, Publ. No. 37, Am. Seed. Trad. Assoc.

Harada, F., H. J. Cross, F. Kimura, S. H. Chang, S. Nishimura, and U. L. Rajbhandary. 1968. 2-Methylthio N^6-(Δ^2-isopentenyl) adenosine: a component of *E. coli* tyrosine transfer RNA. Biochem. Biophys. Res. Commun. 33:299–306.

Ingle, J. 1979. Plant ribosomal RNA genes: a dynamic system? *In* I. Rubenstein, R. L. Phillips, C. E. Green, and B. G. Gengenbach (eds.), Molecular biology of plants, pp. 139–64. Academic Press, New York.

Jacob, F., and J. Monod. 1961. Genetic regulatory mechanisms in the synthesis of proteins. J. Mol. Biol. 3:318–56.

Jacobsen, H. J. 1983. Biochemical mechanism of plant hormone activity. *In* D. A. Evans, W. R. Sharp, P. V. Ammirato, and Y. Yamada (eds.), Handbook of plant cell culture, 1:672–95. Macmillan, New York.

Johnson, K. D., and W. F. Purves. 1970. Ribonucleic acid synthesis by cucumber chromatin. Plant Physiol. 46:581–88.

Johri, M. M., and J. E. Varner. 1968. Enhancement of RNA synthesis in isolated pea nuclei by gibberellic acid. Proc. Natl. Acad. Sci. USA 59:269–76.

Kende, H., and J. E. Tavares. 1968. On the significance of cytokinin incorporation into RNA. Plant Physiol. 43:1244–48.

Larkin, P. J., and W. R. Scowcroft. 1981. Somaclonal variation: a novel source of variability from cell cultures for plant improvement. Theor. Appl. Genet. 60:197–214.

Lewin, B. 1974. Gene expression. I. Bacterial genomes. John Wiley and Sons, New York.

Lewin, B. 1974. Gene expression. II. Eucaryotic chromosomes. John Wiley and Sons, New York.

Liu, M. C. 1981. *In vitro* methods applied to sugarcane improvement. *In* T. A. Thorpe (ed.), Plant tissue culture methods and applications in agriculture, pp. 299–323. Academic Press, New York.

Marcus, A. 1970. Tobacco mosaic virus ribonulceic acid-development amino acid incorporation in a wheat embryo system *in vitro*. J. Biol. Chem. 245:962–66.

Mans, R. J., C. O. Gardner, Jr., and T. J. Walter. 1979. Selective transcription and processing in the regulation of plant growth. *In* I. Rubenstein, R. L. Phillips, C. E. Green, and B. G. Gengenbach (eds.), Molecular biology of plants, pp. 165–95. Academic Press, New York.

Millikan, D. F., and D. Ghosh. 1971. Changes in nucleic acids with maturation and senescence in *Hedera helix*. Physiol. Plant. 24:10–13.

Pardue, M. L. 1975. Evolution of DNA content in higher plants. Genetics 79:159–70.

Rogler, C. E., and M. E. Dahmus. 1974. Gibberellic acid induced phase change in *Hedera helix* as studied by deoxyribonucleic acid-ribonucleic acid hybridization. Plant Physiol. 54:88–94.

Sharp, W. R., M. R. Sondahl, L. S. Caldas, and S. B. Maraffa. 1980. The physiology of *in vitro* asexual embryogenesis. *In* J. Janick (ed.), Horticulture reviews, 2:268–310. AVI Publishing, Westport, Conn.

Sunderland, N. E. 1973. Nuclear cytology. *In* H. E. Street (ed.), Plant tissue and cell culture, pp. 161–90. Blackwell Scientific Publishing, Boston.

Thimann, K. V. 1972. The natural plant hormone. *In* F. C. Steward (ed.), Plant physiology: a treatise, 6B:3–145. Academic Press, New York.

Thompson, W. R., and R. Cleland. 1972. Effect of light and gibberellin on ribonucleic acid species of pea stem tissues as studied by deoxyrobonulceic acid-ribonucleic acid hybridization. Plant Physiol. 50:289–92.

Van't Hoff, J., C. A. Bjerknes, and J. H. Clinton. 1979. Replications of chromosomal DNA fibers of root meristem cell of higher plants. *In* I. Rubenstein, R. L. Phillips, C. E. Green, and B. G. Gengenbach (eds.), Molecular biology of plants, pp. 73–91. Academic Press, New York.

Varner, J. E., and D. H. H. Tuan. 1976. Hormones. *In* J. Bonner and J. E. Bonner (eds.), Plant biochemistry, pp. 713–70. Academic Press, New York.

Walbot, V. 1979. Genome organization in plants. *In* I. Rubenstein, R. L. Phillips, C. E. Green, and B. G. Gengenbach (eds.), Molecular biology of plants, pp. 31–72. Academic Press, New York.

Walker, P. M. B. 1971. "Repetitive" DNA in higher organisms. *In* J. Butler and A. V. Novel (eds.), Progress in biophysics and molecular biology, pp. 147–90. Academic Press, New York.

Watson, J. D., and F. H. C. Crick. 1973. Genetical implications of the structure of deoxyribonucleic acid. Nature 171:964–69.

Widholm, J. M. 1974. Selection and characteristics of biochemical mutants of cultured plant cells. *In* H. E. Street (ed.), Tissue culture and plant tissue, pp. 287–30. Academic Press, London.

Widholm, J. M. 1980. Selection of plant cell lines which accumulate certain compounds. *In* E. J. Staba (ed.), Plant tissue culture as a source of biochemicals, pp. 99–113. CRC Press, Boca Raton, Florida.

Yamamoto, Y., R. Mizugushi, and Y. Yamada. 1982. Selection of a high and stable pigment producing strain in cultured *Euphorbia millii* cells. Theor. Appl. Genet. 61:113–16.

E. C. COCKING

Protoplast Fusion and Gene Transfer

13

Protoplast fusion should not be oversold because other methods of gametic fusion have been splendidly successful in furthering the development of agriculture for many years. Until recently gene flow between plants was essentially restricted to sexual crossings. More recently we have seen the development of protoplast technology. It was used for other things than fusion for many years, including studies on wall formation and virus interactions, which incidentially laid the foundation for a lot of the current work on plasmid manipulation (Cocking, 1981). Then came the realization that because these cells were naked cells, they might fuse together. All this was against the background of many years' work on plant cell cultures in which people were beginning to realize, in an increasing number of species, albeit rather hit and miss, that one could get regeneration of whole plants from single cells. The vista opened up that the plant biologist might be able to do more with this type of somatic fusion system than the animal cell biologists had been able to do. Of course, the animal cell biologists had been splendidly successful, and still are very successful in relation to the use of fusion in the hybridoma field.

Although I will discuss the plant breeding implications of this somatic fusion work, one should not forget that in relation to biotechnology the use of plant cell fusion at the cell culture level, and the production of novel plant secondary products and improved yields, is a major neglected area in this field of endeavor.

As time went on, it became clear that not only could one fuse such naked cells but also that one could devise methods for selecting out the fusion products, and that one could grow these fused cells into various types of hybrid plants. Hence, one saw that one had an extra method of hybridization, with certain differences from sexual hybridization, but also the vista opened up a wider range of possible hybridizations. Of course, this must be viewed against the advances that have been made, over many years, in wider sexual hybridizations in plants. For example, the long, painstaking work on the triticale interspecies hybrid has been especially successful. What I am going to try to do here is to give some personal thoughts on this subject, and try to present to you some perspective as well. The general subject area can easily be oversold, but the reality that I want to present to you is a little about what protoplasts are and what can be done with them.

As time has gone on in this subject, one has begun to realize certain special attributes of protoplast fusion, namely, an ability to carry out transfers not only at the nuclear level but also at the cytoplasmic level. There are certainly some difficulties with application of protoplasts, especially in getting plants from protoplasts in our mainline cereals, and in certain other important crop species. I think the important challenge is to identify some of the needs in the subject area. If the needs are identified, one will break through in these difficult areas. If one is just dabbling with a range of species, and then one takes refuge in good old tobacco or even now in good tomato, one will not do much to advance the subject. That is not to say that tobacco and the tomato are not important. There are myriads of questions that will be posed, using protoplast fusion, and very few of them will be answered adequately unless one can handle a wide range of species. It does take a little bit of imagination and dexterity to resolve some of these unanswered questions. If we look at the detailed situation relating to fusion we are dealing with protoplasts because they are naked cells, which have not got a cell wall. The basic technology has been essentially resolved, and one can induce virtually any one cell to fuse with any other cell these days. One can fuse dicotyledonous protoplasts with monocotyledonous protoplasts. One can fuse plant protoplasts with animal cells. And so on and so forth. One can also get heterokaryons somewhat equivalent to animal cell heterokaryons. The interesting question will be what happens to heterokaryons. There is a more or less equal contribution of the two cytoplasms, contrasting with the situation often encountered in sexual reproduction. At this stage the nuclei are separate. One is interested in whether the nuclei will fuse to produce a hybrid cell that will go on to produce a hybrid plant. It is possible to lose one of the nuclei and finish out with a mixed cytoplasm and the nucleus of one of the species and obtain a cybrid. Alternatively, it is possible to exchange the cytoplasms between different species so that the cytoplasm of one species then goes with the nucleus of the other species. Finally, it is possible to get varying degrees of chromosome elimination. And, of course, if this were better understood and controllable, then one could get a much more rapid introgression of genes from certain species, including species that are sexually isolated from your species of interest (Cocking et al., 1981). One can obtain protoplasts from leaves in abundance, from monocotyledonous leaves or from dicotyledonous leaves. One can also obtain protoplasts readily from cell suspension cultures. Most protoplasts carry a charge at their surface, usually a negative one, and that has to be altered for fusion to take place. A fusogen is needed to fuse them together. This is usually a chemical fusogen such as calcium ions at high pH, or polyethylene glycol. One can obtain protoplasts from various other parts of the plants, as well as the leaf, including roots and cotyledons. Moreover protoplasts can be isolated from roots with varying degrees of vacuolation (Davey, 1983). There are two main types of protoplasts, one with chloroplasts and one without chloroplasts, green and nongreen. This is a very useful marker because if one is fusing things, one wants to be able to identify fusion products. It was a wonderful development when we could get mutants to complement, to come together by fusion, thereby enabling the hybrid product from two protoplasts fusing to be selected out and to grow up into hybrid plants. We have used *petunia* in part because it has got about nine genes that control flower color. It has also got within the genus a very useful range of sexual incompatibilities. Using the range of species within the *Petunia* genus, it was possible to investigate the question of the extent to which sexual incompatibility within the genus could be overcome by somatic fusion.

What this subject needs, and is beginning to get, is the ability to handle large numbers of such heterokaryons without having to rely on mutants; to go directly to wild types, whether it be in *Petunia* or whether it be in the forage legumes or, as I hope in the future, in rice. Fusion, as such, is no longer a central problem; a carrot protoplast can even be fused with *Xenopus* cells using calcium ions at high pH (Cocking, 1984).

In many sexual hybridizations, the contribution on the male side at the cytoplasmic level is very small, and one has maternal inheritance. In somatic fusions one has more or less equal contributions from the two cytoplasms. Moreover, one can subdivide protoplasts into a number of so-called microplasts, very small units of cytoplasm, which one can use for fusions so that one can shunt around cytoplasms between the different species (Cocking, 1981). Also French and Israeli workers have become particularly adept at inactivating one of the nuclei, using irradiation, to get essentially an enucleate protoplast that then fuses with the other species of protoplast, to transfer cytoplasmically based male sterility. In the case of the French work, this was into *Brassica napus* and brought along with it also cytoplasmically based herbicide (atrazine) resistance (Pelletier et al., 1983). For the effective implementation of this technology, one needs to be able to get plants from protoplasts in an increasing number of species. One needs protoplasts that form a wall and then divide to form a callus (or embryoid) and then form a shoot, root, and then a plant (Davey, 1983). *Petunia parviflora* has a trailing habit, very small flowers, and very small leaves, and it can be regenerated from protoplasts into plants. It also has 2n = 18, rather than 2n = 14, and is sexually isolated from all of the other *Petunia* species within the genus. We thought that this was a good target to assess whether we could obtain a somatic hybrid, a flowering somatic hybrid plant that was an amphidiploid between those two species, either by using two albino mutants that complement to give green, or using some method of half selection in which an albino was fused with the wild type, but the wild type would stop growing because of a difference in the culture media requirement. Any green colonies growing through would then be hybrid cells. Colleagues here in the United States have tried to cross *Petunia parviflora* with the other *Petunia* species sexually, by all ramifications of tissue culture applied to sexual hybridization, and have failed. By taking protoplasts of these two species and fusing them and using a complementation selection based on the use of an albino mutant of *Petunia parviflora* with a wild type of *Petunia parodii*, we obtained a somatic hybrid between these two incompatible species (Power et al., 1980). It has a number of chromosomes of 32 and is thus the amphidiploid. Currently, we are attempting to cross this with *P. hybrida* to maintain the hanging basket habit, but with a larger flower.

Recently we turned our attention to two main areas of further investigation to enchance our selection capability. One is to have a mutant that is a double mutant, such as one that is both nitrate reductase-minus and streptomycin-resistant. It was obtained both sexually (Hamill et al., 1983) and somatically between two different tobaccos, one of which was nitrate reductase-minus and the other streptomycin-resistant; and because it has got this double mutant character, it is currently being used to probe for the extent of incompatibility with a wide range of other species. And it is a very powerful method because we do not need this double mutant and the wild species. And we have gone on to utilize this double mutant of *Nicotiana tabacum* by fusions with *Nicotiana rustica*, as the wild type, and to obtain somatic hybrids between *N. tabacum* and *N. rustica*. Although the hybrid can be produced sexually with difficulty, it has a very low fertility; but surprisingly the hybrids produced somatically have very good fertility, and this could be of importance in the general area of tobacco improvement because *N. rustica* has several genes of interest for transfer into *N. tabacum.*

I think that the subject in this area has not significantly advanced during the past couple of years because people have not been able to actually isolate their fusion products. There is great interest in finding out what happens in fusions between a range of species, particularly in heterokaryons between species that are disparate. We are interested in why sometimes in very disparate species somatic hybridization fails to produce anything. One might expect that there would not be complete harmony between the two chromosome sets for a number of reasons and one would expect to get progressive chromosome elimination. The real trouble is that if one is not careful, one can have a complementation selection that will eliminate anything that is in any way like either of the parents and only encourage the balanced hybrid to grow through: as a result any partial hybrids will probably be eliminated. And so our attention has turned to the direct isolation of heterokaryons and their propagation. We need more biochemical and molecular biological probes to prove hybridity. We are getting plants in interspecies fusions in the forage legumes with some hybrid characters in the selected plants, but at present we cannot be certain. We started off by directly isolating heterokaryons using an inverted microscope and capillary pipette and just picking them up (Patnaik et al., 1982). Particularly if one uses a fluorescence-inverted microscope, one can readily identify heterokaryons. We labelled the *N. rustica* protoplasts with fluorescein isothiocyanate, which does not impair the division of the protoplasts. We then fused them with mesophyll protoplasts of *N. tabacum* and obtained what we thought were heterokaryons, using bright field optics. With fluorescence microscopy one can actually be sure because of their double fluorescence, from the red autofluorescence of the chloroplasts and the green fluorescence of the FITC. And we can pick them up and nurse them along and recover plants. This subject now needs automation, thousands and thousands of them, because, particularly in fusions between sexually incompatible species, only a very small percentage of heterokaryons divide to produce hybrids.

I do not wish to review all the ramifications of protoplast fusion at the cytoplasmic level. It is attracting great interest. There are certain fundamental questions still not answered about somatic hybridization. For instance, is there any special heterosis associated with somatic hybridization? This is a difficult question to answer, particularly in view of the extra ramifications of somaclonal variation arising from tissue culture, but an important one that we are trying to assess in alfalfa (Cocking, 1983a). There is also great interest in the extent to which chloroplast recombination will occur. It may be that we do not know enough about fusion. The advent of electrofusion may enable us to not only think about the fusion of protoplasts but also whether we can induce chloroplast to fuse within the fused protoplast system. There is some evidence for mitochondrial recombinants (Belliard et al., 1979).

We have been pleasantly surprised at the readiness with which the forage legumes are behaving at the protoplast level. During the last two years, all the main forage legumes have been shown to regenerate reproducibly, and I would emphasize reproducibly and at high frequency, from protoplast to plant (Davey, 1983). In fact *Lotus corniculatus* regenerates plants virtually 100% from protoplast-derived callus. We have identified with breeders the need to transfer bloat resistance factors from species such as *Lotus corniculatus* and sainfoin into the mainstream forage legumes in the United Kingdom and Europe, the Trifoliums and the Medicagos (Cocking, 1983b). We have got the culture side, the regeneration side, and we have got the fusion capability. What we have not had up to now is the selection capability because we have not had suitable mutants. And so we are now going to large-scale fluorescence-activated cell sorting, which looks distinctly promising in this particular field. If we also have suitable development of markers for the detection of tannin synthesis (anti-bloat factors), we may be able to follow this through adequately in this type of system. There is also great interest in getting whole plant regeneration from soybean protoplasts. Regeneration in the wild species of the *Glycine* genus looks very promising indeed. If one is interested in, say, the transfer of disease-resistance genes from these wild species into *Glycine max* (soybean), then it may not be necessary to have regeneration of *Glycine max* protoplasts. Our work, and that of other workers in the field, demonstrates that regeneration is a sort of dominant character that can be added from wild species. Indeed, this concept could be very important in the case of tomato, where there is a real need to obtain *L. esculentum* lines. And if one is interested in crosses with, say, *L. peruvianum*, which regenerates readily from protoplasts, then one may not have a major problem. So strategy, based on a foundation of knowledge, is very important.

The importance of plant genetic manipulations in relation to cereals is obvious particularly in the developing countries. Through the Overseas Development Administration in the United Kingdom, we have identified with the International Rice Research Institute (IRRI) some of the main objectives in rice-breeding improvement that could utilize protoplasts. I should add that fusion of protoplasts of the water fern *Azolla* could also be exploited rapidly in relation to improved nitrogen fixation in rice paddies. But in relation to rice itself, the need for the transferance of male sterility factors directly is of primary consideration, and also the improvement required in salinity tolerance. Rice is very saline-sensitive, and there are wild species in Bangladesh, for instance, that are desired for a source of useful genes, but sexually isolated. The use of protoplast fusion has been suggested, but I would emphasize that it is not going to be possible to do this type of work until one has the regeneration of plants from protoplasts of rice or the related wild species. In my own country, the evening primrose, which is one of the few plants that produces gammalinolenic acid, has attracted the attention of plant genetic manipulators. Gammalinolenic acid is very important: if one has a weak heart, one cannot synthesize his own gammalinolenic acid. The interesting thing biochemically is that there is only one enzyme difference between linoleic and gammalinolenic acid (Cocking, 1983b). Hence, this is a very good target, not only for fusion efforts between the evening primrose and even tobacco, for instance, but certainly we would like *Brassica napus* not only to produce the normal oils that it produces but perhaps produce some gammalinolenic acid as well. It is also a very good target, and a needed one, for assessments of transformation in plants because one needs examples where there are single gene differences. There has been a lot of interest recently in the extent to which irradiation of pollen with very high doses of irradiation can lead to the transfer of a few genes between species sexually. Our interest is now the extent to which one could imitate this process at the somatic level and extend the range of gene flow using sexually incompatible species. Is it possible to take protoplasts and irradiate them with suitable levels of irradiation to facilitate the transfer of a few genes? Clearly we need to have the ability to select such gene-flow products and to characterize them adequately, and this is a major challenge. Recently we and others have been utilizing nitrate reductase-minus mutants in this respect, and it looks very promising in that we get colonies growing through after irradiation and fusion with a nitrate reductase-minus mutant. However, it is not certain that these colonies are not revertants. I think that this subject area will advance when we have better methods of characterizing possible hybrids. What we need is a better understanding of the stability of the two genomes when they are brought together.

I hope that I have indicated that protoplast fusion has something to offer to extend the range of gene flow between species. It must be viewed against a very successful baseline of conventional plant breeding. The challenge in the field is to identify those areas where conventional plant breeding has failed or is very time-consuming or inefficient, and where this additional mode of gene flow could be advantageous and beneficial.

REFERENCES

Belliard, G., F. Vedel, and G. Pelletier. 1979. Mitochondrial recombination in cytoplasmic hybrids of *Nictotiana tabacum* by protoplast fusion. Nature (London) 281:401–3.

Cocking, E. C. 1981. Opportunities from the use of protoplasts. Phil. Trans. Roy. Soc. Lond. B. 292:557–68.

Cocking, E. C. 1983a. Hybrid vigour in alfalfa. *In* J. Nugent and M. O'Connor (eds.), Better crops for food. Ciba Foundation symposium 97, pp. 142–43. Pitman, London.

Cocking, E. C. 1983b. Applications of protoplast technology to agriculture. *In* I. Potrykus, C. T. Harms, A. Hinnen, R. Hutter, P. J. King, and R. D. Shillito (eds.), Protoplasts 1983: lecture proceedings 6th International Protoplast Symposium, Basel, August 1983, pp. 123–26. Burkhauser Verlag, Basel.

Cocking, E. C. 1984. Factors influencing plant protoplast–animal cell fusions. *In* R. F. Beers and E. G. Bassett (eds.), Cell fusion: gene transfer and transformation, pp. 139–44. Raven Press, New York.

Cocking, E. C., M. R. Davey, D. Pental, and J. B. Power. 1981. Aspects of plant genetic manipulation. Nature (London) 293:265–70.

Davey, M. R. 1983. Recent developments in the culture and regeneration of plant protoplasts. *In* I. Potrykus, C. T. Harms, A. Hinnen, R. Hutter, P. J. King, and R. D. Shillito (eds.), Protoplasts 1983: lecture proceedings 6th International Protoplast Symposium, Base, August 1983, pp. 19–29. Birkhauser Verlag, Basel.

Patnaik, G., E. C. Cocking, J. Hamill, and D. Pental. 1982. A simple procedure for the manual isolation and identification of plant heterokaryons. Plant Sci. Lett. 24:105–10.

Pelletier, G., C. Primard, F. Vedel, P. Chetrit, R. Remy, P. Rousselle, and M. Renard. 1983. Intergeneric cytoplasmic hybridization in Cruciferae by protoplast fusion. Mol. Gen. Genet. 191:244–50.

Power, J. B., S. F. Berry, J. V. Chapman, and E. C. Cocking. 1980. Somatic hybridization of sexually incompatible petunias: *Petunia parodii, Petunia parviflora*. Theor. Appl. Genet. 57:219–27.

C. E. FLICK AND D. A. EVANS

Isolation of Biochemical Mutants by Plant Cell Culture

14

INTRODUCTION

Both *in vitro* and *in vivo* mutant selection has been used successfully in higher plants. *In vivo* selections have been limited primarily to screening populations for morphological variants. Because of the lack of precise control of *in vivo* culture conditions, only limited success has been achieved in isolation of biochemical mutants using *in vivo* selection. In contrast, an extensive array of mutant types have been isolated from *in vitro* cell cultures ranging from autotrophic cell lines to herbicide-resistant mutants that may prove of agricultural interest.

Several aspects of *in vitro* culture simplify the isolation of mutants. By the use of suspension cultures and/or protoplasts, very large populations of cells can be screened in a single experiment. Isolation of mutants from protoplasts insures that mutants will be derived from single cells and reduces the probability of recovering chimeral cell lines or plants. The verification of mutant phenotypes *in vitro* is simplified by the ease of manipulation of cultural conditions enabling mutant selection schemes *in vitro* that are not possible *in vivo*, e.g., selection of mutants resistant to antimetabolites.

There are many developmental mutant phenotypes that are expressed *in vitro* but not *in vivo*, e.g., some instances of resistance to amino acid analogs. Mutations that are expressed *in vitro* but not *in vivo* may be uniquely advantageous as cellular selective markers, such as in protoplast fusion. This is particularly true if the mutant phenotype has no adverse effects on the agronomic characteristics of the regenerated plant. In some instances, *in vivo* expression of a mutation may not be necessary for a mutant to be useful. For example, *in vitro* characterization of a mutant on the cellular level may be sufficient to ascertain usefulness of a cell line in a fermentation-type production of secondary metabolites. There are, however, obvious disadvantages to isolation of mutants from cell cultures. In some instances the lack of expression of a mutant phenotype *in vivo* may preclude its usefulness. For example, cell lines selected for resistance to an herbicide must retain this phenotype in regenerated plants to be agriculturally beneficial. Prolonged culture of cells *in vitro* may produce genetic modifications beyond the desired mutation, i.e., an accumulation of abnormalities in both chromosome number and structure. Some phenotypes cannot be selected *in vitro*, e.g., phenotypes that depend on the structural or functional organization of a whole plant. In this category would fit many instances of disease resistance, for example, those involving vascular wilts. In these cases one must be careful that *in vivo* and *in vitro* effects of a pathotoxin are identical. Other phenotypes may be selectable at the cellular level only under specific cultural conditions, e.g., the use of photo-mixotrophic cell cultures for selection for resistance to herbicides such as metribuzin that function to inhibit photosynthesis during a particular plant developmental stage. However, despite some limitation, plant cell culture offers unique opportunities for isolation of a wide array of biochemical mutants that cannot be readily isolated using *in vivo* methods.

MUTAGENESIS

Both chemical and physical mutagens have been successfully used in higher plants to induce mutants. Physical mutagens are advantageous in some respects, e.g., after mutagenesis the cells do not have to be manipulated to remove the mutagen. Sidorov and colleagues (1981) successfully isolated auxotrophs from haploid *Nicotiana plumbaginifolia* protoplasts following gamma-irradiation. Unfortunately, physical mutagens usually require specialized equipment that is not always readily available, e.g., a gamma-ray source. Ultraviolet light (uv) is one physical method that is convenient for mutagenesis. Equipment is relatively inexpensive, and mutagenesis can be carried out under sterile conditions, e.g., using a portable uv lamp mounted in a laminar flow hood. However, after uv mutagenesis, cells must be incubated in the dark in order to reduce photo-inducible DNA repair.

When using chemical mutagens, it is necessary to wash the cells extensively following mutagenesis to remove all traces of the mutagen. This process may prove as deleterious to the cells as the mutagen treatment, especially if the cells are fragile, e.g., protoplasts. This problem may be circumvented partially by using mutagens that quickly decompose in aqueous solution.

Table 1 summarizes the use of mutagens in higher plant systems based on data previously presented in Flick (1983). It is quite apparent that in more than 50% of the instances reviewed, no mutagen was used. Of those mutants isolated using mutagens, most were isolated following treatment with chemical mutagens (79%). Of those isolated following chemical mutagenesis, EMS was used for isolation of 26.7% of the total number of mutants and was the most common mutagen. Twenty-one percent of the cell lines were isolated following use of physical mutagens, with uv and gamma-rays used equally often. The use of x-rays was reported only once. It seems apparent that use of a particular mutagen depends on the lab in question as much as the biological system being mutagenized.

In some selection procedures, e.g., selection of amino acid analogue-resistant cell lines, it is quite apparent that both mutagenized and unmutagenized cells can be used successfully (see Flick, 1983). Following mutagenesis, a high frequency and a broad spectrum of mutations are produced in a cell. It is possible that in a mutated cell in addition to the mutation of interest other genetic changes can occur. Indeed, some mutagens may cause changes in chromosome structure or gene amplification that lead to extensive changes in gene expression. These secondary mutations arising through mutagenesis may cause effects such as reduced viability, interference with plant regeneration, induction of pollen sterility, and innumerable others. On the other hand, spontaneous mutants are likely to have only a single genetic change per cell.

Before *in vitro* selection for the desired trait, cells should be grown for several generations after mutagenesis. (1) Since the mutation is a physical change in the DNA, the altered gene must be expressed prior to application of selective pressure for the desired phenotype to be apparent. (2) Segregation of the mutation to daughter cells through mitosis must be assured if a mutation is to express a stable phenotype. (3) Mitotic recombination of recessive mutations in diploid cell lines may allow for identification of recessive phenotypes in diploid cell cultures. In the absence of a post-mutagenic growth period, a mutant cell line may be selected against before identification.

The necessity of using mutagens when isolating mutants *in vitro* in higher plants has not been satisfactorily demonstrated. Auxotrophic mutants are usually isolated following mutagenesis. In the only selection from reportedly unmutagenized cells, a BUdR enrichment was used, but BUdR

TABLE I

Use of Mutagens in Isolation of Mutants *in vitro*

Mutagen	Amino Acid Analog Resistant	Base Analog Resistant	Antibiotic Resistant	Agriculturally Useful	Auxotroph	Misc.	Total
None	60.7	43.8	76.9	71.4	8.3	57.1	56.7
EMS	21.6	37.5	23.1	14.3	66.7	14.3	26.7
Azide..........	5.8	0	0	0	0	14.3	3.3
uv..........	9.8	0	0	0	0	0	4.2
X-ray	0	0	0	4.8	0	0	0.8
gamma-ray	0	0	0	9.5	16.7	14.3	4.2
Nitrosoguanidine	2.0	18.8	0	0	8.3	0	4.2

Expressed as percentage of total mutants isolated as reported in Flick (1983).

has mutagenic properties. Both chemical and physical mutagens appear effecttive in plant cell cultures for isolation of auxotrophs. Ethylmethane sulfonate, nitrosoguanidine, and gamma-irradiation have all been used with success. Ranch and colleagues (1983) recently reported a sixfold increase in the frequency of 5-methyltryphtophan-resistant cell lines after nitrosoguanidine mutagenesis of haploid *Datura innoxia.* Other mutagens tested showed no effect in this system. On the other hand, mutagenesis has not been used for *in vitro* isolation of antibiotic-resistant mutants in most instances. Since many of the antibiotic-resistance traits are cytoplasmically inherited, the conventional mutagens may not be as effective on organelle DNA as on nuclear DNA. Perhaps with the use of a mutagen that acts more effectively on organelle DNA than on nuclear DNA, mutagenesis would enhance the success of isolation of cytoplasmic mutants.

SELECTION METHODS

Ultimately the success of *in vitro* mutant isolation will depend on the suitability of the selection scheme in use. The selection depends, of course, on the future use of the isolated mutants. Mutants with direct application are the most simple to isolate. For example, in order to study tryptophan biosynthesis, Widholm (1972a, 1972b) isolated 5-methyltryptophan-resistant cell lines. Herbicide, e.g., picloram (Chaleff and Parsons 1978a), resistance may be directly selected in cell culture. In these instances, one need only identify the problem to be addressed and choose the appropriate selective agent. Most mutants that have been isolated *in vitro* in higher plants have been by direct selection.

A large number of compounds are toxic to higher plant cells *in vitro*. In several instances these are compounds that may produce mutants of agricultural interest. For example, resistance to substituted 2, 6-dinitroaniline herbicides, sulfonylurea or imidizolonone herbicides or metribuzin could be of immediate commerical value. Tolerance of heavy metals, e.g., zinc, copper, and aluminum or NaCl, may increase the possibility of growing crops under conditions previously unsuitable for plant growth. A large number of similar compounds toxic to plant cells may be used as selective agents for *in vitro* mutant isolation.

MUTANT ISOLATION AND CHARACTERIZATION

Several types of biochemical mutants can be isolated *in vitro*, basically falling into two broad classes: (1) auxotrophs, i.e., nutrient-requiring, and (2) antimetabolite-resistant mutants. Auxotrophs have proved difficult to isolate because selection and/or enrichment systems have not been available. Only nitrate reductase-deficient mutants can be isolated by a direct positive selection, i.e., chlorate resistance. On the other hand, antimetabolite-resistant mutants have been easier to isolate. Mutants resistant to amino acid and nucleic acid base analogs and antibiotics have been commonly isolated. Those mutants involving phenotypes of potential agricultural usefulness that have been isolated from cell culture are based on resistance to antimetabolites. These agriculturally important mutants have been difficult to isolate, mostly because the phenotypes must be expressed both in the cultured cells and in the regenerated plant without other deleterious side effects. For example, although maize mutants were isolated from cell cultures of T-type cytoplasmic male sterile lines for resistance to the toxin from *Heminthosporium maydis*, these lines were also found to have reverted to male fertility (Gengenbach and Green, 1975). The most promising results involving antimetabolite mutants of agricultural usefulness have been obtained with herbicide-resistant cell lines (e.g., Chaleff and Parsons, 1978a). In this case cells resistant to normally deleterious herbicides can be isolated in culture, and regenerated plants continue to express the herbicide resistance. Both of these classes of mutants are discussed below.

Auxotrophs

Few auxotrophic mutants have been isolated in higher plants. The best-characterized auxotrophs are mutants involving nitrate reductase activity. In the absence of nitrate reductase activity, chlorate is not converted to toxic chlorite; hence, nitrate reductase-deficient cell lines exhibit chlorate resistance. Chlorate resistance was first used to select nitrate reductase-deficient cell lines from haploid *Nicotiana tabacum* (Mendel and Muller, 1976; Muller and Grafe, 1978). Of nine original chlorate-resistant cell lines, one class, designated cnx, is also deficient in xanthine dehydrogenase. Since nitrate reductase and xanthine dehydrogenase use the same molybdenum cofactor, the cnx allele is believed to affect the molybdenum cofactor but not the nitrate reductase apoprotein. Partial nitrate reductase activity has been restored to cnx lines by addition of 1-10 mM molybdenum (Mendel et al., 1981).

A second class of nitrate reductase mutants, designated nia, lack nitrate reductase activity but retain xanthine dehydrogenase activity, indicating that the mutation affects the apoprotein (Mendel and Muller, 1979). Evidence has substantiated the premise that nia and cnx affect different subunits of nitrate reductase. (1) cnx and nia lines complement *in vitro* in crude extracts to restore nitrate reductase activity (Mendel and Muller, 1978). (2) Hybrids produced by protoplast fusion between cnx and nia cell lines grow on nitrate. *In vivo* complementation implies that both cnx and nia mutations are recessive. Only leaky nia mutants (with reduced levels of FADH2 and viologen dye–dependent nitrate reductase activity, but no NADH-dependent activity) have regenerated shoots (Muller and Grafe, 1978). However, shoots were produced from cnx + nia cell hybrids. Nitrate reductase-deficient cell lines have been isolated *in vitro* in four additional species: *D. innoxia* (King and Khanna, 1980), *Rosa damascena* (Murphy and Imbrie, 1981), *Hyoscyamus muticus* (Straus et al., 1981), and *N. plumbaginifolia* (Marton et al., 1982).

Other auxotrophs have been isolated from higher plants. The first auxotrophs were isolated by Carlson (1970) in haploid *N. tabacum*. Carlson isolated six that required biotin, para-aminobenzoic acid, arginine, hypoxanthine, lysine, and proline. Plants were successfully regenerated from all but the proline auxotroph. Genetic analysis of these mutants showed that the biotin, para-aminobenzoic acid, and arginine auxotrophs resulted from changes in single recessive nuclear alleles, i.e., 3:1 segregation of dominant to recessive in the R_1 generation (Chaleff 1981). The hypoxanthine auxotroph was more complex in that it resulted from mutations in two recessive nuclear alleles, i.e., a segregation ratio of 9:6:1 (dominant:heterozygous:recessive) (Chaleff 1981). Although plants were regenerated from a lysine auxotroph and plant growth was shown to be improved by addition of lysine (Carlson et al., 1973), no genetic analysis was reported.

Malmberg (1979, 1980) has isolated temperature-sensitive (ts) mutants of haploid *N. tabacum*. One such mutant (ts 4) has been biochemically analyzed. The mutation is pleiotropic and decreases activities of ornithine decarboxylase, S-adenosylmethionine decarboxylase, and nitrate reductase but increases the activity of phenylalanine ammonia lyase. The ts 4 mutant also has a decreased level of chlorophyll. A second site revertant of ts 4 (Rt 1) was isolated and has normal levels of ornithine decarboxylase and chlorophyll (Malmberg 1979).

Other auxotrophic mutants isolated in cell culture include a histidine auxotroph of *H. muticus* (Shimamoto et al., 1982), and adenine (King et al., 1980) and calcium pantothenate (Savage et al., 1979) auxotrophs of *Datura innoxia*. Recently, Gebhardt and colleagues (1981) screened a large number of calli recovered from protoplasts of *H. muticus* and successfully isolated several mutants with nutritional requirments. Histidine, nicotinamide, and an undefined temperature-sensitive auxotroph have been regenerated into plants, but not yet analyzed. A tryptophan auxotroph did not regenerate. Autotrophy following protoplast fusion of a trp- and his-auxotroph indicates that these are independent recessive alleles (Gebhardt et al., 1982). A nitrate-reductase mutant, dependent on casein hydrolysate for growth, and three auxotrophs also have recently been isolated from gamma-irradiated protoplasts isolated from leaf mesophyll tissue of haploid *N. plumbaginifolia* (Sidorov et al., 1981). An isoleucine auxotroph has been most extensively studied. This cell line is deficient in threonine deaminase, the first enzyme in the isoleucine pathway of biosynthesis. Plants have been regenerated from the isoleucine auxotroph, but genetic analysis has not yet been reported. A uracil auxotroph (Sidorov et al., 1981) and a leucine auxotroph (Sidorov and Maliga, 1982) also were isolated, but neither has been analyzed or regenerated.

The value of auxotrophic mutants in somatic hybridization has been shown by Sidorov and Maliga (1982). The recessive nature of uracil, isoleucine, and leucine auxotrophs was clearly demonstrated by protoplast fusion-complementation (Sidorov and Maliga 1982). In addition, only following protoplast fusion could uracil and leucine auxotrophs of *N. plumbaginifolia* be regenerated into plants, i.e., the homozygous recessive auxotroph does not regenerate. Fusion-complementation of chlorophyll-deficient mutants was also demonstrated by Sidorov and Maliga (1982). In addition, the successful use of auxotrophs as recipients in vector-mediated DNA transformation will be essential.

Resistance to Antimetabolites

Selection for resistance to antimetabolites is the most direct method of *in vitro* screening for mutants. Several factors influence the success of a selection system. (1) The action of the antimetabolite must be irreversible, i.e., the agent should be lethal, not just growth-inhibitory. (2) The selective agent should not be degraded in the culture medium either by factors in the medium itself or by wild-type cells. If the compound is degraded, sensitive cells must be killed before its degradation. In the case of a labile-selective agent, cells may recover and begin to grow after the antimetabolite has been degraded. Although some of the cell colonies isolated will be resistant variants, many will also be sensitive. Some mechanisms of resistance to antimetabolites and reported examples are summarized in table 2.

Cytoplasmic mutants, isolated as antibiotic-resistant mutants, have been used extensively in cytoplasmic transfer. The streptomycin-resistant mutant SR1 has been most extensively studied as a selective marker. Menczel and colleagues (1981) demonstrated chloroplast transfer from *N. tabacum*

TABLE 2

MECHANISMS OF RESISTANCE TO ANTIMETABOLITES

Mechanism of Resistance	Phenotype	Species	Reference
Decreased feedback inhibition	Lysine + threonine-resistant	*Z. Mays*	Hibberd et al., 1980.
Increased production secondary metabolite	5-methyltryptophan-resistant p-Fluorophenylalanine-resistant	*N. tabacum* *N. tabacum*	Widholm, 1972a. Berlin & Widholm, 1977.
Altered enzyme activity	S-aminoethyl cysteine resistant	*D. carota*	Matthews et al., 1980.
Increased free pool of natural compound	Lysine + threonine-resistant	*Z. Mays*	Hibberd et al., 1980.
Uptake-deficient	S-aminoethyl cysteine-resistant	*N. tabacum*	Widholm, 1976.
Detoxification of Antimetabolite	Cycloheximide-resistant	*D. carota*	Gresshoff, 1979.

to *Nicotiana knightiana* using the SR 1 mutant. In addition, SR 1 was transferred from *N. tabacum* to *Nicotiana sylvestris* without nuclear hybridization (Medgyesy et al. 1980). Cytoplasmically inherited lincomycin-resistant cell lines have also been isolated (Cseplo and Maliga 1982) in *N. sylvestris* and *N. plumbaginifolia*. Lincomycin resistance is cytoplasmically inherited and can be transferred through protoplast fusion. Cytoplasmic markers expressed at the cellular level can be used to select for other cytoplasmic markers that are not expressed in cultured cells. For example, the transfer of cytoplasmic male sterility (cms) in *Nicotiana* using protoplast fusion can be accomplished by cellular selection for the cytoplasmic SR 1 marker. Segregation of cytoplasmic markers can be observed, as demonstrated in *Nicotiana* somatic hybrids by Belliard and colleagues (1978). In this instance chloroplast DNA type as determined by restriction enzyme analysis was shown to segregate from cms.

The potential usefulness of biochemical mutants in an understanding of biosynthetic pathways has not been fully realized. However, a number of important characteristics of antimetabolite-resistant mutants have been elucidated. (1) Resistance to amino acid analogues can lead to decreased feedback inhibition of an enzyme (Widholm 1972a; Widholm 1974; Hibberd et al. 1980). (2) Unregulated enzyme activity can lead to increased intercellular pools of the naturally occurring compound, e.g., a 3– to 23-fold increase in free tryptophan was found in 5-methyltryptophan (5MT)-resistant *D. innoxia* cells (Ranch et al. 1983). (3) Single mutations can produce pleiotropic effects. For example, S-aminoethylcysteine-resistant cell lines of *Daucus carota* showed altered activities of both aspartokinase and dihydropicolinic acid synthase (Matthews et al. 1980). (4) Antimetabolite resistance expressed *in vitro* may not be expressed in the regenerated plant. Widholm (1974) described a 5MT-resistant cell line that as a regenerated plant is 5MT sensitive, but callus reinitiated from the plant is 5MT-resistant. (5) Antimetabolite resistance may even vary within the regenerated plant. 5MT-resistant *D. innoxia* could be maintained as shoot cultures on 5MT, but rooting was inhibited in the presence of 5MT (Ranch et al., 1983). In none of these instances has a systematic analysis of a biosynthetic pathway been accomplished using a variety of analogues as in microorganisms.

In vitro mutant isolation may be useful for isolation of other agriculturally important mutants. These include mutants for disease resistance, stress tolerance, and herbicide tolerance. Behnke (1979) isolated *Phytophthora infestans* toxin (early blight)-resistant cell lines of *Solanum tuberosum* (potato). Regenerated plants as well as callus initiated from these plants were resistant to the toxin. Subsequently Behnke (1980a) demonstrated that toxin-resistant plants are 25% more resistant to the late blight caused by *P. infestans* than parental lines. A genetic basis for this resistance has not yet been established since *S. tuberosum* is commercially propagated asexually. Similarly, Behnke (1980b) isolated potato cell lines resistant to a culture filtrate from *Fusarium oxysporum*, which causes a wilt in many species. Resistance to the culture filtrate was retained in regenerated plants as well as callus initiated from these plants.

Resistance of *Zea mays* to *H. maydis* (southern leaf blight) has been extensively studied *in vitro*. T-type cytoplasmic male sterile (T-cms) lines are sensitive to *H. maydis* whereas fertile and other cms (N, S, C) types are resistant. Gengenbach and Green (1975) selected *in vitro* for resistance of T-type cms cell lines to a culture filtrate from *H. maydis*. All of the regenerated plants (52/65) resistant to *H. maydis* have also reverted to male fertility (Gengenbach et al., 1977). Male sterile lines resistant to *H. maydis* do not have the T-type cytoplasm (Gengenbach et al., 1981). Kemble and colleagues (1982) showed that fertile plants regenerated from T-type cms lines with *H. maydis* resistance do not have normal fertile N cytoplasm-type mitochondrial DNA. As both *H. maydis* resistance and T-cms are maternally inherited, it is apparent that two phenotypes are tightly linked genetically and cannot be separated using these methods (Gengenbach et al., 1981). No plants have been produced of T-cms type with *H. maydis* resistance.

Both *Phoma lingam* fungal growth and toxin resistance were screened in haploid *Brassica napus* culture (Sacristan 1982). Only screening for toxin resistance, however, produced resistant plants. Technical difficulties, e.g., fungal growth on plant culture media, precluded use of the fungus itself. Resistant plants also arose among a nonselected population of regenerated plants. Somaclonal variation in disease resistance among regenerates has also been reported for sugar cane eyespot and Fiji disease resistance (Heinz et al., 1977) and for resistance of protoplast-derived potato clones to *Alternaria solani* (Matern et al., 1978).

Numerous instances of *in vitro* selection of NaCl-tolerant cell lines have been reported (see Flick, 1983). In several of these reports, plants have been regenerated. Sexual transmission of NaCl-tolerance was first reported in *N. tabacum* (Nabors et al., 1980), but segregation frequencies were not Mendelian, as expected. Regenerated plants and callus initiated from regenerated plants of NaCl-tolerant cell lines of *D. innoxia* were resistant to NaCl (Tyagi et al. 1981). In addition, plants of *Kickxia ramosissima* are resistant to NaCl up to 120 mM (Mathur et al. 1980). NaCl-tolerant cell lines have also been isolated in *N. sylvestris, Capsicum annuum* (Dix and Street, 1975; Dix and Pearce 1981), and *Medicago sativa* and *Oryza sativa* (Rains et al., 1979).

Numerous examples of herbicide resistance have been reported in higher plants. The picloram-resistant mutants of *N. tabacum* described by Chaleff (Chaleff and Parsons, 1978a; Chaleff, 1980) are the best analyzed. Of seven resistant cell lines isolated, plants, could not be regenerated from one. One cell line, although stably resistant in cell culture, regenerated only sensitive plants. Only sensitive callus could be initiated from these plants. In another cell line, callus initiated from plants was picloram resistant although seedlings were sensitive. Indeed, only 4/7 cell lines regenerated plants in which stable heritable resistance was present. These four picloram-resistant lines were analyzed genetically; in three lines, picloram resistance is inherited as a dominant allele. In the fourth line, picloram resistance is inherited as a semidominant allele. These four picloram-resistant cell lines define three linkage groups (Chaleff, 1980). Analysis of these mutants did not pinpoint the biochemical lesion.

Cross-resistance to an unrelated antimetabolite was reported in mutants isolated for picloram resistance. Chaleff (1981) selected cell lines of *N. tabacum* resistant to hydroxyurea, which inhibits ribonucleotide reductase in animal cells and microorganisms. These mutants behaved as dominant nuclear alleles (Chaleff, 1981). In at least one instance, an allele for hydroxyurea resistance was found to be tightly linked to a previously isolated allele for picloram resistance (Chaleff and Kiel, 1981). A large proportion of Chaleff's picloram-resistant mutants are also hydroxyurea-resistant (Chaleff and Kiel, 1981). This phenomenon has not been adequately explained. In addition, the recovery of hydroxyurea resistance in so many picloram-resistant mutants indicates that there may have been some selective pressure (Chaleff and Kiel, 1981). This is the only reported instance of seemingly unrelated mutant phenotypes arising simultaneously in a selected cell line of higher plants.

MUTANT STABILITY

The stability of mutant phenotypes in culture will determine their usefulness. Only a portion of those variants isolated will be stable mutants. For example, of 250 p-fluorophenylalanine (PFP)-resistant cell lines isolated in *N. tabacum* Su/Su, only 21 were stably resistant to the phenylalanine analogue (Flick et al., 1981). The stability of variants can only be ascertained after prolonged subculturing in the absence of the selective agent followed by subculturing onto media with the selective agent. Variant cell lines isolated *in vitro* with any selection system should be verified as mutants. Several characteristics can be used to define a mutant. (1) Mutants arise at low frequency, e.g., at a frequency of 10^{-5} to 10^{-6} in the absence of mutagenesis. Up to a hundred-fold higher mutation frequency may be possible following mutagenesis. In those instances where a selected phenotype arises at a very high frequency, e.g., 10^{-2} to 10^{-3} a mutational basis for the variant is unlikely and physiological adaptation should be suspected. (2) A mutant must be stable in the absence of selection. Physiological adaptation may involve induction of certain biochemical functions in the presence of the antimetabolite. In the absence of selective pressure, variants may revert to sensitivity. Variant cell lines that are mutants will retain the selected phenotype after prolonged culture in the absence of selective pressure. The preceding two criteria for defining a mutant can be observed in cell culture; the following criteria require plant regeneration. (3) A mutant cell line must remain stable after plant regeneration. Although plants may or may not show the selected phenotype, callus initiated from the regenerated plants must express it. (4) Sexual transmission of a selected phenotype is the ultimate test to assure its identity as a mutant. For example, Chaleff and Parsons (1978a) demonstrated sexual transmission of picloram resistance and observed Mendelian segregation of sensitive and resistant phenotypes. Since in many instances plants regenerated from mutants are aneuploid and/or sterile, genetic analysis may be complicated.

Stable epigenetic changes may be selected in cell cultures. This has been a persistent problem in isolation of some agriculturally interesting variants. For example, cell lines with increased cold tolerance have been isolated in *N. sylvestris* (Dix and Street, 1976; Dix, 1977) and *D. carota* (Templeton-Somers et al., 1981). Although plants were regenerated from three lines of *N. sylvestris* that expressed cold tolerance (Dix 1977), the phenotype was not transmitted sexually. Templeton-Somers and colleagues (1981) reported that although cold-tolerant callus of *D. carota* was isolated, this trait was not expressed in embryos. This instance emphasizes the problem that not all traits selected in unorganized callus cultures will be expressed in the organized differentiated plant or represent true mutations.

In other instances expression of stable mutant phenotypes may be dependent on the state of differentiation. Chaleff and Parsons (1978b) have described glycerol-utilizing mutants that can only be detected in callus cultures. Although the regenerated plant and its sexual progeny show no mutant phenotype, callus initiated from these plants maintains the ability to utilize glycerol. A similar case has been described by Chaleff for hydroxyurea-resistant mutants. Widholm (1972a) showed that whereas a feedback insensitive anthranilate synthetase could be detected in callus of 5-methyl-tryptophan-resistant cells of *N. tabacum*, regenerated plants are feedback sensitive. Callus initiated from regenerates expresses the feedback-insensitive mutant phenotype. On the other hand, some phenotypes are expressed in plants but not callus, e.g., sensitivity to the herbicide metribuzin, which inhibits photosynthesis (Ellis 1978).

Concluding Remarks—Use of Mutants

The wide range of mutants isolated *in vitro* are potentially useful in both basic and applied research areas. As in other organisms, mutants are powerful tools in understanding biochemical pathways and their regulation. In addition, these mutants can be used as selective markers for identification of somatic hybrids and contribute to an understanding of genome interrelationships in interspecific or intergeneric hybrids. A wide range of agricultural applications of mutants are possible. Somatic hybrids are useful for transfer of desired genes between sexually incompatible species. The transfer of cytoplasmic male sterility between cell lines via protoplast fusion is an exciting possibility, only possible when mutants are available as selective markers. A wide range of agriculturally useful crops may be developed through *in vitro* mutant isolation. Pathotoxin resistance may enable direct one-step selections to create new forms of disease resistance. Herbicide-resistant crops developed *in vitro* are possible. These may be especially useful in the development of new and more effective herbicides. *In vitro* mutant isolation in combination with traditional breeding techniques may lead to a wide variety of plants more adapted to their environment.

REFERENCES

Behnke, M. 1979. Selection of potato callus for resistance to culture filtrates of *Phytophthora infestans* and regeneration of resistant plants. Theor. Appl. Genet. 55:69–71.

Behnke, M. 1980a. General resistance to late blight of *Solanum tuberosum* plants regenerated from callus resistant to culture filtrates of *Phytophthora infestans*. Theor. Appl. Genet. 56:151–52.

Behnke, M. 1980b. Selection of dihaploid potato callus for resistance to the culture filtrate of *Fusarium oxysporum*. Z. Pflanzenzuchtg. 85:254–58.

Belliard G., G. Pelletier, F. Vedel, and F. Quetier. 1978. Morphological characteristics and chloroplast DNA distribution in different cytoplasmic parasexual hybrids of *Nicotiana tabacum*. Molec. Gen. Genet. 165:231–37.

Berlin, J., and J. M. Widholm. 1977. Correlation between phenylalanine ammonia lyase activity and phenolic biosynthesis in p-fluorophenylalanine-sensitive and -resistant tobacco and carrot tissue cultures. Plant Physiol. 59:550–53.

Carlson, P. S., R. D. Dearing, and B. M. Floyd. 1973. Defined mutants in higher plants. *In* A. M. Srb (ed.), Genes, enzymes, and populations, pp. 99–107. Plenum Publishing, New York.

Chaleff, R. S. 1980. Further characterization of picloram-tolerant mutants of *Nicotiana tabacum*. Theor. Appl. Genet. 58:91–95.

Chaleff, R. S. 1981. Genetics of higher plants: applications of cell culture. Cambridge University Press, New York.

Chaleff, R. S., and R. L. Keil. 1981. Genetic and physiological variability among cultured cells and regenerated plants of *Nicotiana tabacum* Mol. Gen. Genet. 181:254–58.

Chaleff, R. S., and M. F. Parsons. 1978a. Direct selection *in vitro* for herbicide-resistant mutants of *Nicotiana tabacum*. Proc. Natl. Acad. Sci. USA 75:5104–7.

Chaleff, R. S., and M. F. Parsons. 1978b. Isolation of glycerol-utilizing mutant of *Nicotiana tabacum*. Genetics 89:723–28.

Cseplo, A., and P. Maliga. 1982. Lincomycin resistance, a new type of maternally inherited mutation in *Nicotiana plumbaginifolia*. Current Genetics 6:105–10.

Dix, P. J. 1977. Chilling resistance is not transmitted sexually in plants regenerated from *Nicotiana sylvestris* cell lines. Z. Pflanzenphysiol. 84:223–26.

Dix, P. J., and R. S. Pearce. 1981. Proline accumulation in NaCl-resistant and sensitive cell lines of *Nicotiana sylvestris* Z. Pflanzenphysiol. 102:243–48.

Dix, P. J., and H. E. Street. 1975. Sodium chloride resistant cultured cell lines from *Nicotiana sylvestris* and *Capsicum annuum*. Plant Sci. Lett. 5:231–37.

Dix, P. J., and H. E. Street. 1976. Selection of plant cell lines with enhanced chilling resistance. Ann. Bot. 40:903–10.

Ellis, B. E. 1978. Non-differential sensitivity to the herbicide Metribuzin in tomato cell suspension cultures. Can. J. Plant Sci. 58:775.

Flick, C. E. 1983. *In vitro* mutant isolation. *In* D. A. Evans, W. R. Sharp, P. V. Ammirato, and Y. Yamada (eds.), Handbook of plant tissue culture, 1:393–441. Macmillian, New York.

Flick, C. E., R. A. Jensen, and D. A. Evans. 1981. Isolation, protoplast culture, and plant regeneration of PFP-resistant variants of *Nicotiana tabacum* Su/Su. Z. Pflanzenphysiol. 103:239–45.

Gebhardt, C., H. Fankhauser, and P. J. King. 1982. Amino acid auxotrophs of *Hyoscyamus muticus* and their characterization. *In* A. Fujiwara (ed.), Plant tissue culture 1982, pp. 461–64. Jap. Assoc. for Plant Tissue Culture, Tokyo, Japan.

Gebhardt, C., V. Schnebli, and P. J. King. 1981. Isolation of biochemical mutants using haploid mesophyll protoplasts of *Hyoscyamus muticus*. Planta 153:81–89.

Gengenbach, B. G., and C. E. Green. 1975. Selection of T-cytoplasm maize callus cultures resistant to *Helminthosporium maydis* race T pathotoxin. Crop Science 15:645–49.

Gengenbach, B. G., J. A. Connelly, D. R. Pring, and M. F. Conde. 1981. Mitochondrial DNA variation in maize plants regenerated during tissue culture selection. Theor. Appl. Genetics. 59:161–67.

Gengenbach, B. G., C. E. Green, and C. M. Donovan. 1977. Inheritance of selected pathotoxin resistance in maize plants regenerated from cell cultures. Proc. Natl. Acad. Sci. USA 74:5113–17.

Gresshoff, P. M. 1979. Cycloheximide resistance in *Daucus carota* cell cultures. Theor. Appl. Genet. 54:141–43.

Heinz, D. J., M. Kirshnamurthi, L. G. Nickell, and A. Maretzki. 1977. Cell, tissue, and organ culture in sugarcane improvement. *In* J. Reinert and Y. P. S. Bajaj (eds.), Plant cell, tissue, and organ culture, pp. 3–16. Springer-Verlag, New York.

Helgeson, J. P., G. T. Haberlach, and C. D. Upper. 1975. A dominant gene conferring disease resistance to tobacco plants is expressed in tissue cultures. Phytophathology 66:91–96.

Hibberd, K. A., T. Walter, C. E. Green, and B. G. Gengenbach. 1980. Selection and characterization of a feedback-insensitive tissue culture of maize. Planta 148:183–87.

Kemble, R. J., R. B. Flavell, and R. I. S. Brettell. 1982. Mitochondrial DNA analysis of fertile and sterile maize plants derived from tissue culture with the Texas male sterile cytoplasm. Theor. Appl. Genet. 62:213–17.

King, J., R. B. Horsch, and A. D. Savage. 1980. Partial characterization of two stable auxotrophic cell strains of *Datura innoxia* (Mill.). Planta 149:480–84.

King, J., and V. Khanna. 1980. A nitrate reductase-less variant isolated from suspension cultures of *Datura innoxia* (Mill.). Plant Physiol. 66:632–36.

Malmberg, R. L. 1979. Temperature sensitive variants of *Nicotiana tabacum* isolated from somatic cell culture. Genetics 92:215–21.

Malmberg, R. L. 1980. Biochemical, cellular, and developmental characterization of a temperature-sensitive mutant of *Nicotiana tabacum* and its second site revertant. Cell 22:603–9.

Marton, L., T. M. Dung, R. R. Mendel, and P. Maliga. 1982. Nitrate reductase deficient cell lines from haploid protoplast culture of *Nicotiana plumbaginifolia*. Molec. Gen. Genet. 182:301–4.

Matern, U., G. Strobel, and J. Shephard. 1978. Reactions to phytotoxins in a potato population derived from mesophyll protoplasts. Proc. Nat. Acad. Sci. USA 75:4935–37.

Mathur, A. K., P. S. Ganapathy, and B. M. Johri. 1980. Isolation of sodium chloride-tolerant plantlets of *Kickxia ramosissima* under *in vitro* conditions. Z. Pflanzenphysiol. 99:287–94.

Matthews, B. F., S. C. H. Shye, and J. M. Widholm. 1980. Mechanism of resistance of a selected carrot cell suspension culture to S(2-aminoethyl)-L-cysteine. Z. Pflanzenphysiol. 96:453–63.

Medgyesy, P., L. Menczel, and P. Maliga. 1980. The use of cytoplasmic streptomycin resistance: chloroplast transfer from *Nicotiana tabacum* into *Nicotiana sylvestris*, and isolation of their somatic hybrids. Molec. Gen. Genet. 179:693–98.

Menczel, L., F. Nagy, Zs. R. Kiss, and P. Maliga. 1981. Streptomycin resistant and sensitive somatic hybrids of *Nicotiana tabacum* × *Nicotiana knightiana*: correlation of resistance to *N. tabacum* plastids. Theor. Appl. Genet. 59:191–95.

Mendel, R. R., Z. A. Alikulov, N. P. Lvov, and A. J. Muller. 1981. Presence of the molybdenum-cofactor in nitrate reductase-deficient mutant cell lines of *Nicotiana tabacum*. Mol. Gen. Genet. 181:395–99.

Mendel, R. R., and A. J. Muller. 1976. A common genetic determinant of xanthine dehydrogenase and nitrate reductase in *Nicotiana tabacum*. Biochem. Physiol. Pflanzen. 170:538–41.

Mendel, R. R., and A. J. Muller. 1978. Reconstitution of NADH-nitrate reductase *in vitro* from nitrate reductase-deficient *Nictotiana tabacum* mutants. Molec. Gen. Genet. 161:77–80.

Mendel, R. R., and A. J. Muller. 1979. Nitrate reductase-deficient mutant cell lines of *Nicotiana tabacum*. Mol. Gen. Genet. 177:145–53.

Muller, A. J., and R. Grafe. 1978. Isolation and characterization of cell lines of *Nicotiana tabacum* lacking nitrate reductase. Molec. Gen. Genet. 161:67–76.

Murphy, T. M., and C. W. Imbrie. 1981. Induction and characterization of chlorate-resistant strains of *Rosa damascena* cultured cells. Plant Physiol. 67:910–16.

Nabors, M. W., S. E. Gibbs, C. S. Bernstein, and M. F. Meis. 1980. NaCl-tolerant tobacco plants from cultured cells. Z. Pflanzenphysiol. 97:13–17.

Ranch, J. P., S. Rick, J. E. Brotherton, and J. M. Widholm. 1983. Expression of 5-methyltryptophan resistance in plants regenerated from resistant cell lines of *Datura innoxia*. Plant Physiol. 71:136–40.

Rains, D. W., T. P. Croughan, and S. J. Stavarek. 1979. Selection of salt-tolerant plants using tissue culture. *In* D. W. Rains, R. C. Valentine, and A. Hollaender (eds.), Genetic engineering of osmoregulation. pp. 279–92. Plenum Publishing, New York.

Sacristan, M. D. 1982. Resistance response to *Phoma lingam* of plants regenerated from selected cell and embryogenic cultures of haploid *Brassica napus*. Theor. Appl. Genet. 61:193–200.

Savage, A. D., J. King, and O. L. Gamborg. 1979. Recovery of a pantothenate auxotroph from a cell suspension culture of *Datura innoxia* (Mill.). Plant Sci. Lett. 16:367–76.

Shimamoto, K., G. Aeschbacher, and P. J. King. 1982. A histidine requiring clone of *Hyoscyamus muticus* isolated by negative selection using 5-bromodeoxyuridine. *In* A. Fujiwara (ed.), Plant tissue culture 1982, pp. 465–66. Jap. Assoc. for Plant Tissue Culture, Tokyo, Japan.

Sidorov, V. A., and P. Maliga. 1982. Fusion complementation analysis of auxotrophic and chlorophyll-deficient lines isolated in haploid *Nicotiana plumbaginifolia* protoplast cultures. Molec. Gen. Genet. 186:328–32.

Sidorov, V., L. Menczel, and P. Maliga. 1981. Isoleucine-requiring *Nicotiana* plant deficient in threonine deaminase. Nature 294:87–88.

Strauss, A., F. Bicher, and P. J. King. 1981. Isolation of biochemical mutants using haploid mesophyll protoplasts of *Hyoscyamus muticus*. Planta 153:75–80.

Templeton-Somers, K. M., W. R. Sharp, and R. M. Pfister. 1981. Selection of cold-resistant cell lines of carrot. Z. Pflanzenphysiol. 103:139–48.

Tyagi, A. K., A. Rashid, and S. C. Maheshwari. 1981. Sodium chloride resistant cell line from haploid *Datura innoxia* (Mill.). A resistance trait carried from cell to plantlet and vice versa *in vitro*. Protoplasma 105:327–32.

Widholm, J. M. 1972a. Cultured *Nicotiana tabacum* cells with an altered anthranilate synthetase which is less sensitive to feedback inhibition. Biochim. Biophys. Acta 261:52–58.

Widholm, J. M. 1972b. Anthranilate synthetase from 5-methyltryptophan-susceptible and -resistant cultured *Daucus carota* cells. Biochim. Biophys. Acta 279:48–57.

Widholm, J. M. 1974. Cultured carrot cell mutants: 5-methyltryptophan-resistant trait carried from cell to plant and back. Plant Sci. Lett. 3:323–30.

Widholm, J. M. 1976. Selection and characterization of cultured carrot and tobacco cells resistant to lysine, methionine, and proline analogs. Can. J. Bot. 54:1523–29.

Recombinant DNA of Plant Genes

15

Conventional plant-breeding techniques have, over the last few decades, made dramatic increases in crop plant productivity, and this will continue to be the major way forward into the future. However, the plant breeder is anxious now to use molecular and somatic genetics to help overcome present restrictions in the extent to which he can transfer genes in breeding programs. This paper deals with various aspects of recombinant DNA of plant genes. Plant tissue culture as a foundation for genetic engineering, and the use of vectors for inserting plant genes, are the subjects of other papers in this volume.

Recombinant DNA techniques include isolation of plant DNA, its restriction by endonucleases, ligation to various vectors, and cloning. The re-isolated recombinant DNA will eventually be used to transform either bacteria, plant cultures, or plants.

Plant genes may be obtained directly by restricting the genomic DNA if selective procedures are available. Normally, however, mRNA has to be identified, partially purified, and DNA complementary to it made; this cDNA can then be restricted and cloned, of the cDNA a particular protein gene selected, and used as a probe to isolate the gene from digests of genomic DNA. If genomic DNA is to be used to transform and if plant genes have a modified structure in the "switched-off" condition, e.g., increased methylation as has been recently suggested for animal genes (Naveh-Many and Cedar, 1981), it may be necessary for their subsequent expression that genomic genes be isolated from tissues in which the protein product is being actively synthesized.

Copy DNAs and mRNAs of several plant proteins are now available, e.g., those of the small subunit of ribulose-1, 5-bisphosphate carboxylase (Bedbrook et al., 1980) leghaemaglobin (Verma, 1979), and several seed storage proteins (Burr and Burr, 1979; Hall et al., 1979; Croy et al., 1981). In this paper I shall concentrate on describing work done at Durham on the construction, cloning, characterization, and sequence of cDNAs encoding *Pisum sativum* (pea) storage protein precursors, particularly legumin.

Legumin is the major storage protein in many legume seeds and is a hexameric molecule of M_r 360,000 to 400,000, consisting of six subunit pairs, each of which is constituted with a 40,000 M_r polypeptide disulfide-linked to a 20,000 M_r polypeptide (Derbyshire et al., 1976). As synthesized in pea, the 60,000 M_r is a covalently unbroken precursor polypeptide (Croy et al., 1980).

Initially, polyadenylated mRNA was purified from polysomes, isolated from developing seed cotyledons that were actively synthesizing legumin, by oligo dT-chromatography and sucrose gradient centrifugation (fig. 1). The 18s poly (A)-Rich polysomal RNA fraction so isolated contained many seed protein messages as shown by hybridization kinetics (data not shown), but was greatly enriched with storage protein messages as shown by its translation in the reticulocyte lysate translation system. Double-stranded, blunt-ended, open, complementary DNAs were synthesized from this partially purified messenger RNA according to the protocol set out in figure 2.

Synthetic Bam H1 oligonucleotide linkers were then attached and the DNAs size-fractionated on Sepharose 6B. The longest lengths of cDNA were cloned into the Bam H1 site, of pBR 322 (Sippel et al., 1978) (fig. 3).

Phosphate-treated plasmid DNA was used in order to prevent self-ligation and nonrecombinant transformation. Transformants were selected on ampicillin plates and tetracycline-sensitive, ampicillin-resistant clones were further selected by colony hybridization on nitrocellulose filters using 32 P-labeled mRNA and cDNA as probes (fig. 4). Hybridizations were performed in the presence of polyadenylic acid. Those colonies that strongly hybridized were selected and plasmid minipreps made according to Kline and colleagues (Klein et al., 1980). They were then restricted with Bam H1 to excise the cDNA inserts, sized on agarose gels. Southern "blots" from these gels strongly hyridized mRNA and cDNA probes.

Plasmids with cDNA inserts greater than 500 bp in length were purified, cDNA sequences excised and bound to DBM paper (Smith et al., 1979; Alwine et al., 1977). Storage protein clones were then identified by hybrid selected translation using the reticulocyte lysate system (Croy et al., 1981). Initially, twelve clones were selected; these included several with cDNAs encoding vicilin 50,000 M_r and 47,000 M_r subunits, two encoding the le-

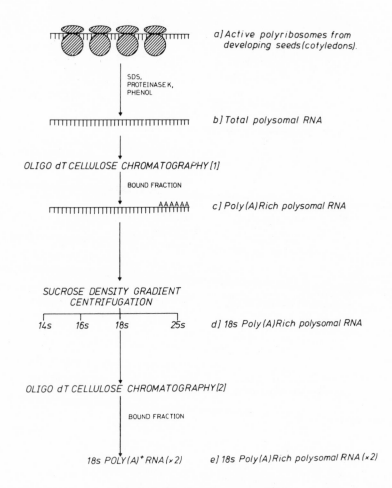

a] *Active polyribosomes from developing seeds(cotyledons).*

SDS,
PROTEINASE K,
PHENOL

b] *Total polysomal RNA*

OLIGO dT CELLULOSE CHROMATOGRAPHY [1]

BOUND FRACTION

c] *Poly (A)Rich polysomal RNA*

SUCROSE DENSITY GRADIENT CENTRIFUGATION

14s   16s   18s   25s   d] *18s Poly (A)Rich polysomal RNA*

OLIGO dT CELLULOSE CHROMATOGRAPHY [2]

BOUND FRACTION

18s POLY(A)⁺ RNA (×2)   e] *18s Poly(A)Rich polysomal RNA (×2)*

Fig. 1 Purification of messenger RNA.

gumin 60,000 M_r precursor and others coding unidentified polypeptide products (Croy et al., 1981).

The legumin cDNA pRC 2.2.4 has been completely sequenced and corresponds to a 638 nucleotide sequence from the 3′ end of the legumin mRNA. It encodes the legumin basic subunit as is evident from the correspondence of its predicted amino acid sequence with the N-terminal sequence published by Casey (Casey et al., 1981) and also with those of five major peptides isolated from tryptic digests of the legumin basic subunit

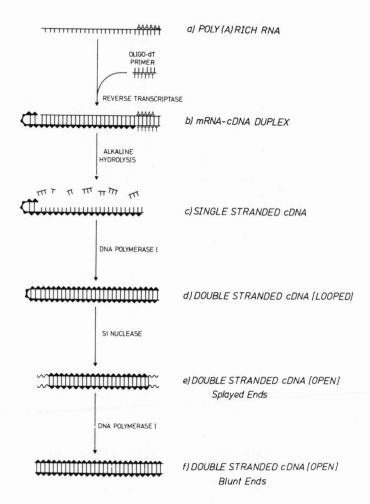

a) POLY (A) RICH RNA

OLIGO-dT
PRIMER

REVERSE TRANSCRIPTASE

b) mRNA-cDNA DUPLEX

ALKALINE
HYDROLYSIS

c) SINGLE STRANDED cDNA

DNA POLYMERASE I

d) DOUBLE STRANDED cDNA [LOOPED]

S1 NUCLEASE

e) DOUBLE STRANDED cDNA [OPEN]
   Splayed Ends

DNA POLYMERASE I

f) DOUBLE STRANDED cDNA [OPEN]
   Blunt Ends

Fig. 2 Synthesis of complimentary DNA.

(Croy et al., 1981). These data locate the basic (20,000 M_r) subunit at the 3′ end of the precursor. A larger clone extends the sequence upstream at the equivalent of 30 amino acid residues, and this nucleotide sequence contains no initiator or termination codons, confirming the results from cell-free translation experiments that demonstrated a large legumin precursor. Northern "blots" of total and polyadenylated RNA of developing cotyledons probed with 32P-labeled legumin cDNAs showed the presence of a singe legumin mRNA band corresponding to a 19S mRNA (2.2×10^3

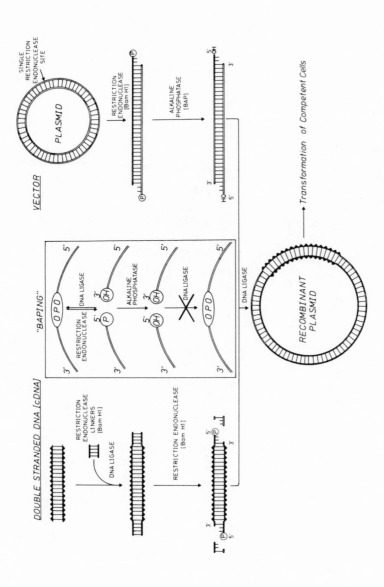

Fig. 3 Cloning strategy 1.

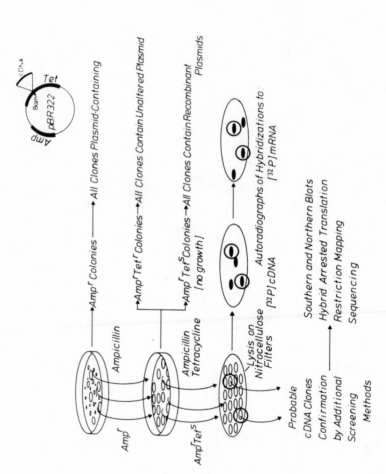

Fig. 4 Selection strategy.

bases), i.e., of sufficient size to code a 60,000 M_r polypeptide (Croy et al., 1981).

The cDNAs have also been used to probe for legumin genomic genes. Pea genomic DNA digested to completion with Eco R1 restriction enzyme and probed with legumin cDNA showed the presence of four Eco R1 genomic single-copy fragments hybridizing to legumin cDNA (Croy et al., 1981).

The plant breeder is particularly interested in transforming agronomically good crop varieties with genetic information that would improve yield and confer resistance to major pests and diseases. These particular characteristics are in the main polygenic, and at present their genetic basis is unknown. Therefore, one of the most important tasks for the future is to develop our basic understanding of developmental physiology in plants in order that we may identify those genes that play an important role in these key processes.

Another realistic possibility would be to attempt to improve the nutritional quality of a crop, especially the profile of its major protein, e.g., seed storage protein, by mutagenesis *in vitro* of isolated genes that would then be reintroduced into the plant using a suitable vector.

Realization of these possibilities will depend on a better understanding of the organization of the plant genome and on the isolation of particular genes, investigations in which recombinant DNA techniques will play the major role. Also required will be the development of suitable transfer vectors for higher plants, but that is the subject of other papers in this volume.

ACKNOWLEDGMENT

I would like to thank Dr. R. R. D. Croy for help in the preparation of the manuscript.

REFERENCES

Alwine, J. C., J. D. Kemp, and G. R. Stark. 1977. Method of detection of specific RNAs in agarose gels by transfer to diazobenzyloxymethyl-paper and hybridization with DNA probes. Proc. Nat. Acad. Sci. USA 74:5350–54.

Bedbrook, J. R., S. S. Smith, and R. J. Ellis. 1980. Molecular cloning and sequencing of cDNA encoding and precursor to the small subunit of chloroplast ribulose 1,5 biphosphate carboxylase. Nature 287:692–97.

Burr, F. A., and B. Burr. 1979. The cloning of Zein sequences and an approach to Zein genetics. *In* C. J. Leaver (ed.), Genome organisation and expression in plants, pp. 227–31. Plenum, New York.

Casey, R., J. F. March, and E. Sanger. 1971. N-terminal amino acid sequence of β-subunits of legumin from *Pisum sativum*. Phytochem. 20:161–63.

Croy, R. R. D., J. A. Gatehouse, J. N. Yarwood, and D. Boulter. 1981. Characterisation of the storage protein subunits synthesized in vitro by polyribosomes and RNA from developing pea (*Pisum sativum* L.). Planta 148:49–56.

Croy, R. R. D., G. W. Lycett, J. A. Gatehouse, J. N. Yarwood, and D. Boulter. 1981. Cloning and analysis of cDNAs encoding plant storage protein precursors. Nature 295:76–79.

Derbyshire, E., D. J. Wright, and D. Boulter. 1976. Legumin and vicilin, storage proteins of legume seeds. Phytochemistry 15:3–24.

Hall, T. C., S. M. Sun, B. U. Buchbinder, J. W. Pyne, F. A. Bliss, and J. D. Kemp. 1979. Bean seed globulin mRNA: translation, characterization, and its use as a probe towards genetic engineering of crop plants. *In* C. J. Leaver (ed.), Genome organisation and expression in plants, pp. 259–72. Plenum, New York.

Klein, R. D., E. Selsing, and R. D. Wells. 1980. A rapid microscale technique for isolation of recombinant plasmid DNA suitable for restriction enzyme analysis. Plasmid 3:88–91.

Naveh-Many, T., and H. Cedar. 1981. Active gene sequences are under-methylated. Proc. Natl. Acad. Scie. USA 78:4246–50.

Sippel, A. E., H. Land, W. Lindenmaier, M. C. Nguyen-Huu, T. Wurts, K. N. Timmis, K. Giesecke, and G. Shutz. 1978. Cloning of chicken lysozynae structural gene sequences synthesized *in vitro*. Nucleic Acids Res. 5:3275–94.

Smith, D. F., P. E. Searle, and J. G. Williams. 1979. Characterisation of bacterial clones containing DNA sequences derived from *Xenopus laevis* vitellogenin mRNA. Nucleic Acids Res. 6:487–506.

Verma, D. P. S. 1979. Expression of host genes during symbiotic nitrogen fixation. *In* C. J. Leaver (ed.), Genome organisation and expression in plants, pp. 437–52. Plenum, New York.

M. L. R. AGUIAR—PERECIN

Molecular Organization of Plant Chromosomes and Perspectives for Genetic Engineering

16

ABSTRACT

New potential techniques for plant breeding involve regeneration of entire plants from tissue culture. Cytogenetic analysis should be a basic routine in such technologies. On the other hand, new techniques have pointed out interesting aspects concerning chromosomal organization and function that if better investigated and understood certainly could contribute greatly to technologies involving genetic manipulation.

INTRODUCTION

In the early 1970s, new techniques in the biological sciences—the recombinant DNA methodology—were used to demonstrate the possibility of taking pieces of DNA from an organism and joining them using biochemical methods to pieces of DNA from another organism. Geneticists and biochemists were quick to see the applicability of these methods to answering some of the most important questions about how in living cells, particularly the complex cells of higher organisms, the genetic information in their chromosomes is organized and the expression of this information regulated.

Apart from scientific applications toward the greater understanding of the nature and mode of regulation of eukaryotic genes, several broad areas of application of recombinant DNA technology have been recognized in industry, agriculture, and medicine. Various claims have been made regarding possible contributions of genetic engineering to plant breeding. These potential technologies would involve many kinds of techniques, whose perspectives have been discussed (Rubenstein et al., 1977; Jackson and Stick, 1979; Sprague et al., 1980). Recombinant DNA technologies coupled with known techniques for growing entire plants from tissue culture cells could make possible the development of new plants that would represent hybrids of different species. At present the problems concerning the expression of foreign DNA in plant cells (transferred through recombinant DNA methodology, or otherwise incorporated into plant cells) are practically unknown, being still in an initial phase of investigation. On the other hand, techniques involving plant tissue culture have progressed in the last few decades. The intent has been to manipulate and select for favorable phenotypes at the cellular level and then regenerate complete plants that could be used in plant breeding programs. Methods such as anther culture and protoplast fusion of different species have been investigated widely as tools for plant breeding.

Among the several problems to be considered concerning techniques involving regeneration of entire plants from tissue culture, chromosome instability has been observed to occur in cell cultures and respective regenerated plants of several species. On the other hand, in the case of protoplast fusion, hybrid materials will be recovered. In this context it is hoped that cytogenetic analysis will be very important. Furthermore, since protoplast fusion will produce plants with chromosomes of different origins within their cells, a better understanding of plant chromosome organization and function will be very important in this area (Evans and Reed, 1981). In this article some important aspects will be discussed concerning the knowledge of plant chromosome organization and investigations concerning the linear differentiation of plant chromosomes. The organization of plant chromosomal DNA sequences and the importance of this information in the context of plant genetic manipulation will be particularly emphasized.

DNA CONTENT IN PLANTS

Eukaryotes in a general sense and particularly higher plants have genomes much larger than would be expected to code and regulate the expression of the genes required during development. Higher plants have not been investigated as much as animals, but an analysis of available data shows that plants frequently have much more repetitive DNA than most animals. In his regard plants have been considered as particularly favorable material for studies on the origin and evolution of repeated sequences (McCarthy, 1969; Thompson et al., 1980).

In a general sense, variation in genome size in plants is also more pronounced among related species than it is among animals, and there is little correlation between DNA content and organismic complexity. In a compliation of 753 measurements of angiosperms, DNA per genome varied between 3 pg in *Utrica urens* (Urticaceae) and 127. 14 pg in *Fritillaria assyriaca* (Liliaceae) (Bennet and Smith, 1976). Of course, on the broad scale of evolution from virus to higher plants there are levels of complexity that demand more genes and more DNA for their fulfillment. On the other hand, in higher plants there are cases where it is difficult to explain how DNA amounts can vary considerably without raising or lowering the general level of intricacy of organization. For instance, species of *Drosophyllum* that have 15 pg per genome, compared with species of *Dorsera* with 0.3-0.9 pg, although distinguished as distinct species, have nothing in their phenotypes to explain these differences of DNA content (Sparrow and Nauman, 1976). In a compilation of data on the DNA amount per genome in about 2,400 species, it was shown that within each major phylogenetic group a multimodal distribution was evident. This was interpreted as a series of doublings of minimum genome size in each taxonomic group (Sparrow and Nauman, 1976).

The variations of quantity and kinds of DNA have been investigated through cytological and biochemical techniques

THE LINEAR DIFFERENTIATION OF PLANT CHROMOSOMES

Banding techniques have made possible the visualization of a linear differentiation along metaphase chromosomes of animals and plants (Hsu, 1973; Comings, 1978).

In one of these methods, quinacrine, which may produce regions that fluoresce under ultraviolet light, is employed. Fluorescent bands have been shown to contain AT-rich DNA. Other methods involve different kinds of pretreatment followed by staining with Giemsa, which, according to the procedure used, reveals the so-called G- or C-bands. G-bands have been evidenced in animals after various treatments, as for example with trypsin. In plants G-bands have not been observed (Greiluhber, 1977). C-bands are revealed after exposure of fixed chromosomes to an alkaline agent followed by treatment with a saline-citrate solution (2XSSC) at relatively high temperature. With C-banding various kinds of constitutive heterochromatin in plants and animals have been distinguished. Since this was first shown (Pardue and Gall, 1970; Yunis and Yasmineh, 1970), evidence that C-bands are frequently associated with repetitive DNA has accumulated.

In plants, apart from evidence that heterochromatin associated with nucleolus organization stains by C-banding procedures, it has been shown that tandem arrays of identical repeats frequently seem to be localized in heterochromatic chromosomal regions (Flavell, 1980).

C-banding has been the most used banding method in plant cytogenetics. Species so far investigated have had variable aspects. Dark staining bands have been found in centromeric, intercalary, and telomeric regions of plant chromosomes. Patterns are rather constant in some species and extremely variable in others. The significance of such large heterochromatic blocks in chromosome organization and chromosome action has not yet been understood fully and has been widely discussed. First, the contributions that light microscopists have made to their understanding by observing their occurrence and variation both within and between species will be commented on briefly.

Since we have been studying maize somatic chromosomes, using a C-banding procedure, some results will be reported, for maize represents a species particularly interesting in studies of the structure of heterochromatic regions. Stocks of different knob constitutions were analyzed. In all the stocks, distal bands were observed in metaphase chromosomes from root tip metistems. The number of such bands varied among stocks, and a strict correspondence between these bands and knobs was seen in pachytene. Other heterochromatic regions had different degrees of staining, such as NOR heterochromatin, pericentromeric heterochromatin, and satellite of chromosome 6. Heterochromatin blocks of B chromosomes did not have differential staining (Aguiar-Perecin, 1985). In a general sense, the results agree with data published by Ward (1980), with divergences concerning degree and variation in staining of heterochromatic regions other than knobs. It was observed also that banded chromosomes are larger than chromosomes without bands. It appears that knobs are additional material in the chromosome (Aguiar-Perecin and Vosa, 1985). The analysis of interphase nuclei stained by the C-banding procedure also showed a correlation between the number of distal bands and chromocenters. This correspondence is visualized more easily in stocks with a lower number of bands. In materials with a high number of bands, there is a higher chance of occurrence of fusion among chromocenters (Aguiar-Perecin, 1985).

These observations in maize may well illustrate the usefulness of C-banding methodology in cytogenetic analyses of plant cells *in vitro* or of cells and plants originated from protoplast fusion. In this connection the validity of chromocenter counting for ploidy assessment, *in vitro*, was tested using recently initiated haploid tissue cultures of corn. It was found that chromocenter counting may be a convenient method for ploidy assessment in mixed dividing or resting plant cell culture populations (Dhaliwal and King, 1979).

Maize represents a species in which a high degree of chromosomal polymorphism is observed, but other species so far examined have different situations, from near-absence of bands to cases where similar patterns of C-bands can be recognized among species. Some cases representative of these aspects can be found among cereals and Liliaceae.

Wheat and rye species have been investigated widely (Yosa, 1974; Gill and Kimber, 1974a,b; Verma and Rees, 1974; Weimark, 1975; Iordansky et al., 1978). The karyotype of some polyploid wheat species consists of "constant" chromosomes occurring in almost all the species investigated and a group of "variable" chromosomes that are either specific for one species in question or occur in a few species (Zurabishvili et al., 1978). Giemsa-banded chromosomes of *Secale cerale* (rye) have a characteristic banding pattern that greatly facilitates their identification, particularly in triticale lines (Sapra and Stewart, 1980). On the other hand, extensive polymophism was found in the chromosomes of six inbred lines of cultivated rye. Variations were observed not only with regard to the presence and size of thin, intercalary bands, but also of the characteristic telomeric bands (Lelley et al., 1978).

Some groups of *Scilla* species (Liliaceae) have been extensively investigated (Grielhuber and Speta, 1976, 1978; Greilhuber, 1977a, b). By analysis of DNA content and C-banded caryotypes, some interesting aspects were shown. In a group of cultivated species of the *Scilla siberica*, group different patterns of bands were found, but chromosome structure and "banding style" also permitted systematic grouping of closely related species. By quantitative analysis of C-banded caryotypes, it was shown that C-bands do not "replace" but are added to euchromatin. In another group of related species, *S. drunensis* (2n = 36.4×, 1 C = 12.8 pg), *S. bifolia* (2n = 18.2×, 1 C = 6.2 pg), and *S. vindobonensis* (2n = 18.2 ×, 1 C = 9.4 pg), it was demonstrated that these two last species differed by an excess of euchromatic DNA, although the higher DNA content in *S. vindobonensis* was also combined with higher heterochromatin content.

The cases mentioned give an idea of the variability in C-band patterns found in many groups of plants. Such observations have raised questions regarding the significance of these bands in chromosomal organization and function. Some interesting evidence seems to correlate them with the duration of mitotic cycle and annual or perennial life cycles of plants (Nagi, 1974). Besides these cytological methods, biochemical techniques that permit the study of the nature and disposition of chromosomal DNA sequences have revealed important aspects of animal and plant chromosome organization.

REPETITIVE AND UNIQUE DNA SEQUENCE ORGANIZATION
IN PLANT CHROMOSOMES

Investigations in this area are based on the techniques of DNA denaturation and reassociation, described by Britten and Kohne (1969), that permit one to calculate the proportion of classes of highly and moderately repetitive as well as single-copy DNA in a genome. It is also possible to measure relative lengths of the various categories of sequences and so classify the DNA content of a genome in a very precise manner. Furthermore, their relative positions can be found from which models of organization of regions of a chromosome can be constructed. Although plants have not yet been investigated so much as animals, an extensive literature has grown around this subject (Flavell, 1980; Thompson et al., 1980; Hake and Walbot, 1980; Bendich and Ward, 1980; Bedbrook et al., 1980) giving important information on the organization of DNA sequences in plant chromosomes. As was mentioned in the introduction to this article, large amounts of repetitive DNA are found in several species of plants, within many taxonomic groups. The importance of these investigations is emphasized, not only in the sense of understanding plant chromosome organization in relation to gene control, but also that such studies should be done in a context that includes evolutionary as well as developmental perspectives. In other words, a better understanding of the origin of repetitive sequences through cycles of amplification should be very important to the understanding of plant chromosome organization and evolution (Thompson et al., 1980). Some data will be discussed briefly to illustrate the general situation of the knowledge of molecular organization of plant chromosomes.

The proportion of repeated sequences in a wide range of angiosperms is quite variable, from 30% to 95%. When data of reassociation kinetics are compared, they suggest that in the smallest plant genomes, there is only a very small proportion of repeated DNA sequences and that much of what is present is likely to be ribosomal DNA (Flavell, 1980).

An interesting illustration of the different proportions of repetitive sequences, found in related species with different contents of DNA, can be appreciated in a study on *Pisum sativum* (pea) that contains 4.6–4.8 pg of

DNA and *Vigna radiata* (mung bean) with 0.48–0.53 pg of DNA (Thompson et al., 1980). In this comparative study, it was demonstrated that pea DNA differed from mung bean DNA in the amount of repetitive sequences, both in maximum repetition frequency and in the number of different sequences. The data are consistent with the idea that different groups of families arose during cycles of amplification events in the evolutionary history of a species (Thompson et al., 1979). Students of the linear arrangement of repetitive and single-copy DNA (6, 14) have demonstrated extensive interspersion of short (ca. 300 nucleotides) repetitive sequences at intervals among single-copy sequences (ca. 2,000 nucleotides) in the DNA of a wide variety of animals and plants (Flavell, 1980; Davidson et al., 1975). Variations of this pattern have been found particularly in plants. Data from higher plants so far analyzed show that a substantial fraction of the genome consists of short lengths (200 to 4,000 base pairs) of single-copy DNA, interspersed with short lengths (50 to 2,000 base pairs) of repeated sequences of nucleotides. Other chromosomal segments contain long stretches of single-copy DNA or reverse repeats. The remaining segments comprise substantial fractions of the plant chromosome. In these fractions two kinds of arrangements can be distinguished: those that consist of very similar repeating units tandemly arranged, and those that are constituted of unrelated short repeats interspersed with each other in different permutations (Thompson et al., 1980; Hake and Walbot, 1980).

Here again the comparative study of pea and mung bean (Thompson et al., 1980) has interesting aspects. The excess of DNA found in pea is constituted of repetitive sequences (shorter and longer than 1,200 base pairs) and short single copies (ca. 300–400 base pairs). The data are consistent with the hypothesis that such single-copy sequences in pea DNA are members of ancient interspersed repetitive families that have accumulated so many mutations that one can no longer recognize them as repetitive.

As these investigations progress, the authors recognize that few data are available still on such problems as genic regulation in plants, structure of genes, or significance of the arrangements of DNA (Flavell, 1980; Bendich and Ward, 1980). Recently molecular cloning technology has provided new ways of studying eukaryotic chromosomes. In plants such techniques already have enabled the study of several aspects such as: localization of tandem arrays of identical repeats in telomeric heterochromatin of *Secale cereale* (Bedbrook et al., 1980), cloning of 18 S, 5.8 S and 25 S ribosomal RNA gene units from wheat and barley (Gerlach and Bedbrook, 1979), and of zein sequences (Burr and Burr, 1980), and studies of their structure.

REFERENCES

Aguiar-Perecin, M. L. R. de. 1985. C-banding in maize. I. Band patterns. Caryologia 38:23–30.
Aguiar-Perecin, M. L. R. de and C. G. Vosa. 1985. C-banding in maize. II. Identification of somatic chromosomes. Heredity 54:31–43.
Bedbrook, J., W. Gerlach, S. Smith, J. Jones, and R. Flavell. 1980. Chromosome and gene structure in plants: a picture deduced from analysis of molecular clones of plant DNA. *In* C. J. Leaver (ed.), Genome organization and expression in plants, pp. 49–62. Plenum Press, New York.
Bendich, A. J., and B. L. Ward. 1980. On the evolution and functional significance of DNA sequence organization in vascular plants. *In* C. J. Leaver (ed.), Genome organization and expression in plants, pp. 17–30. Plenum Press, New York.
Bennett, M. D., and J. B. Smith. 1976. Nuclear DNA amounts in angiosperms. Phil. Trans. Roy. Soc. London B274:227–74.
Britten, R. J., and D. E. Kohne. 1969. Repetition of nucleotide sequences in chromosomal DNA. *In* A. Lima-de-Faria (ed.), Handbook of molecular cytology, pp. 21–36. North-Holland, Amsterdam.
Burr, F. A., and B. Burr. 1980. The cloning of zein sequences and an approach to zein genetics. *In* C. J. Leaver (ed.), Genome organization and expression in plants, pp. 227–31. Plenum Press, New York.
Comings, D. E. 1978. Mechanisms of chromosome banding and implications for chromosome structure. Ann. Rev. Genet. 12:25–46.
Davidson, E. H., G. A. Galau, R. C. Angerer, and R. J. Britten. 1975. Comparative aspects of DNA organization in mitazoa. Chromosoma (Berlin) 51:253–59.
Dhaliwal, H. S., and P. J. King. 1979. Ploidy analysis of haploid-derived tissue cultures of *Zea mays* by chromocentre counting. Maydica 24:103–12.
Evans, D. A., and S. M. Reed. 1981. Cytogenetic techniques. *In* T. A. Thorpe (ed.), Plant tissue culture, pp. 213–40. Academic Press, New York.
Flavell, R. 1980. The molecular characterization and organization of plant chromosomal DNA sequences. Ann. Rev. Plant Physoil. 31:569–96.
Gerlach, W. L., and R. J. Bedbrook. 1979. Cloning and characterization of ribosomal RNA genes from wheat and barley. Nucleic Acid Res. 7:1869–85.
Gill, B. S., and G. Kimber. 1974a. The Giemsa C-banded karyotype of rye. Proc. Natl. Acad. Sci. USA 71:1247–49.
Gill, B. S., and G. Kimber. 1974b. Giemsa C-banding and the evolution of wheat. Proc. Natl. Acad. Sci. USA 71:4086–90.
Greilhuber, J. 1977a. Nulcear DNA and heterochromatin contents in the *Scilla hohenackeri* group, *S. persica* and *puschkinia scilloides* (Liliaceae). Plant. Syst. Evol. 128:243–57.
Greilhuber, J. 1977b. Why plant chromosomes do not show G-bands. Theor. Appl. Genet. 50:121–24.
Greilhuber, J., and F. Speta. 1976. C-banded karyotypes in the *Scilla hohenackeri* group, *S. persica* and *puschkinia* (Liliaceae). Plant Syst. Evol. 126:149–88.
Greilhuber, J., and F. Speta. 1978. Quantitative analyses of C-banded karyotypes and systematics in the cultivated species of the *Scilla siberica* group (Liliaceae). Plant. Syst. Evol. 129:63–109.
Hake, S., and V. Walbot. 1980. The genome of *Zea mays*, its organization and homology to related grasses. Chromosoma (Berlin) 79:251–70.
Hsu, T. C. 1973. Longitudinal differentiation of chromosomes. Ann. Rev. Genet. 7:153–76.

Iordansky, A. B., T. G. Zurabishvili, and N. S. Bradaev. 1978. Linear differentiation of cereal chromosomes. I. Common wheat and its supposed ancestors. Theor. Appl. Genet. 51:145–52.

Jackson, D. A., and S. P. Stick. 1979. The reombinant DNA debate. Prentice-Hall, Englewood Cliffs, N.J.

Lelley, T., K. Josifek, and P. J. Kaltsikes. 1978. Polymorphism in the Giemsa C-banding pattern of rye chromosomes. Can. J. Genet. Cytol. 20:307–12.

McCarthy, B. J. 1969. The evolution of base sequences in nucleic acids. *In* A. Lima-de-Faria (ed.), Handbook of molecular cytology, pp. 3–20. North-Holland, Amsterdam.

Nagl, W. 1974. Role of heterochromatin in the control of cell cycle duration. Nature 249:53–54.

Pardue, M. L., and J. G. Gall. 1970. Chromosome localization of mouse satellite DNA. Science 168:1356–58.

Rubenstein, I., R. L. Phillips, C. E. Green, and R. J. Desnick. 1977. Molecular genetic modification of eucaryotes. Academic Press, New York.

Sapra, V. T., and M. D. Stewart. 1980. Staining of heterochromatin bands and the determination of rye chromosomes in triticale (wheat × rye hybrid). Euphytica 29:497–509.

Sparrow, A. H., and A. F. Nauman. 1976. Evolution of genome size by DNA doublings. Science 192:524–29.

Sprague, G. F., D. E. Alexander, and J. W. Dudley. 1980. Plant breeding and genetic engineering: a perspective. Bioscience 30:17–21.

Thompson, W. F., M. G. Murray, and R. E. Cuellar. 1980. Contrasting patterns of DNA sequence organization in plants. *In* C. J. Leaver (ed.), Genome organization and expression in plants, pp. 1–15. Plenum Press, New York.

Verma, S. C., and H. Hees. 1974. Giemsa staining and the distribution of heterochromatin in rye chromosomes. Heredity 32:118–22.

Vosa, C. G. 1974. The basic karyotype of rye (*Secale cereale*) analysed with Giemsa and fluorescence methods. Heredity 33:403–8.

Ward, E. J. 1980. Banding patterns in maize mitotic chromosomes. Can. J. Genet. Cytol. 22:61–67.

Weimark, A. 1975. Heterochromatin polymorphism in the rye karyotype as detected by the Giemsa C-banding technique. Hereditas 79:293–300.

Yunis, J. J., and W. G. Yamineh. 1970. Satellite DNA in constitutive heterochromatin of the guinea pig. Science 168:263–65.

Zurabishvilli, T. G., A. B. Iordansky, and N. S. Badaev. 1978. Linear differentiation of cereal chromosomes. II. Polyploid wheats. Theor. Appl. Genet. 51:497–509.

Virus RNA *In Vitro* and *In Vivo*

17

INTRODUCTION

This work deals with purification studies of the protein and nucleic acid of pepper ringspot virus (VAP), one of the Rattle virus or Tobravirus group. By several tests and properties such as absorption, electron microscope examination, virus crystalization, serological reaction, and infectivity, the purified VAP could be used to obtain nucleic acid. The molecular weight of two nucleoproteins (L and S) particles were 37×10^6 D and 9.0×10^6 D, respectively. By radioautographic examination it was found that the RNA of the two particles replicates in the nucleolar area of the infected cell.

The amount and purification grade of the L and S particles of the VAP obtained by the method used (Kitajima, 1967) and the radioautographic information (Silva, 1965) about the nuclear replication of this virus seem to be good areas for research in genetic engineering as well as starting material to develop a vehicle for introduction of foreign genetic information into higher plants.

Several aspects of the VAP are now well known. Besides permitting its inclusion among the special group named Tobravirus (Harrison and Robinson, 1978), they include some interesting biological properties that deserve the attention of experts in genetic engineering. The Tobravirus group includes plant viruses that occur in more that 100 plant species in Solanaceae, Compositae, Cruciferae, and Leguminosae. The viruses in this group frequently contain two or more kinds of nucleoprotein particles.

The population of Brazilian VAP possesses two kinds of rods: long (L) 2,000 A and short (S) 550 A (Cadman and Harrison, 1959; Oliveira, 1967). Both types of rods are nucleoproteins, but only the L rods are infective. In general, the L particles are associated with mitochondria, but in some cases, an association between the mitchondria and S particles was found in our laboratory (Silva and Nogueria, 1979).

Lister (1966, 1967) working with the Brazilian ringspot virus named Rattle-CAM showed that the S particles RNA have the genetic code of the capsomeres found in both L and S rods. The nucleic acid of L particles also contains the gene for proteins such as the replicase and some others for several symptomatological characteristics.

So the VAP should be considered an infective system containing mainly two components in a real symbiosis relation. The RNA obtained by the common purification procedure most frequently is composed of single strips of about 2,000 nucleotides. The short RNA, with 2,000 nucleotides, as with the short nucleoproteic particles, is relatively easy to introduce into a host cell and could become a convenient vector for plant genetic engineering.

VIRUS PURIFICATION

Inoculated tobacco leaves with typical symptoms of the VAP were frozen, crushed, and filtered through cheesecloth. The juice was treated with n-butanol at 8% of the total volume and centrifuged at 5,000 rpm for 15 min. The supernatant, filtered through a cotton pad, was ultracentrifuged at 30,000 rpm for 75 min. The pellets obtained were suspended with phosphate buffer (0.1 M, pH 7.0) and submitted to 5,000 rpm centrifugation for 15 min. The supernatant was again centrifuged for 75 min. at 30,000 rmp. The pellets were suspended in phosphate buffer and centrifuged at low and high rotation as before.

The pellets were suspended in phosphate buffer and submitted to a sucrose gradient density centrifugation as described by Brakke (1951). Two layers were observed in the tubes and after careful extraction of each layer (with L or S particles, respectively) were ultracentrifuged at 45,000 rpm for 65 min. The suspended pellets were examined with an electron microscope, and the particle suspensions were kept in the refrigerator to be used in the next stages.

QUALITY CONTROL OF THE VAP PURIFICATIONS

The preparations kept in phosphate buffer 0.1 M, pH 7.0 were examined with a spectrophotometer and the absorption relation at 2800 and 2600 A was used as an index of the preparation's quality. Electron microscope examinations were used in several stages of the earliest purified preparations in order to follow the purification procedure. Finally, bioassays were used at certain stages of VAP purifications were used to follow the infectivity.

These tests were made by mechanical inoculations on *Chenopodium amaranticolor* leaves with crude and purified preparations. In order to check the specific infectivity, the protein content of each sample to be inoculated was obtained by a microjeldahl determination. These data related to the number of local lesions per unit area allowed calculation of the specific infectivity value of the L particles and the contamination in the S particles preparations.

Crystals of the L rods were obtained as needles in ammonium sulphate at 60% and kept overnight in a refrigerator, and by electron microscope examination many particles were observed that were broken in water solution.

PROPERTIES OF THE L OR S PARTICLES

Antiserum Test

Antisera were prepared by rabbit injection against the two sizes of VAP particles. The antisera with high titre (about 2040) were tested against the two kinds of particles by the double diffusion test described by Ouchterlone (1962).

Fingerprint of VAP-protein Treated with Trypsin

Virus samples of 10–15 mg were submitted to acetic acid treatment after the Fraenkel-Conrat method (Fraenkel-Conrat, 1977; Rushizky and Knight, 1960). The protein precipitates obtained after 2–3 days dialysis against water were dissolved in phosphate buffer 0.1 M, pH 7.0 and ultracentrifuged to be virus-free. To the pellet suspended in the same buffer at pH 5.0 was added a trypsin solution with 1% final concentration and kept in water bath at 35°C for 2 hr. The pH of the reaction medium was kept at 7.5 by a pH-stat apparatus. Aliquots of 40–45 μl were collected and put on Whatman paper no. 1 in order to make an electrophoresis in the first direction. After drying the Whatman paper, the sample was run with n-butanol, acetic acid, water (4:1:5) in the perpendicular direction to the first electrophoresis. The development of the oligopeptides was made with *ninydrin* and/or starch-iodine.

Estimates for the L and S Particles Molecular Weight

The L and S particle weight estimates were made by the Paul method (Paul, 1959) where the b_1 parameter (relation between the pure absorbances at 2600 and 2800 A radiations) is directly proportional to the molecular weight of the studied particles. The estimated values of the molecular weight was calculated from the following Paul relation:

$$b_v = \frac{E_v\,2800 - T_v\,2800}{E_v\,2600 - T_v\,2600}$$

where $E_v\,2800$ and $E_r\,2600$ stand for absorbances at 2800 and 2600 A, $T_v\,2800$ and $T_v\,2600$ represent the turbidity when radiated with 2600 A and 2800 A.

Nucleic Acid from the VAP and
Infectivity Test for L and S Particles

The nucleic acid from both particles of the VAP preparations was obtained by the Gierer and Scharam method (Gierer and Scharam, 1956; Rushizky and Knight, 1960). The nucleic acid was prepared for electron microscope examinations as described by Evenson and colleagues (Caro and Van Tubergen, 1962) and was submitted to serological tests (double diffusion on agar) to check presence of virus protein.

Infectivity tests were performed by mechanical inoculation of the following preparations on *C. amaranticolor* leaves: VAP-nucleic acid + ribonulcease and VAP-nucleoprotein without ribonuclease.

In order to avoid the presence of the nucleoprotein particles in the nucleic acid, preparations they were ultracentrifuged at 4,000 rpm for 60 min., and the supernatant was used for inoculation tests.

Radioautography at Electron Microscope Level to
Study the Intracellular VAP-RNA Replication

The following method was used to prepare the radioactive specimen for electron microscope examination. Small pieces (1×2 mm) from healthy or diseased *Nicotiana tabacum* leaves were immersed in actinomycin D solution (75 μg/ml) for 16 hr. The healthy and diseased samples without actinomycin D were run as controls. All samples were treated with uridine [3]H (specific activity 29.2 Ci/mM) for 5 hr. The samples were washed three times with phosphate buffer 0.15 M, pH 7.2 and left for 48 hr. The samples from the four treatments were prepared for electron microscope as follows: glutaraldehyde and osmic acid fixation gradative dehydration with ketone infiltration and inclusion of the tobacco cells with Epon 82 at 50, 70, and 100%, keeping the specimen at 38, 45, and 60°C during 8, 16, 24 hr., respectively. The ultrathin sections (60–90 nm) obtained in Porter-Blum ultramicrotome were put on copper grids (260 mesh) previously covered with collodion and carbon membranes. The grids with thin sections of all treatments were covered with Ilford L4 emulsion after Caro and Van Tubergen method (Reynolds, 1963), modified by Steves. These grids were kept at 4°C in dried atmosphere for 2–3 months. Grids with samples covered with emulsion were withdrawn at various intervals, developed with Microdol, and fixed with sodium thiosulfate at 22°C. After washing with double-distilled water, they were washed again with acetic acid solution at 4%, 37°C. After gelatin removal the leaf cuts were stained with uranil acetate at 2.5% washed and stained with lead citrate (Nadler, 1971). The stained sections were examined at electron microscope Elmiskop 1A-Siemens; electron photomicrographs of each treatment were obtained and used to make statistical analysis by the Nadler (1971) method.

RESULTS

The purified preparations by the method described were considered by several kinds of tests to have reasonable purity when on grids. The optical absorption in the range 2200–3000 A had one maximum at $\lambda = 2600$ A and one minimum at $\lambda = 2400$ A. The absorbance relation maximum: minimum was always close to 1.23. In figure 1(I), I show the absorption curve obtained. Serological tests could not distnguish L from S particles.

The purified preparations as examined at electron- microscope had reasonable purity grids. In figure 2 (I, II), I show that L and S particles after purification.

By the experiments to test the infectivity of the two kinds of particles, I showed that L particles were infective, and that the low number of lesions produced by the S particles should be due to contamination with L particles. The results concerning infectivity of the two kinds of particles can be seen in figures 1 (II).

Particles from purified preparations treated with ammonium sulfate produced crystals with needle form and at the same time had some purity grades too. In figure 2 (III), I show the mentioned crystals.

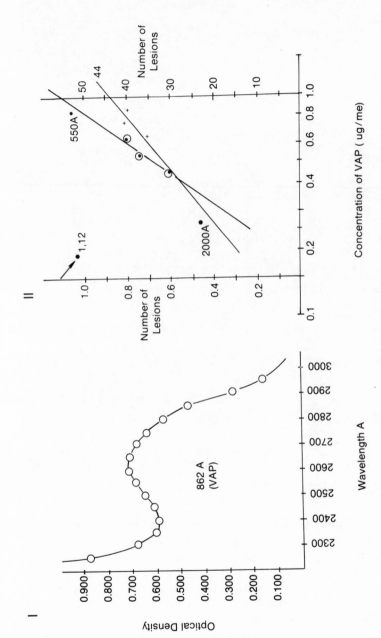

Fig. 1. Absorption spectrum of VAPs purified preparation. II. Specific lesion number on *Chenopodium amaranticolor* for L particles (2000 Å). The S (550 Å) straight line shows results due to L particle contamination.

Fig. 2. Electronphotomicrographs. I and II show L and S particles of VAPs purified preparation. III. Crystals of VAP examined with optical microscope.

The molecular weight estimates of Brazilian VAP particles found by the Paul method were 37×10^6 D and 9×10^6 D for the L and S particles, respectively.

The polypeptides produced by tryptic digestion and detected by the fingerprint method could not reveal differences between polypeptides obtained from L and S particles. Figure 3 (1) contains the common fingerprint obtained with those protein particles.

The spectrum absorption of nucleic acid preparation of L and S particles

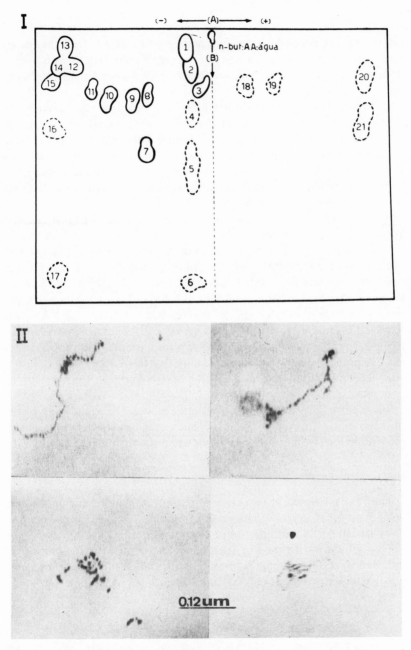

Fig. 3. I. Oligopeptide map obtained with tryptic digestion of VAP protein. II. Electronmicrograph of VAP RNA.

that gave negative serological results had a peak at 2600 A. Only the nucleic acid from L particles was infective. The control inoculations with nucleic acid treated with ribonulcease was inactive, as shown by the infectivity tests.

Pictures of the active nucleic acid prepared for electron microscope by a drop diffusion method contained after 16 hours short particles and many RNA filaments with length estimate about 2,000 nucleotides (fig. 3 [II]). This kind of particle represents more or less one-fourth of the L particles with 7,500 nucleotides, but as far as is known at this time, these RNA filaments should not be similar to the S particle's RNA since only 500 nucleotides are common between its two RNAs.

By experiments in our laboratory using radioautography at electron microscope level, I showed that the RNA from L and S particles should replicate in the nucleolus (fig. 4 [II]). Figure 4 (II) is of silver filaments on nucleus, nucleolus produced by VAP's RNA labeled with uridine-^{3}H. Recently Bisaro and Siegel (1981) showed by using hybridization techniques among cDNA for the two viral RNAs of VAP and five subgenomic particles of the same virus RNA species that the RNA-3, RNA-4, RNA-6, and RNA-7 were able to hybridize to cDNA probes prepared to sRNA. Only RNA-5 was able to hybridize to cDNA probes prepared to lRNA. The L and S genomes by being processed as subgenoic RNA act as mRNA. Bisaro and Siegel inferred that either the sRNA or lRNA contains an overlapping gene set.

Short particles associated with mitochondria are not common at least at the beginning of the infection period, but sometimes they are found in ultrathin sections from leaf tissue, as shown in the figure 4 (I).

DISCUSSION AND CONCLUSIONS

The VAP is formed by two kinds of nucleoprotein (200 and 55 nm) and belongs to tobacco Rattle virus or Tobravirus. The VAP is classified as serotype III, without afinity to I or II serotypes.

Through the purification of diseased tobacco juice with organic solvent followed by differential and sucrose gradient ultracentrifugations, it was possible to get two kinds of VAP particles as inferred from:

1. Some contamination of long particles in purified preparations of the short ones, explaining why some lesions were produced by the S particle preparation. S particle contamination was found in L particle preparation.

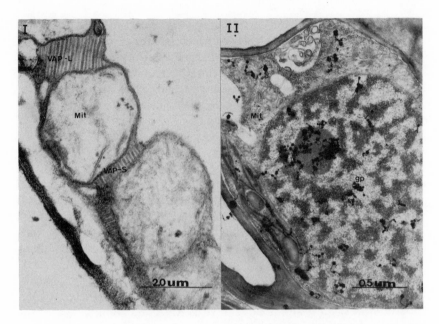

Fig. 4. I. Electronphotomicrographs showing L and S particles associated with mitochondria II. Tobacco cell containing RNA viral with uridine[3]H.

2. The absorbances of the purified preparations were similar and had one minimum at 240 nm and one maximum at 260 nm as expected from nucleoprotein preparations.
3. The obtaining of cystals with needles form, from VAP, is one more evidence of purity of the virus preparation grids.

Comparative research with the S and L purfied preparations included:

1. Serological evidences that the virus protein subunits in S and L particles are the same.
2. That the tryptic digests from S or L proteins give similar fingerprints is more evidence of the similarities of the protein cap of the two kinds of particles.
3. Infectivity differences between the particles; only the L particles were infective.

From these results it is reasonable to accept that the Brazilian VAP belongs to the Tobravirus and is a special system among the multicomponent viruses. The proteins of the tobacco necrosis virus components are not similar as in the VAP. The cowpea mosaic virus on the other side is a multicomponent virus (three components) that requires all the components to be infective, whereas the VAP does not need the S particle to be infective.

Preliminary research concerning the VAP nucleic acid showed that its RNA had a low concentration of uridilic acid and a high concentration of guanilic acid.

By the radioautographic program the electron microscope level carried out with the objective to get information about the virus-RNA replication, it can be concluded that the nucleolus is the incorporation site of tritiated uridine in the virus RNA. These results support that the VAP RNA synthesis is in the nucleolus in contrast to a preliminary hypothesis based on the common association of mitochondrial with viruses seen with the electron microscope. The radioautographic studies substantiate that the S particles were synthesized in the same intracellular places as the L particles.

The mitochondria 5 hr after introduction of labeled uridine showed that at least 78% of silver grains are due to virus RNA and about 22% are produced by the mitochondrial RNA.

Taking into account that:

1. The VAP is relatively easy to purify;
2. The production normally is around 50 mg/kg of tobacco leaf; and
3. The L and S particles contain 2.0 and 7.5 kb and can be cloned after obtaining cDNA through reverse transcriptase (Lage, 1980),

it seems reasonable to conclude that VAP is a good candidate to be used as a gene vector for plant cells.

REFERENCES

Bisaro, M., and A. Siegel. 1981. Viral RNA species in tobacco rattle virus. Abstract. Fifth International Congress of Virology, Strasburg, France.
Brakke, M. K. 1951. Density gradient centrifugation: a new separation technique. J. Amer. Chem. Soc. 73:1847–48.
Cadman, C. H., and B. D. Harrison. 1959. Studies on the properties of soil-borne viruses of the tobacco rattle type occurring in Scotland. Ann. Appl. Biol. 47:542–56.
Caro, L. G., and Van Tubergen. 1962. High resolution autoradiography. I. Methods. J. Cell. Biol. 15:173–88.

Fraenkel-Conrat, H. 1957. Degradation of tobacco mosaic virus with acetic acid. Virology 4:1−4.

Gierer, A., and G. Scharam. 1956. Infectivity of ribonucleic acid from tobacco mosaic virus. Nature 177:702−4.

Harrison, B. D., and D. J. Robinson. 1978. The Tobravirus. *In* M. A. Lauffer, F. B. Bang, K. Maramorosh, and K. S. Smith (eds.), Advances in virus research, 23:25−77. Academic Press, New York.

Kitajima, E. W. 1967. Morfologia das particulas do virus do anel do pimentao e ultraestructura dos tecidos infectados. Ph.D. diss., ESALQ/USP. 54 pp.

Lage, G. 1980. Autoradiografia quantitativa a nivel de microscopia electronica em celulas de fumo (*Nicotiana tabacum* L.) infectadas com o virus do anel do pimentao. M.S. thesis, ESALQ/Univ. Sao Paulo, Piracicaba, S.P., Brasil. 108 pp.

Lister, R. M. 1966. Possible relationship of virus specific products of tobacco rattle virus infections. Virology 28:350−53.

Lister, R. M. 1967. A symptomatological difference between some unstable and stable variants of pea early browning virus. Virology 31:739−42.

Nadler, N. J. 1971. Interpretation of grains count in electron microscope autoradiography. J. Cell. Biol. 49:877−82.

Oliveira, A. R. 1967. Serologia aplicada ao estudo do virus do anel do pimentao. Ph.D. diss., ESALQ/USP, Piracicaba. 40 pp.

Ouchterlone, O. 1962. Diffusion-in-gel methods for immunological analysis. II Prog. Allergy 6:30−154.

Paul, H. L. 1959. Der Zusammenhang zwischendem Molekulargewicht pflanzlicher viren und der spektrophotometric bestimmbaren Grousse b-v. Z. Naturforsch Band. 14b. 7:432−33.

Reynolds, E. S. H. 1963. The use of lead citrate at high pH as an electron-opaque stain in electron microscopy. J. Cell Biol. 17:208−12.

Rushizky, G. W., and C. A. Knight. 1960. An oligonucleotide mapping procedure and its use in the study of tobacco mosaic virus nucleic acid. Virology 11:237−49.

Silva, D. M., and N. L. Nogueria. 1979. Ocorrencia de particulas pequenas do virus do anel pimentao em becido foliar de fumo. Revista da Agricultura LIV 3:137−38. (Piracicaba.)

Silva, D. M. 1965. Estudos sobre purificacao, composicao e algunas propriedades do virus do anel do pimentao. Ph.D. diss. ESALQ/USP, Piracicaba. 106 pp.

A. P. RUSCHEL AND S. M. T. SAITO

Bioenergetics of Nitrogen Utilization in Symbiotic Systems

18

ABSTRACT

Energy costs of nitrogen fixation on N-incorporation from soil as well as the economy of N_2-fixation efficiency of legumes on the tropics are reviewed and discussed. Studies concerning the efficiency of N-ultization in soybean and *Phaseolus* beans are included. By using an isotope dilution method with ^{15}N as ammonium and nitrate fertilizers, the nitrogen derived from fertilizer (Ndff) and from the soil (Ndfs) was measured. Nitrogen derived from fixation ($NdfN_2$) was measured by comparing nodulated plants with non-nodulated for soybean, and with wheat for *Phaseolus* beans. Distribution of N from different sources (soil, fertilizer, and fixation) was observed in aerial parts of plants, root, pods, and nodules. Plants grown in soil supplemented with ammonium sulphate did not have $NdfN_2$ in root; however, if nitrate was supplied, the roots had $NdfN_2$. Soybean pods always have higher $NdfN_2$ than aerial parts, independent of the N-source of fertilizer used. Forty-seven percent of energy used by nitrogenase is due to H_2-evolution during the N_2-fixation process in *Phaseolus*-beans symbiosis; 80% and 66% of nodule nitrogen of soybeans and *Phaseolus*-beans respectively is derived from N_2-fixation, indicating that 20% to 34% of energy in nodules goes into obtaining N from sources other than fixation. Energy expended in N-utilization by the plant was estimated by the ratio of plant parts to pods as of weight and total-N of plants.

INTRODUCTION

Theoretically, the cost of N_2-fixation is mainly measured as the ATP requirement by nitrogenase, which includes at the same time N_2 and H^+

reduction, diverting 75% and 25% energy costs of nitrogen and hydrogen, respectively (Schrauzer, 1977). The energy expended in the nitogen fixation process, when considering 2 moles of ATP hydrolysed per electron by nitrogenase, is summarized by the following reactions (Pate et al., 1981):

$$N_2 + 6e^- + 12ATP + 8H^+ \quad 2NH^+_4 + 12ADP + 12 \text{ Pi}$$

$$2H_2 + 2e^- + 4ATP \quad H_2 + 4ADP + 4 \text{ Pi}$$

Assuming 38 moles ATP/mol of glucose used and 6.3 moles ATP/mol CO_2 and $P/2e^-$ of 3, the cost of fixation would be 21 ATP/N_2 reduced and 3.3 CO_2/N_2 (reaction 1), as well as 1/3 of that energy in H_2 evolution by nitrogenase, which means 7ATP/H_2 reduced (reaction 2). However, recovery of energy occurs by oxidation of H_2 evolved by nitrogenase, assuming $P/2e^-$ of 3 during nitrogen fixation, indicating that 3/4 of ATP, or 43% of energy, will thus be retained with a final consumption, after H_2 recovery, of 25 ATP/N_2 reduced.

On the other hand, plants expend energy in obtaining nitrogen (NO^-_3) from soil. Following the same calculations and taking into consideration that reduction of nitrate to ammonia requires $P/2e^-$ of 4, then NADPH or NADH equivalent to 25 moles of ATP go into NO^-_3 reduction, which is similar to the energy that goes into N_2-fixation, if H_2 uptake process (Hup^+) is recycling part of the ATP lost during N_2 reduction.

Since costs of N-assimilation from soil or from air are almost the same, some revelant information from recent studies on energy required in symbiotic nitrogen fixation is considered, and data obtained at CENA laboratories of N-utilization in *Phaseolus*-beans and soybeans are discussed. Partitioning of nitrogen from soil, from fertilizer (NH^+_4 and NO^+_3), and from fixation in legume plants is studied. Suggestions are made to enhance biological nitrogen fixation (BNF) in legumes.

ECONOMY OF NITROGEN FIXATION × NITRATE REDUCTION

The controversy over the energy cost of both ways of N-assimilation is, at present, almost cleared up. At the end of the 1960s, several investigators studied the cost of biological N_2-fixation. Many attempts to measure energy consumption were made using respiratory/ATP consumption, or simply through growth or yield parameters. Two different results came from these studies, one reporting little difference in energy of the two processes (Gibson, 1966; Minchin and Pate, 1973); the other, the high loss of carbohydrate (photosynthates) through respiration in nodulated legumes

(Mahon, 1977; Silsbury, 1977). These findings were reinforced by the Australian research group (Atkins et al., 1978, 1979; Herridge and Pate, 1977; Layzell et al., 1979; Pate and Herridge, 1978; Pate et al., 1977, 1979).

Calculation, on a theoretical basis (Bergerson, 1971) showed that N_2-fixation required 12 ATP per molecule of fixed N_2, i.e., the amount of ATP necessary to reduce nitrate. Later this amount of ATP was confirmed (Ching et al., 1975).

The theorectical approach to the estimation of BNF costs on an ATP basis was recently studied with batch cultures. It was found (Evans et al., 1980) that values much higher than those of nitrate reduction are necessary (20−30 ATP per mole of fixed N_2).

These conclusions are not related to other aspects than net balance of energy consumed. Many factors involving the physiology of the growing legume, which could minimize the effects of the high requirement of N_2-fixation, must be taken into account. All efforts to increase the C supply to the legume (a C_3 plant) have been intensively explored since the early 1930s. However, there are few papers considering the energy balance based on the growth phases of the plant and how this extra C benefits or becomes only another C chain of the vegetative part and, therefore, not enhancing seed yield. An interesting comparison between efficiency of nodulated and N-dependent legumes in an overall functioning system has been made (Neves, 1981). Although CO_2 losses due to respiration are increased in nodulated roots (Silsbury, 1977; Ryle et al., 1978, 1979; Neves et al., 1981; Pate et al., 1979), this effect is partially compensated by the high respiration in below-ground parts of plants supplemented with mineral N, due to a more developed root system, an effect more easily found with vegetative growth (Neves, 1981). Since mineral N can depress respiration in nondulated roots (Mahon, 1979), one has to consider that even for species that have nitrate-reductase also in the roots, this respiratory difference is not pronounced (e.g., in *Pisum sativum*). On the other hand, in plants having the enzyme mainly in the leaves (e.g., *Vigna unguiculata, Glycine max*), the nitrate assimilation can be more efficiently processed, as additional energy can be obtained from photophosphorylation (Ryle et. al., 1977, 1978). Although these findings are relevant to the economy of the growing plant, there is evidence that the extra C and N remain in the vegetative parts of the plant, not being efficiently translocated to the pods (Summerfield et al., 1977; Minchin et al., 1980). In table 1 we show the difference between the CO_2 output per N assimilated in nodulated and nitrate supplemented *Lupin* (at flowering) and *V. unguiculata* (at seed formation). However, little difference between the nodulated and the 200 pm N-supplemented *Vigna* can be

TABLE 1

RELATIVE ECONOMY OF C AND N IN NODULATED AND N-SUPPLEMENTED
LEGUMES DURING VEGETATIVE AND REPRODUCTIVE STAGES

	FLOWERING*			GRAIN RIPENING†		
	NODULATED		NON-NODULATED (5mM KNO₃)	NODULATED		NON-NODULATED 200 ug N/ml
	Root	Nodule	Root	Root	Nodule	Root
1. Estimated CO_2 output						
N_2 assimilation (mg C/plant)	-	299	-	1504	2165	-
NO_3 reduction (mg C/plant)	-	-	170	-	-	1939
2. Losses of CO_2 (respiration and growth (mg C/plant))	702	55	454	12179	2686	11750
3. Belowground parts						
Mesured CO_2 output (mg C/plant)	-	1056.0‡	624	15848	4851	13689
N assimilated (mg N/plant)	-	103.0‡	76.9	-	1989	3024
4. CO_2 output per unit N assimilated (mg C/mg N)	-	10.3‡	8.1	8.0	2.4	4.5
5. Net photosynthate consumed to N from rooting medium (mgC:mgN) ratio	-	28.6‡	32.2	-	-	-
6. Seed weight (g/plant)	-	-	-	31.5	-	54.0
7. Seed weight: total plant dry weight ratio	-	-	-	0.44	-	0.48
8. Seed weight: pod weight ratio	-	-	-	3.6	-	4.1

* Measured in *Lupinus albus* L. cv Ultra, *Rhizobium* W U 425 (Pate et al., 1979).
† Measured in *Vigna unguiculata* cv TVu 150 3, *Rhizobium* R 5028 (Neves et al., 1981).
‡ Root + nodule data

observed when comparing seed and total plant dry weight ratio, corresponding to 0.44 and 0.48, respectively, in the nodulated and non-nodulated N-dependent plants. This suggests N limitation on yield rather than C starvation.

EFFICIENCY OF N_2-FIXATION

N_2-fixation is a biological process with a high requirement of energy, and, therefore, loss of energy should be minimized to enhance its efficiency. A survey (Schubert and Evans, 1976) among many legume-*Rhizobium* associations revealed that few associations are capable of processing N_2 reduction without any diversion of energy, followed by high plant growth or high plant nitrogen content, or both (Carter et al., 1978; Albrecht et al., 1979). Results indicate that hydrogen evolution, by nitrogenase, is a major factor affecting the efficiency of nitrogen fixation by many legumes, including soybean and pea, which could divert 40–60% of the electron flow to the air. It was also demonstrated that some-leguminous plants, such as *Alnus rubra*, and other legumes, such as *Vigna unguiculata*, were apparently capable of reutilizing the evolved H_2 through a recycling mechanism. The Hup^+ was demonstrated to be associated with a unidirectional reversible hydrogenase, not dependent on ATP and found in some nodule-bacteria (Dixon, 1972). An extra gain of 2–3 ATP was shown to be associated with ths physiological process (Dixon, 1972, 1978). Considering that the average theoretical cost of BNF is 28 ATP (Pate et al., 1981), this could represent an economy of around 10–15% of energy from the process. The ability to synthesize the hydrogenase system was shown to be associated with the fixing organism and is considered a desirable characteristic for strain selection (Albrecht et al., 1979).

Tropical Legumes

In most tropical legumes, there is an efficient system of storing fixed N, not frequently found in temperate legumes, that is associated with the N transport in the plant. In the first group, the solutes from nodules are exported in great proportion as ureides and also as amides, amino acids, and small peptides (Minchin and Pate, 1973; Neves et al., 1981; Matsumoto et al., 1977). Ureides, represented by allantoin or allantoic acid, have a low C:N ratio, i.e., each molecule can transport 4 atoms of N, whereas an amide can transport only 2. This high economy of C might represent an increase in efficiency of 50% in terms of carbon for agronomically important legumes, such as *V. ungiculata, G. max,* and *Phaseolus vulgaris* (Minchen et al.,

1980; Kushizaki et al., 1964; Cookson et al., 1980) in relation to legumes exporting mainly amide, such as *P. sativum* and *Lypinus albus* (Minchin and Pate, 1973; Pate and Herridge, 1977). This transport, however, was proved to be highly dependent on an efficient supply of water because of the low solubility of ureides. The high transportation rate found in tropical legumes could be responsible for the efficient transport of ureides in the plant (Goi and Neves, 1980).

Respiration in Legumes

The efficiency of C and N metabolism can be measured using both parameters. Factors affecting each or both metabolisms may alter the C:N relationship and affect the whole plant energy balance. Some inefficient systems normally include a high respiratory loss of carbon under adverse environmental conditions, such as high temperature (Minchen et al., 1980; Halliday, 1975). This could perhaps be a major factor in the high respiratory cost of N_2-fixation found in *P. vulgaris* (Mahon, 1977; Tjepkema and Winship, 1980). Another desirable characteristic is the carboxylation process by PEP carboxylase in the nodules, which could partially recapture the CO_2 lost through respiration as found in *P. vulgaris, L. albus,* and *Vicia faba* nodules (Jackson and Coleman, 1959; Lawrie and Wheeler, 1975; Christeller et al., 1977).

There is a lack of information, especially from tropical regions where the system has a high potential for energy use. In these cases research should aim at better combinations, for instance, of plant genotypes and *Rhizobium* strains more adapted to adverse environments (high temperature and problems related to soil acidity). Such information is of vital importance to any program with the objective of improving legume-*Rhizobium* association.

DISCUSSION

In a study (Phillips et al., 1981) of the carbon and nitrogen limitations of symbiotically grown soybean seedlings, using three levels of mineral nitrogen (NH_4NO_3) to assay the efficiency of N-utilization and apparent photosynthesis, it was concluded that these seedlings were primarily N-limited rather than C-limited and that any calculation of the energy retained could be in error if the growth period had not been considered. Such conclusion indicates that, in the field, soil-N and fertilizer have a very important effect on symbiotically fixed nitrogen. Experiments to study the distribution of N derived from fertilizer (Ndff), from soil (Ndfs), and from fixation ($NdfN_2$)

in soybean using the isotope dilution method and $(^{15}NH_4)_2SO_4$ have been carried out (Ruschel et al., 1979). It was observed that the roots did not contain nitrogen from fixation, and it was assumed that fixed nitrogen was primarily distributed to the aerial parts of the nodulated plants. Nitrogen in the pods had 68% $NdfN_2$ and indicated the importance of fixation in the supply of seed protein (table 2). However, when NO_3 $(K^{15}NO_3)$ was applied (Ruschel and Azevedo, 1981) to field-grown soybeans, high Ndff in roots as well as in other parts of the plant was noticed. The same effect of NH^+_4 and NO^-_3 on nitrogen partitioning was observed in a greenhouse experiment (Saito et al., 1980) (table 3). The reasons for this effect are not understood since plants grown on NH^+_4 had Ndfs and Ndff as well as from fixation in shoots, pods, and nodules. Further research needs to be done to study the effect of sulphate and discrimination of fixed nitrogen in roots grown with ammonium sulphate. In soybean the $NdfN_2$ in pods is always greater than in aerial parts independent of N-source used (nitrate or ammonium), which seems to indicate that fixed nitrogen is transferred right away from nodules to pods. This is not the case in bean (table 4) since aerial parts and pods have the same level of $NdfN_2$.

TABLE 2

Distribution of N Derived From Soil (Ndfs), Fertilizer (Ndff),
and Fixation ($NdfN_2$) in Non-nodulated and Nodulated
Soybeans grown under Greenhouse Conditions with 25 kg N/ha ($^{15}NH_4$).

	Shoots	Roots	Pods	Nodules
Non-nodulated				
%Ndfs	83.1	87.5	84.2	
%Ndff	16.9	12.5	15.8	
%NdfN$_2$	0	0	0	
Nodulated				
%Ndfs	62.3	87.2	26.1	16.3
%Ndff	10.3	12.8	4.9	2.4
%NdfN$_2$	27.4	0	69.0	81.3

Source: Ruschel et al., 1979

It is interesting to note in tables 2, 3, and 4 that 80% of the nitrogen in soybean nodules and 66% in *Phaseolus* nodules is derived from N_2-fixation, indicating that 20% and 34%, respectively, of the energy used in the nodules originated from sources other than fixation. This again indicates the importance of soil nitrogen on N_2-fixation in bean and soybean. The differences obtained in both species studied indicate that energy used

TABLE 3

DISTRIBUTION OF N DERIVED FROM SOIL (Ndfs), FERTILIZER (Ndff) AND FIXATION
(NdfN₂) IN NON-NODULATED AND NODULATED SOYBEANS GROWN UNDER
FIELD CONDITIONS WITH 20 kg N/ha (K^{15}NO₃ Azevedo, 1981).

	Shoots	Roots	Pods	Nodules
Non-nodulated				
%Ndfs	77.8	79.5	80.4	
%Ndff	22.1	20.8	19.6	
%NdfN₂	0	0	0	
Nodulated				
%Ndfs	36.7	38.6	22.3	15.3
%Ndff	10.4	10.6	5.3	4.1
%NdfN₂	52.8	50.2	73.0	80.6

SOURCE: Ruschel and Azevedo, 1981.

TABLE 4

DISTRIBUTION OF N DERIVED FROM SOIL (Ndfs)
FERTILIZER (Ndff), AND FIXATION (NdfN₂)
IN *PHASEOLUS* BEANS GROWN IN SOIL AMENDED WITH
20 kg N/ha AS (^{15}NH₄)₂SO₄ AND K^{15}NO₃

	Shoots	Roots	Pods	Nodule
(^{15}NH₄)₂SO₄				
%Ndfs	22.2	95.53	22.3	42.8
%Ndff	1.9	4.47	1.9	1.48
%NdfN₂	75.9	-	75.8	55.8
K^{15}NO₃				
%Ndfs	25.34	56.85	20.46	32.03
%Ndff	2.36	5.55	2.16	1.63
%NdfN₂	72.3	37.6	77.38	66.34

SOURCE: Saito et al., 1981.

in fixation by *Phaseolus*-beans nodules are greater than in soybean nodules. This can be explained by the high energy spent in nitrogenase in
hydrogen reduction in *Phaseolus* nodules (Saito et al., 1981) (table 5), indicating that 47% goes into H₂-evolution in *Phaseolus* nodules during whole
plant growth (Bethlenfalvay and Phillips, 1977). The studies also indicate
that Hup$^+$ is very low in *Rhizobium phaseoli–Phaseolus* symbiosis, and
that PEP carboxylase is probably not very active in nodules.

By studying the two ratios (weight of aerial parts or roots to pods and
total-N of roots to pods), the energy of N-utilization in nodulated and non-
nodulated soybeans (table 6) grown in field soil amended with 20 kgN/ha

TABLE 5

ENERGY USED IN *PHASEOLUS - RHIZOBIUM*
SYMBIOSIS (SAITO ET AL., 1980).

N_2 - fixed (μmoles N/g nodules.h)	13.99
H_2 - evolved (μmoles H_2/g nodules.h)	37.43
Electron cost of nitrogenase ($3N_2 + H_2$)	79.40
% Energy of N_2-fixation	52.8
% Energy of H_2-evolution	47.2

SOURCE: Saito et al., 1980.

TABLE 6

ENERGY EXPENDED IN YIELD OF NON-NODULATED AND
NODULATED SOYBEAN (IN VARIOUS PLANT PARTS)

Part of Plant	Plant Weight kg/ha	Total-N kg/ha	YIELD ENERGY RATIO: AERIAL PARTS (OR ROOTS)/PODS	
			Weight	N-total
Non-nodulated				
Aerial	3,524	36.5	9.0	2.2
Root	612	3.1	1.6	0.2
Pods	390	16.5	1.0	1.0
Nodulated				
Aerial	4,444	97.0	4.3	2.0
Root	682	5.3	0.6	0.1
Pods	1,031	47.5	1.0	1.0

labeled with ^{15}N was observed. The accumulation of carbon is aerial parts of non-nodulated plants was higher than in nodulated plants, but the ratio for total nitrogen being the same for both plants studied, confirms results already discussed for nodulated soybean seedlings (Gibson, 1966) and suggesting that non-nodulated plants were N-limited rather than C-limited.

Improvement of symbiosis, aiming at a better energy efficiency could be achieved, (1) by using *R. phaseoli* strains capable of recycling H_2 evolved by nitrogenase; (2) efficient and competitive strains to match the legume variety used; (3) varieties that can have both good nitrate reductase and nitrogenase; (4) varieties with high seed N-utilization; (5) varieties with PEP carboxylase in nodules that could recycle respiratory CO_2.

REFERENCES

Albrecht, S. L., R. J. Maier, F. J. Hanus, S. A. Russel, D. W. Emerich, and H. J. Evans. 1979. Hydrogenase in *Rhizobium japonicum* increases nitrogen fixation by nodulated soybeans. Science 203:1255–57.

Atkins, C. A., D. F. Herridge, and J. S. Pate. 1978. The economy of carbon and nitrogen in nitrogen-fixing annual legumes. Experimental observations and theoretical observations. *In* Isotopes in biological dinitrogen fixation, IAEA, Vienna, pp. 211–42.

Atkins, C. A., J. S. Pate, and D. B. Layzell. 1979. Assimilation and transport of nitrogen in nonnodulated (NO_3-grown) *Lupinus albus* L. Plant Physiol. 64:1078–82.

Bergersen, F. J. 1971. The central reactions of nitrogen fixation. Plant Soil, Special Volume, pp. 511–23.

Bethlenfalvay, G. F., and D. A. Phillips. 1977. Photosynthesis and symbiotic nitrogen fixation in *Phaseolus vulgaris* L. *In* A. Hollaender et al.(eds.), Genetic engineering for nitrogen fixation, pp. 401-9. Plenum Press, New York.

Carter, K. R., N. T. Jennings, J. Hanus, and H. J. Evans. 1978. Hydrogen evolution and uptake by nodules of soybeans inoculated with different strains of *Rhizobium japonicum*. Can. J. Microb. 24:307–11.

Ching, T. M., S. Hedtke, S. A. Russel, and H. J. Evans. 1975. Energy state and dinitrogen fixation in soybean nodules of dark-grown plants. Plant Physiol. 55:796–98.

Cookson, C., H. Hughes, and J. Coombs. 1980. Effects of combined nitrogen on anapleurotic carbon assimilation and bleeding sap composition in *Phaseolus vulgaris* L. Planta. 148:338–45.

Christeller, J. T., W. A. Laing, and W. D. Sutton. 1977. Carbon dioxide fixation by lupin root nodules. Plant Physiol. 60:47–50.

Dixon, R. O. D. 1972. Hydrogen in legume root nodule bacteroids: occurrence and properties. Arch. Microb. 85:195–201.

Dixon, R. O. D. 1978. Nitrogenase-hydrogenase interrelationships in rhizobia. Biochimie 60:233–36.

Evans, H. J., D. W. Emerich, T. Ruiz-Argueso, R. J. Maier, and S. L. Albrecht. 1980. Hydrogen metabolism in the legume-*Rhizobium* symbiosis. *In* W. E. Newton and W. H. Orme-Johnson (eds.), Nitrogen fixation, 2:69–86. University Park Press, Baltimore.

Gibson, A. H. 1966. The carbohydrate requirements of symbiotic nitrogen fixation: a "whole-plant" growth analysis approach. Aust. J. Biol. Sci. 19:499–515.

Gibson, A. H. 1976. The energy requirements of symbiotic nitrogen fixation. *In* W. J. Broughton and C. K. John (eds.), Proceedings of soil microbiology and plant nutrition symposium, pp. 1–14. University of Malaya Press.

Goi, S. R., M. V. P. Neves, and J. Dobereiner. 1980. Teor de ureidos, tipo de nodulo e atividade da nitrogenase de leguminosa forrageiras, florestais e de graos. *In* X reunion Latinoamericana de *Rhizobium*, September, Maracay, Venezuela, pp. 299–312.

Halliday, J. 1975. An interpretation of seasonal and short term fluctuations in nitrogen fixation. Ph.D. thesis, University of Western Australia. 141 pp.

Herridge, D. F., and J. S. Pate. 1977. Utilization of net photosynthate for nitrogen fixation and protein production in an annual legume. Plant Physiol. 60:759–64.

Jackson, W. A., and N. T. Coleman. 1959. Fixation of dioxide by plant root through phosphoenolpyruvate carboxylase. Plant Soil 11:1–16.

Kushizaki, M., J. Ishizuka, and F. Akamatsu. 1964. Physiological studies on the nutrition of soybean plants. 2. Effects of nodule formation on nitrogenous constituents of soybeans. J. Sci. Soil Manure 35:323–27.

Lawrie, A. C., and C. T. Wheeler. 1975. Nitrogen fixation in the root nodules of *Vicia faba* L. in relation to the assimilation of carbon. II. The dark fixation of carbon dioxide. New Phytol. 74:437–45.

Layzell, D. B., R. M. Rainbird, C. A. Atkins, and J. S. Pate. 1979. Economy of photosynthate use in nitrogen-fixing legume nodules. Plant Physiol. 64:888–91.

Mahon, J. D. 1977. Respiration and energy requirement for nitrogen fixation in nodulated pea roots. Plant Physiol. 60:817–21.

Mahon, J. D. 1979. Environmental and genotypic effects on the respiration associated with symbiotic nitrogen fixation in peas. Plant Physiol. 63:892–97.

Matsumoto, T., M. Yatazawa and Y. Yamamoto. 1977. Incorporation of ^{15}N into allantoin in nodulated soybean plants supplied with ^{15}N$_2$. Plant Cell Physiol. 18:459–62.

Minchin, F. R., and J. S. Pate. 1973. The carbon balance of a legume and the functional economy of its root nodules. J. Exp. Bot. 24:259–71.

Minchin, F. R., R. J. Summerfield, and M. C. P. Neves. 1980. Carbon metabolism, nitrogen assimilation and seed yield of cowpea (*Vigna unguiculata* L. Walp.) grown in an adverse temperature regime. J. Exp. Bot. 31:1327–45.

Neves, M. C. P. 1981. Interdependencia fisiologica entre os componentes do sistema simbiotico *Rhizobium*-Leguminosas. R. Bras. Ci. Solo, 5:79–22.

Neves, M. C. P., F. R. Minchin, and R. J. Summerfield. 1981. Carbon metabolism, nitrogen assimilation and seed yield of cowpea (*Vigna unguicalata*) plant dependent on nitrate-nitrogen or one of two strains of *Rhizobium*. Trop. Agric. 58:115–32.

Pate, J. S., D. B. Layzell, and C. A. Atkins. 1979. Economy of carbon and nitrogen in a nodulated and nonnodulated (NO$_3$-grown) legume. Plant Physiol. 64:1083–88.

Pate, J. S., and D. F. Herridge. 1978. Partitioning and utilization of net photosynthate in a nodulated annual legume. J. Exp. Bot. 29:401–12.

Pate, J. S., D. B. Layzell, and D. L. McNeil. 1977. Modelling the transport and utilization of carbon and nitrogen in a nodulated legume. Plant Physiol. 63:730–37.

Pate, J. S., P. J. Sharkey, and C. A. Atkins. 1977. Nutrition of a developing legume fruit functioning economy in terms of carbon, nitrogen, water. Plant Physiol. 59:506–10.

Pate, J. S., C. A. Atkins, and R. M. Rainbird. 1981. Theoretical and experimental costing of nitrogen fixation and related processes in nodules of legumes. *In* A. H. Gibson and W. E. Newton (eds.), Current perspectives in nitrogen fixation, pp. 105–16. Australian Academy of Science, Canberra.

Phillips, D. A., T. M. DeJong and L. E. Williams. 1981. Carbon and nitrogen limitations on symbiotically-grown soybean seedling. In. A. H. Gibson and W. E. Newton (eds.), Current perspectives in nitrogen fixation, pp. 117–20. Australian Academy of Science, Canberra.

Ruschel, A. P., P. B. Vose, R. L. Victoria, and E. Salati. 1979. Comparison of isotope techniques and nonnodulating isolines to study the effect of ammonium fertilizer on dinitrogen fixation in soybean, *Glycine max*. Plant Soil. 53:513–25.

Ruschel, A. P., and M. E. Azevedo. 1981. Contribuicao de N do solo, fertilizante, e fixacao simbiotica na planta e solo cultivado com soja. Reuniao Regional da Soja, Agosto, Londrina-Parana, Brasil.

Ryle, G. J. A., C. E. Powell, and A. J. Gordon. 1977. The respiration costs of nitrogen fixation in soybean, cowpea and white clover. II. Comparisons of the cost of nitrogen fixation and the utilization of combined nitrogen. J. Exp. Bot. 30:145–53.

Ryle, G. J. A., C. E. Powell, and A. J. Gordon. 1978. Effect of source of nitrogen fixation on the growth of fiskeby soybean: the carbon economy of whole plants. Ann. Bot. 42:637–48.

Saito, S. M. T., E. Matsui and E. Salati. 1981. $^{15}N_2$ fixation, H_2 evolution and C_2H_2 reduction relationships in *Phaseolus vulgaris*. Physiol. Plant 49:37–42.

Saito, S. M. T., R. L. Victoria, and A. P. Ruschel. 1981. Avaliacao da fixacao simbiotica de N_2 em *Phaseolus vulgaris* por diluicao isotopica com $^{15}NH_4+$ e $^{15}N_3$. *In* Regional Seminar on Use of Nuclear Techniques in Soil-Plant-Atmosphere Relationship Studies, April, Piracicaba-SP, Brasil.

Schrauzer, G. M. 1977. Nitrogenase model systems and the mechanism of biological nitrogen reduction. *In* W. E. Newton, J. R. Postgate, and C. Rodrigues-Barrueco (eds.), Recent developments in nitrogen fixation, pp. 109-19. Academic Press, New York.

Schubert, K. R., and H. J. Evans. 1976. Hydrogen evolution: a major factor affecting the efficiency of nitrogen fixation in nodulated symbionts. Proc. Natl. Acad. Sci. USA 73:1207–11.

Silsbury, J. H. 1977. Energy requirments for symbiotic nitrogen fixation. Nature 267:149–50.

Summerfield, R. J., P. J. Dart, P. A. Huxley, A. R. J. Eaglesham, F. R. Minchin, and J. M. Day. 1977. Nitrogen nutrition of cowpea (*Vigna unguiculata*). I. Effects of applied nitrogen and symbiotic nitrogen fixation on growth and seed yield. Expl. Agric. 13:129–42.

Tjepkeme, T. D., and L. J. Winship. 1980. Energy requirement for nitrogen fixation in actinorhizal and legume root. Science 209:279–81.

R. J. WHITAKER

Opportunities for Genetic Engineering in *Pseudomonas*

19

INTRODUCTION

For many years microbiologists have marveled at the biochemical versatility and vast diversity of bacterial species belonging to the *Pseudomonadaceae*. The taxonomy of this extremely heterogeneous group of microorganisms historically has been limited to morphological and nutritional analysis (Doudoroff and Palleroni, 1974), but more recent classifications have been based on nucleic acid hybridization studies and the enzymological patterning of aromatic amino acid biosynthesis (Palleroni et al., 1973; Byng et al., 1980; Whitaker et al., 1981a, b). A summary of the current phylogenetic organization of pseudomonad species can be found in table 1. Although each of the five major pseudomonad homology groups contain enormously interesting and important species in terms of genetic and biochemical potential as well as pathogenicity to both plants and animals, certainly the fluorescent species of Group Ib (table 1) are the best known and most broadly studied members of the *Pseudomonadaceae*. *Pseudomonas aeruginosa* has been the focus of much attention in the areas of clinical medicine and biochemical genetics (Jacoby, 1979), whereas *P. putida* is renowned for its plasmid-encoded biochemical versatility (Chakrabarty, 1976). *P. syringae* represents a number of pathovars (table 1) that are of vital importance to the agriculture industry.

The various methodologies that fall under the umbrella of genetic engineering have long been viewed with a sense of enthusiasm for their potential applications to biochemical and genetic research at both the basic and applied levels. Indeed, the traditional approaches of auxotrophic mutant

TABLE 1

PHYLOGENY OF PSEUDOMONAD SPECIES

REPRESENTATIVE SPECIES[a]	RRNA HOMOLOGY GROUP[b]	TYROSINE PATTERNING GROUP[c]	PHENYLALANINE PATTERNING GROUP[d]
P. stutzeri P. mendocina P. alcaligenes			Ia

P. aeruginosa
P. putida
P. fluorescens
P. syringae
  P. coronafaciens
  P. atrofaciens
  P. glycinea
  P. morsporunorum
  P. mori
  P. atptata
  P. helianthi
  P. savastanoi
  P. tomato
  P. tabaci
P. cichorii
	I	I	Ib

P. cepacia P. marginata P. mallei P. caryolphlii	II	II	IIa
P. pickettii P. solanacearum			IIb
P. acidovorans P. testosteroni			IIIa
P. delafieldii P. facilis	III	III	IIIb
P. saccharophila			IIIc
P. diminuta P. vesicularis	IV	IV	IV
P. maltophilia X. campestris	V	V	V

[a]Some of the pathovars of P. syringae are listed separately.
[b]Palleroni et al., 1973.
[c]Byng et al., 1980.
[d]Whitaker et al., 1981a, b.

isolation and amino acid analog resistance have proved extremely valuable for defining precise gene-enzyme relationships and patterns of regulatory control in a number of microbial species. In a similar fashion, great strides have been gained in the areas of structural and functional analysis of the genome, the regulation of transcription and translation, selective mutagenesis, complementation analysis, and DNA sequencing as a direct result of molecular cloning. In this way segments of DNA can be spliced onto small, autonomously replicating plasmids and studied in great detail. However, the use of recombinant DNA as a tool for probing many of the basic biological functions of bacterial species has been somewhat limited to *Escherichia coli* K-12 because of the lack of suitable vector systems and limited availability of genetic markers in other species.

This chapter will specifically address itself to the potential applications of the methodologies of genetic engineering for unraveling a few of the most interesting research problems currently under investigation in pseudomonad bacteria. Topics to be considered include: metabolic versatility and the biodegradation of unusual organic compounds, the regulation of complex biosynthetic pathways, and the construction of transformation vectors to be used for studying phytopathogenicity and potentially for plant cell transformation. Each of these areas of pseudomonad research is significant along two fronts: first, a more complete definition of the biochemical and genetic interactions of *Pseudomonas* species with their environment; and second, the potential commercial applications that may be derived from the basic research. Since the fluorescent species of *Pseudomonas* (Group Ib) have been studied in the most detail, these species will be the focus of attention. However, where possible, examples will be cited from other pseudomonad groups.

METABOLIC DIVERSITY AND THE BIODEGRADATION OF UNUSUAL ORGANIC COMPOUNDS

Pseudomonads are notorious for the wide range of unusual and complex organic compounds they can use as sources of carbon and energy. A great deal of the biochemical versatility found in pseudomonads is encoded by the genophore, but the presence of degradative plasmids greatly enhances their metabolic capabilities (Gunsalus et al., 1975). Each type of degradative plasmid possesses a set of genes coding the biodegradation of a specific organic substrate (table 2; Chakrabarty, 1976). Thus camphor is metabolized by a pathway of 15–20 inducible enzymes coded by the self-transmissible CAM plasmid (Palchaudhuri, 1977). The exotic array of or-

TABLE 2

Degradative Plasmids of Pseudomonas

Plasmid	Degradative Pathway	Native Strain[a]	Size[b]	Reference
SAL	Salicylate	P. putida R_1	40,55	Chakrabarty, 1972
CAM	Camphor	P. putida PpG1	150	Palchaudhuri 1977;
				Rheinwald et al., 1973
NAH	Napthalene	P. putida PpG7		Dunn and Gunsalus, 1973
OCT	Alkanes	P. oleovorans		Chakrabarty et al., 1973
XYL	Xylene	Pseudomonas Pxy	10	Friello et al., 1976
TOL	Xylene	P. putida (arvilla)	75	Worsey and Williams, 1975
pJP1	2,4-Dichlorophenoxyacetic acid	A. paradoxus	58	Fisher et al., 1981
pAC25	3-Chlorobenzoic acid	P. cepacia	68	Chatterjee et al., 1981
pAC27	4-Chlorobenzoic acid	P. cepacia	59	Karns et al., 1981
pAC31	3,5-Dichlorobenzoic acid	P. cepacia	59	Karns et al., 1981

[a] Abbreviations: P= Pseudomonas; A = Alcaligenes. A. paradoxus has been placed by rRNA-DNA homology and aromatic biosynthetic patterning as a Group III pseudomonad (Table 1, Byng et. al. 1983).

[b] Millions of daltons.

ganic substrates metabolized by pseudomonads is indicative of the broad spectrum of enzyme species and substrate specificities available in these organisms.

The multitude of enzymatic reactions routinely performed by pseudomonads has potentially far-reaching implications for the chemical industry. Enzymes can be used to replace chemical reactions currently requiring expensive synthetic catalysts. In the most simple circumstance, one would identify the desired enzyme activity and using the current recombinant DNA technology clone the segment of DNA (either chromosomal or plasmid) coding that activity on a high copy number plasmid (fig. 1). This plasmid could then be used to transform a selected bacterial host that, because of the high copy number of the plasmid, would produce dramatically elevated levels of the desired enzyme. The enzyme can then be purified and used as a reagent in a conversional process or possibly immobilized on a chromatography resin and used to catalyze substrates as they are passed through a column. It must be noted that this scenario is most likely oversimplified and one may well need to modify the substrate specificity or other enzymatic parameters in order to obtain the necessary reaction. However, the isolation of the DNA coding a particular enzymatic activity on a plasmid allows a great deal of latitude in the selection of altered enzyme forms. In any event, the use of enzymes for industrial applications represents a vital aspect of biotechnology, and because of their metabolic versatility, *Pseudomonas* species can be viewed as a valuable natural resource.

Another area that takes advantage of the metabolic potential inherent to *Pseudomonas* species is the construction of strains for the dissimilation of toxic chemical pollutants (Saunders, 1977). The natural occurrence of degradative plasmids in pseudomonads offers a pool of genetic information that can be recombined so as to facilitate the breakdown of specific recalcitrant pollutants. One can certainly envision the construction of hybrid degradative plasmids by gene-splicing techniques. In this process, specific segments of DNA from one plasmid could be spliced with DNA fragments from another degradative plasmid. In this way a specific enzyme activity such as aromatic oxidase could be teamed with another enzyme or set of enzymes to yield a bacterial species with modified metabolic capabilities.

However, an intuitively more simple alternative is available. Most degradative plasmids belong to different incompatibility groups so that specific combinations of different degradative plasmids can be introduced into a single host cell. This approach, based on principles of plasmid evolution and interaction, has been termed "plasmid assisted molecular breeding"

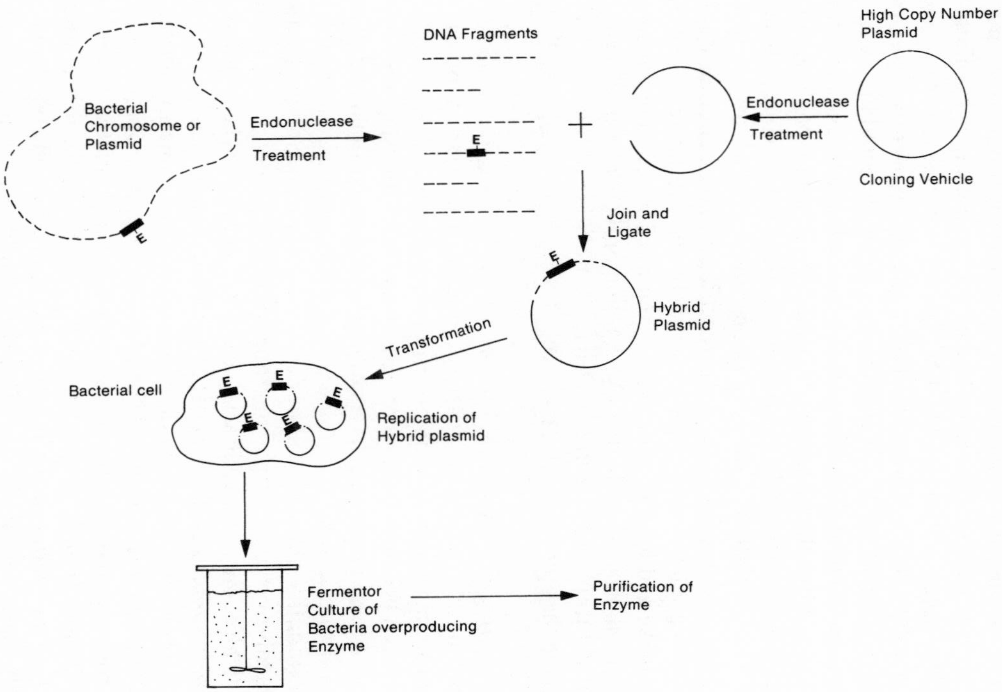

Fig. 1. Molecular cloning for enzyme overproduction. The letter E represents the gene coding enzyme E. See text for explanation.

(Karns et al., 1981; Kellogg et al., 1981). The extension of the natural substrate range of a bacterium by introducing a combination of degradative
plasmids into a single microorganism has resulted in the breeding of a pure
culture of *P. cepacia* that can utilize the problematic chemical pollutant
2,4,5-trichlorophenoxyacetic acid (2,4,5-T) as a sole source of carbon and
energy (Karns et al., 1981). Essentially this culture of *P. cepacia* was created
by combining plasmid pAC25, which encodes a complete and highly specific 3-chlorobenzoate degradative pathway in a cell with a TOL plasmid
that provides a substrate-ambiguous benzoate oxygenase capable of converting 4-chlorobenzoate to 4-chlorocatechol. The simultaneous presence
of TOL and pAC25 allowed the cells to degrade 4-chlorobenzoate to 4-
chlorocatechol by the action of the TOL-specified benzoate oxygenase followed by the degradation of 4-chlorocatechol by the enzymes encoded by
pAC25 (Reineke and Knacknauss, 1980). Continued selection in the presence of other chlorinated aromatics like 3,5-dichlorobenzoate and eventually 2,4,5-trichlorophenoxyacetic acid led to the isolation of cultures capable of utilizing these compounds as sources of carbon and energy (Kellogg
et al., 1981).

The engineering of new plasmid forms by plasmid-assisted molecular
breeding potentially could yield a variety of pseudomonad strains possessing specific capabilities for dissimulation of toxic pollutants. The feasibility
of this approach becomes even more realistic and exciting when one considers the wealth of biochemical versatility (both chromosomal and plasmid borne) already described in *Pseudomonas* and the likelihood that additional degradative plasmids will be described in this exceedingly diverse
and continually evolving group of microorganisms. It is interesting to note
that a 2,4-dichlorophenoxyacetic acid-degrading plasmid has been isolated
in *Alcaligenes paradoxus* (Fisher et al., 1978). The commercial benefits of
molecularly-bred plasmids for the detoxification of pollutants are substantial, but this area of research has also helped support some proposed mechanisms of bacterial and plasmid evolution. Various degradative plasmids
have considerable homology with one another, pointing to their natural
evolution by gene recruitment from other plasmids (Farrell and Chakrabarty, 1979). The isolation of a 2,4,5-T–utilizing microorganisms explicitly
demonstrates gene recruitment as the benzoate oxygenase of the TOL
plasmid was recruited to initiate the degradation of 2,4,5-T. In essence, the
2,4,5-T–utilizing bacterium is an example of "forced" evolution in that this
microorganism was constructed by taking advantage of the natural genetic
diversity present in *Pseudomonas* and the interaction of these genetic elements under strong selective pressure.

ELUCIDATION OF THE ROUTE AND
REGULATION OF BIOCHEMICAL PATHWAYS

It is becoming increasingly clear that the presence of separate enzymatic sequences leading to a major metabolite may be more common in nature than previously suspected. Indeed, alternative routes of isoleucine biosynthesis are known to operate in *E. coli* (Abramsky and Shemin, 1965; Phillips et al., 1972), *Leptospire* (Charon et al., 1974), and *Saccharomyces cerevisiae* (Vollbrecht, 1974). Additionally, biochemical alternatives in the biosynthesis of homocysteine (Datko et al., 1974), serine (Heptenstall and Quayle, 1970), cysteine (Pasyewski and Grabski, 1975), and methionine (Pasyewski and Grabski, 1975) have been reported. In no biochemical system has the simultaneous presence of alternative biosynthetic routes been more extensively studied than in the synthesis of aromatic amino acids in *Pseudomonas* (Byng et al., 1983). In addition to the *E. coli*-like phenylpyruvate and 4-hydroxyphenylpyruvate routes to *L*-phenylalanine and *L*-tyrosine, respectively, *Pseudomonas* species of Groups Ia, IIa, IIb, IIIb, IIIc, and V (table 1) simultaneously possess the *L*-arogenate pathway to both aromatic amino acids (fig. 2). The presence of this alternative pathway is presumed to be at least in part responsible for the native resistance of these species to growth inhibition by aromatic amino acid analogs that have been successfully employed in the selection of regulatory mutants in other bacteria (Fowden et al., 1967 and Cohen, 1965). Similarly auxotrophic mutant isolation is severely hampered because a single mutation in one pathway can be masked effectively by the presence of the alternative route.

The presence of the alternative arogenate route raises several questions as pertains to gene: enzyme relationships and carbon flow in aromatic biosynthesis. The aromatic pathway in *Pseudomonas* takes on additional significance in that it not only carries out the synthesis of the aromatic amino acids (phenylalanine, tyrosine, and tryptophan) but it is also the source of many potentially useful secondary metabolites. Pyrrolnitrin and a number of phenazine compounds have been implicated for their antibiotic activity (see Leisinger and Margraff, 1979). Therefore, the elucidation of the route and regulation of the aromatic pathway is important for both basic and applied science.

Recently the challenge of isolating regulatory mutants of the aromatic pathway in *P. aeruginosa* was met by the approach of limiting carbon flow into the aromatic pathway (Fiske et al., 1983). Presumably the presence of multiple flow routes in the aromatic pathway allows sufficiently large amino acid pools to account for the observed resistance to the antimetabo-

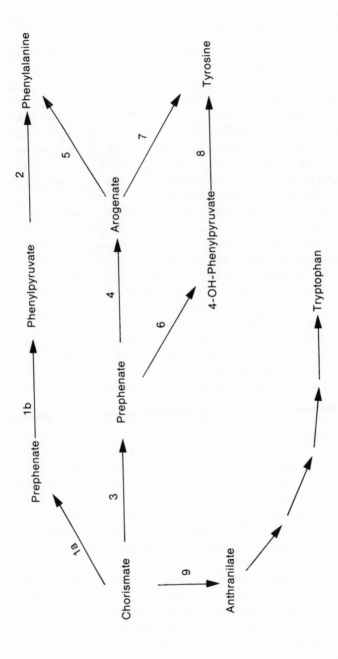

Fig. 2. Aromatic amino acid biosynthesis in *Pseudomonas*. All species of Groups, Ib, IIa, IIb, IIIb, IIIc, and V (table 1) possess each enzyme activity shown. Species of Group Ia lack enzyme 5, as do Group IIIa species. Group IV species lack 1a and 1b activities. Enzymes 1a and 1b, chorismate mutase/-prephenate dehydratase complex; 2, phenylpyruvate aminotransferase; 3, chorismate mutase; 4, prephenate aminotransferase; 5, arogenate dehydratase; 6, prephenate dehydrogenase; 7, arogenate dehydrogenase; 8, 4-hydroxy-phenylpyruvate aminotransferase; and 9, anthranilate synthase.

lite action of aromatic amino acid analogs. The solution to this problem was hidden in the observation that metabolite flow to aromatic amino acids is growth-limiting during culture on fructose as the sole source of carbon and energy (Calhoun and Jensen, 1972). This phenomenon is probably explained by the presence of a phosphoenolpyruvate-dependent phosphotransferase system to transport and phosphorylate fructose (Durham and Phibbs, 1982; and Roehl and Phibbs, 1982). Since biosynthesis of each aromatic amino acid molecule requires two molecules of phosphoenolpyruvate, aromatic pathway biosynthesis is particularly sensitive to intracellular depletion of phosphoenopyruvate. Thus, with growth on fructose, phosphoenolpyruvate is diverted from aromatic synthesis for fructose transport and the organisms are rendered sensitive to analogs of phenylalanine or tyrosine so that regulatory mutants can be directly selected as having a phenotype of analog resistance. Indeed, this approach has resulted in the isolation and enzymological characterization of p-fluorophenylalanine- and m-fluorophenylalanine-resistant mutants in *P. aeruginosa* (Fiske et al., 1983). The analog-resistant mutants obtained fall into two classes. The first class lacked feedback sensitivity of prephenate dehydratase (fig. 3) and was easily recognized by dramatic excretion of *L*-phenylalanine; the second possessed a completely feedback-resistant 3-deoxy-*D*-arabinoheptulosonate 7-phosphate (DAHP) synthase-Tyr, the major species of two isozymes. Deregulation of DAHP-Tyr resulted in the channeling of most chorismate molecules produced into an unregulated overflow route via the intermediate arogenate to phenylalanine (see fig. 3). Additionally, even in the wild type, where all allosteric regulation is operable, some phenylalanine overflow was detected on glucose media but not on fructose media. These data indicate a potential role for the arogenate route to phenylalanine as an overflow mechanism that only operates under conditions of nutritional wealth. Thus nutritional manipulation of *P. aeruginosa* has facilitated the traditional approach of regulatory mutant isolation so that we may begin to understand the genetic and physiological organization of aromatic amino acid biosynthesis.

Aromatic amino acid biosynthesis in *Pseudomonas* species can be used conveniently to demonstrate yet another approach to defining precisely biochemical routes where traditional approaches are not applicable. As mentioned above, the multiplicity of biosynthetic routes for phenylalanine and tyrosine biosynthesis has largely precluded the isolation of auxotrophs, since single mutations blocking one pathway are presumably masked by the alternative route. Although strategies of sequential muta-

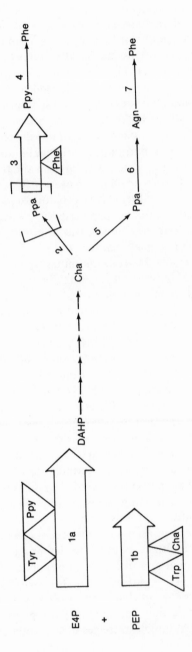

Fig. 3. Allosteric control points along two flow routes leading to L-phenylalanine in *P. aeruginosa*. Heavy arrows indicate the three control points, and the effector molecules cognate to each enzyme molecule are shown symbolically. Regulatory isozymes (1a) and (1b) contribute major and minor fractional output of DAHP as shown schematically. Enzyme (1a) and (1b) are DAHP synthase-tyr and DAHP synthase-trp, respectively. Enzyme activities (2) and (3) are chorisamte mutase and prephenate dehydratase of the bifunctional "P" protein. Enzyme (5) denotes a monofunctional protein possessing a second chorismate mutase activity. Other enzymes: (4), phenylpyruvate aminotransferase; (6), prephenate aminotransferase; and (7), arogenate dehydratase. *Abbreviations:* DAHP, 3-deoxy-D—*arabino*-heptulosonate 7-phosphate; CHA, chorisamte; PPA, prephenate; PPY, phenylpyruvate; PHE, L-phenylalanine; AGN, L-arogenate; Tyr, L-tyrosine; and L-tryptophan; E4P, erythrose-4-phosphate; and PEP, phosphoenolpyrurate.

genesis have been employed to isolate aromatic aminotransferase mutants (Whitaker et al., 1982), this process is extremely arduous and a viable alternative is desirable. This alternative might very well reside in recombinant DNA technology. A vast array of well-characterized phenylalanine and tyrosine auxotrophs are available in *E. coli*. This infinitely more simple system with only single pathways to each aromatic amino acid could therefore be used as a test system for cloned segments of *Pseudomonas* DNA. Fragments of a *Pseudomonas* genophore could be cloned on a broad host range plasmid and used to transform specific *E. coli* phenylalanine and/or tyrosine auxotrophs. By selecting for prototrophy, one could isolate plasmids carrying segments of *Pseudomonas* DNA coding particular aromatic enzymes. This approach could potentially facilitate the assignment of precise gene-enzyme relationships in *Pseudomonas* and define the genetic organization of the aromatic pathway. One could also use transformed, prototrophic *E. coli* lines for mutagenesis so that mutant pseudomonad DNA segments aromatic enzymes could be isolated. These mutant segments could then be used to transform pseudomonad cells and the effect on phenotype ascertained. Although this approach to mutant isolation is speculative in that it has not yet been achieved, it is feasible and potentially very useful. Additionally, broad host range plasmids that move freely between *E. coli* and *Pseudomonas* species have been constructed. Most notable among these is plasmid RSF1010 and its derivatives (Bagdassarian et al., 1981).

PHYTOPATHOGENIC PSEUDOMONADS

The phytopathogenic pseudomonads offer an exciting challenge to the emerging technologies of genetic engineering. These bacteria are an exceedingly large and diverse group of microorganisms that have been implicated as the causative agent in a number of economically important diseases among crop plants. However, the genetic basis of pathogenicity in these organisms is not well understood in most cases. Although plasmids are often associated with phytopathogenic bacteria and have been proposed as a controlling factor in pathogenicity (Piwowarski and Shaw, 1982), this has been unequivocally demonstrated in only a few bacterial species (and only one pseudomonad).

The Ti plasmid of *Agrobacterium tumefaciens* has been well established as coding the genetic information responsible for crown gall tumor formation (Kao et al., 1982). A similar relationship has been observed also between the Ri plasmid of *A. rhizogenes* and hairy root tumors (White et al., 1982). Species of *Rhizobium* have been shown to contain large molecular weight plasmids that encode genetic information for plant host range and nitrogen fixation (Rosenberg et al., 1981). Mega-plasmids, similar to those reported in *R. meliloti* have also been demonstrated in *P. solanacearum*, although a genetic relationship between this plasmid and pathogenicity has not yet been established (Rosenberg et al., 1982). Indeed, the only pseudomonad that has been shown to have plasmid-encoded phytopathogenicity is *P. savastanoi* (Comai et al., 1982). The pIAA1 plasmid codes for the enzymes of indoleacetic acid biosynthesis (Comai and Kosuge, 1982). Indoleacetic acid is a plant hormone produced by *P. savastanoi* resulting in galls or knots in olive and oleander (Comai and Kosuge, 1982).

It is evident that there is a great need for basic biochemical and genetic research on the biology of phytopathogenicity in *Pseudomonas*. It has been argued that the likelihood of plasmids being the exclusive carriers of the genetic determinants of pathogenicity in pseudomonads (or other species) is rather limited (Panopoulos, 1982). In most instances the only available means for studying pathogenicity is by cloning fragments of the bacterial genome and looking for genetic complementation in virulence or host-range mutants of the bacterial strain in question. It is quite likely that complex traits such as host-range and virulence are encoded by a number of specific genes whose expression is very intricately regulated. The most commonly used and widely known vector system is the pBR322 system (Bolivar et al., 1977). However, this narrow host range plasmid will be of limited use in the study of phytopathogenicity in *Pseudomonas*, since functional expression of cloned genes is directly related to the phylogenetic similarity of the organism from which DNA is cloned and the *E. coli* host (Franklin et al., 1981). Indeed, alternative vector systems that will supply the entire range of experimental needs required to unravel phytopathogenicity will need to be developed.

The versatility of the broad host range plasmid RSF1010 has recently attracted attention to its potential as a cloning vector for phytopathogenic pseudomonads. Plasmid RSF1010 is a small (5.5 Md), nonconjugative, naturally occurring plasmid coding resistance to both streptomycin and sulfonamides (Bagdasarian et al., 1979). Plasmid RSF1010 was first shown to be stably maintained and replicated in *P. aeruginosa* and *P. putida* by Nagahari and Sakaguchi (1978). Subsequently RSF1010 has been shown to be capable of autonomous replication in virtually all phytopathogenic pseudomonads examined (Panopoulos, 1981). Therefore, RSF1010 and its derivatives (Bagdasarian et al., 1981) are the most promising candidates as cloning vectors in phytopathogenic pseudomonads. Indeed, RSF1010 has already been used to clone the tryptophan monooxygenase gene, an enzyme involved in the synthesis of indolacetic acid in the olive pathogen *P. savastanoi* (Comai and Kosuge, 1982). The development of RSF1010 and derivative vectors will greatly enhance research efforts in the field of pathogen/host interaction.

A greater understanding of pathogenicity in *Pseudomonas* species has the potential to result in the development of pseudomonad-mediated plant cell transformation vectors. A great deal of effort has been focused on the development of plant cell transformation vectors. Much of the effort has been placed on *A. tumefaciens* and the Ti plasmid (or its derivatives) (Ream and Gordon, 1982). However, the Ti plasmid does have some drawbacks, most notably, the limited host range of *A. tumefaciens*, which may in fact restrict its usefulness as a generalized plant transformation vector. So far many transformation experiments have relied on bacterial infection to introduce foreign DNA. In this respect the broad range of plant species routinely infected by various pseudomonad phytopathogens points to the potential these bacteria represent in the area of plant cell transformation. It should be noted that Ti plasmid DNA has been introduced into plant cells by means of protoplast-bacterial spheroplast fusion (Hasegawa et al., 1981). However, this approach is only practical in those plant species that will regenerate from protoplasts into whole plants. Of course, in order to take advantage of the diversity of pseudomonad phytopathogens, a vector that is transmitted and stably maintained in all pathovars is essential. Once again, RSF1010 or a derivative might be ideal. Naturally, a great number of other factors need to be considered when developing pseudomonad-based plant cell transformation vectors. These include: (1) size of DNA that can be transferred by the transformation vehicle; (2) the ability of the transferred DNA to become stably incorporated into the plant genome (as has been shown for the T-DNAs [Matzbe and Chilton, 1981]); and (3) the availability of an active promotor on the vector so that the transferred genetic sequence can be expressed. Many of these characteristics are fulfilled by the Ti-plasmid, but there is certainly no reason to believe that suitable alternatives or, indeed, improved vectors could not be isolated or constructed using the genetic versatility present in pseudomonad species. As more is learned about the genetic and biochemical capabilities of pseudomonad plasmids, this speculation on pseudomonad plant cell transformation vehicles may indeed find its place in reality.

SUMMARY

Pseudomonad species represent an impressive collection of biochemical and genetic capabilities. With the advent of recent biotechnological methodologies, it may now be possible to harness this vast resource of genetic information for agricultural and industrial purposes. It is evident that much still needs to be learned about the genetic regulation and routing of specific biochemical pathways and the nature of the array of plasmids in *Pseudomonas* on the basic scientific level. The technologies of genetic engineering should prove to be valuable tools in unraveling these complex systems.

ACKNOWLEDGMENTS

The author wishes to thank D. A. Evans and G. S. Byng for their helpful comments and K. M. Selover for typing this manuscript.

REFERENCES

(Key references' authors' names appear in boldface type.)

Abramsky, T., and D. Shemin. 1965. The formation of isoleucine from B-methylaspartic acid in *Escherichia coli* W. J. Biol. Chem. 240:2971–75.

Bagdasarian, M., R. Lunz, B. Ruckert, F. C. H. Franklin, M. M. Bagdasarian, J. Frey, and K. N. Timmis. 1981. Specific-purpose plasmid cloning vectors. II. Broad host range, high copy number, RSD1010-derived vectors, and a host-vector system for gene cloning in *Pseudomonas* Gene 16:237–47.

Bagdasarian, M., N. N. Bagdasarian, S. Coleman, and K. N. Timmis. 1979. New vector plasmids for gene cloning in *Pseudomonas*. *In* K. N. Timmis and A. Puhler (eds.), Plasmids of medical, environmental, and commercial importance, pp. 411–22. Elsevier/North Holland Biomedical Press, Amsterdam.

Bolivar, F., R. L. Rodriquez, P. J. Greene, M. L. Betlack, H. L. Neyneker, H. W. Boyer, J. Crosa, and S. Falkow. 1977. Construction and characterization of new cloning vehicles. II. A multi-purpose cloning system. Gene 2:95–113.

Byng, G. S., R. J. Whitaker, R. L. Gherna, and R. A. Jensen. 1980. Variable enzymological patterning in tyrosine biosynthesis as a means of determining natural relatedness among the *Pseumonadaceaae*. J. Bacteriol. 144:247–57.

Byng G. S., J. L. Johnson, R. J. Whitaker, R. L. Gherna, and R. A Jensen. 1983. The evolutionary pattern of aromatic amino acid biosynthesis and the emerging phylogeny of pseudomonad bacteria. J. Mol. Evol. 19:373–82.

Calhoun, D. J., and R. A. Jensen. 1972. Significance of altered carbon flow in aromatic amino acid synthesis: an approach to the isolation of regulatory mutants in *Pseudomonas aeruginosa*. J. Bacteriol. 109:365–73.

Chakrabarty, A. M. 1976. Plasmids in *Pseudomonas*. Ann. Rev. Genet. 10:7–30.

Chakrabarty, A. M., G. Chou, and I. C. Gunsalus. 1973. Genetic regulation of octane dissimilation plasmid in *Pseudomonas*. Proc. Nat. Acad. Sci. USA 70:1137–40.

Chakrabarty, A. M. 1972. Genetic basis of the biodegradation of salicylate in *Pseudomonas*. J. Bacteriol. 112:815–23.

Charon, N. W., R. C. Johnson, and D. Peterson. 1984. Amino acid biosynthesis in the spirochete *Leptospire*: evidence for a novel pathway of isoleucine biosynthesis. J. Bacteriol. 117:200–211.

Chatterjee, D. K., S. T. Kellogg, S. Hamada, and A. M. Chakrabarty. 1981. Plasmid specifying total degradation of 3-chlorobenzoate by a modified ortho pathway. J. Bacteriol. 146:639–46.

Cohen, G. N. 1965. Regulation of enzyme activity in microorganisms. Ann. Rev. Microbiol. 95:105–26.

Comai, L., and T. Kosuge. 1982. Cloning and characterization of iaa M, a virulence determinant of *Pseudomonas savastanoi*. J. Bacteriol. 149:40–46.

Comai, L., G. Surico, and T. Kosuge. 1982. Relation of plasmid DNA to indoleacetic acid production in different strains of *Pseudomonas syringae* pv. *savastanoi*. J. Gen. Microbiol. 128:2157–63.

Datko, A. H., J. Grovanelli, and S. H. Mudd. 1974. Homocysteine biosynthesis in green plants. J. Biol. Chem. 249:1139–55.

Doudoroff, M., and N. J. Palleroni. 1974. *Pseudomonas*. *In* R. E. Buchanan and N. E. Gibbons (eds.), Bergey's manual of determinative bacteriology, 8th ed., pp. 217–43. Williams and Wilkins, Baltimore.

Dunn, N. W., and I. C. Gunsalas. 1973. Transmissable plasmid coding early enzymes of napthalene oxidation in *Pseudomonas putida*. J. Bacteriol. 114:974–79.

Durham, D. R., and P. V. Phibbs. 1982. Fractionation and characterization of the phosphoenolpyruvate: fructose 1-phosphotransferase system from *Pseudomonas aeruginosa*. J. Bacteriol. 149:534–41.

Farrell, R., and A. M. Chakrabarty. 1979. Degradative plasmids: molecuar nature and mode of evolution. *In* K. N. Timmis and A. Puhler (eds.), Plasmids of medical, environmental, and commercial importance, pp. 97–109. Elsevier/North Holland Biomedical Press, Amsterdam.

Fisher, P. R., J. Appleton, and J. M. Pemberton. 1981. Isolation and characterization of the pesticide-degrading plasmid pJP1 from *Alcaligenes paradoxus*. J. Bacteriol. 135:798–804.

Fiske, M. J., R. J. Whitaker, and R. A. Jensen. 1983. Hidden overflow pathway to *L*-phenylalanine in *Pseudomonas aeruginosa*. J. Bacteriol. 154:623–31.

Fowden, L., D. Lewis, and H. Tristram. 1967. Toxic amino acids: their action as antimetabolites. *In* F. F. Nord (ed.), Advances in enzymology, 29:89–163. John Wiley and Sons, New York.

Franklin, F. C. H., M. Bagdasarian, M. M. Bagdasarian, and K. N. Timmis. 1981. A. molecular and functional analysis of the TOL plasmid pWWO from *Pseudomonas putida* and cloning genes for the entire regulated aromatic ring meta-cleavage pathway. Proc. Nat. Acad. Sci. USA.

Friello, D. A., J. R. Myhoie, D. T. Gibson, J. E. Rodgers, and A. M. Chakrabarty. 1976. XYL, a non-conjugative xylene-degradative plasmid in *Pseudomonas* Pxy. J. Bacteriol. 1727:1217–24.

Gunsalus, I. C., M. Hermann, W. A. Toscano, D. Katz, and G. K. Garg. 1975. Plasmids and metabolic diversity. *In* D. Schlessinger (ed.), Microbiology-1974, pp. 207–12. American Society for Microbiology, Washington.

Hasegawa, S., T. Nagata, and K. Syono. 1981. Transformation of *Vinca* protoplasts mediated by *Agrobacterium* spheroplasts. Mol. Gen. Genet. 182:206–10.

Heptenstall, J., and J. R. Quayle. 1970. Pathways leading to and from serine during growth of *Pseudomonas* AM1 on C_1 compounds or succinate. Biochem. 227:563–72.

Jacoby, G. A. 1979. Plasmids of *Pseudomonas aeruginosa*. In R. G. Dogett (ed.), *Pseudomonas aeruginosa*: clinical manifestation of infection and current therapy, pp. 271–309. Academic Press, New York.

Kao, J. C., K. L. Perry, and C. I. Kado. 1982. Indoleacetic acid complementation and its relation to host range specifying genes on the Ti plasmid of *Agrobacterium tumefaciens*. Mol. Gen. Genet. 188:425–32.

Karns, J. S., J. J. Kilbane, D. K. Chatterjee, and A. M. Chakrabarty. 1981. Laboratory breeding of a bacterium for enhanced degradation of 2,4,5-T. In O. J. Crocomo, F. C. A. Tavares, and D. Sodrzeicski (eds.), Genetic engineering for biotechnology: proceedings of the International Symposium on Genetic Engineering, pp. 37–42. Promocet, Sao Paulo, Brazil.

Kellogg, S. T., D. K. Chatterjee, and A. M. Chakrabarty. 1981. Plasmid-assisted molecular breeding: new technique for enhanced biodegradation of persistent toxic chemicals. Science 214:1133–45.

Leisinger, T., and R. Margraff. 1979. Secondary metabolites of the fluorescent pseudomonads. Microbiol. Rev. 43:422–42.

Matzki, A. J. M., and M.-D. Chilton. 1981. Site specific insertion of genes into T-DNA of the *Agrobacterium* tumor-unducing plasmid: an approach to genetic engineering of higher plant cells. J. Mol. Appl. Genet. 1:39–49.

Nagahari, K., and K. Sakaguchi. 1978. RSF1010 plasmid as a potential cloning vehicle in *Pseudomonas* species. J. Bacteriol. 133:1527–29.

Palchudhuri, S. 1977. Molecular characterization of hydrocarbon degradative plasmids in *Pseudomonas putida*. Biochem. Biophys. Res. Commun. 77:518–25.

Palleroni, N. J., R. Kunisawa, R. Contopoulou, and M. Doudoroff. 1973. Nucleic acid homologies in the genus *Pseudomonas*. Int. J. Syst. Bacteriol. 23:333–59.

Panapoulos, N. J. 1981. Emerging tools for in vitro and in vivo genetic manipulations in phytopathogenic pseudomonads and other nonenteric gram negative bacteria. In N. J. Panopoulos (ed.), Genetic engineering in the plant sciences, pp. 155–86. Praeger Scientific, New York.

Pasyewski, A., and J. Grabski. 1975. Enzymatic lesions in methionine mutants of *Aspergillis nidulans*: role and regulation of an alternative pathway for cysteine and methionine synthesis. J. Bacteriol. 124:893–904.

Phillips, A. T., J. I. Nuss, J. Mossic, and C. Foshay. 1972. Alternative pathway for isoleucine biosynthesis in *Escherichia coli*. J. Bacteriol. 109:714–19.

Piwowarski, J. M., and P. D. Shaw. 1982. Characterization of plasmids from plant pathogens *Pseudomonas*. Plasmid 7:85–94.

Ream, L. W., and M. P. Gordon. 1982. Crown gall disease and prospects for genetic manipulation of plants. Science 218:854–59.

Reineke, W., and H. J. Knacknauss. 1980. Hybrid pathway for chlorobenzoate metabolism in *Pseudomonas* sp. B13 derivatives. J. Bacteriol. 142:467–73.

Rheinwald, J. G., A. M. Chakrabarty, and I. C. Gunsalus. 1973. A transmissable plasmid controlling camphor oxidation in *Pseudomonas putida*. Proc. Nat. Acad. Sci. USA 70:885–89.

Roehl, R. A., and P. V. Rhibbs. 1982. Characterization and genetic mapping of fructose phosphotransferase mutations in *Pseudomonas aeruginose.* J. Bacteriol. 150:402–6.

Rosenberg, C., P. Boistard, J. Denarie, and F. Casse-Delbart. 1982. Genes controlling early and late functions in symbiosis are located on a megaplasmid in *Rhyzobium meliloti.* Mol. Gen. Genet. 184:326–33.

Saunders, J. R. 1977. Degradative plasmids. Nature 269:470.

Volbrecht, D. 1974. Three pathways of isoleucine biosynthesis in mutant strains of *Saccharomyces cerevisiae.* Biochim. Biophys. Acta 362:382–89.

Whitaker, R. J., C. G. Gaines, and R. A. Jensen. 1982. A multispecific quintet of aromatic aminotransferases that overlap different pathways in *Pseudomonas aeruginoas.* J. Biol. Chem. 257:13550–56.

Whitaker, R. J., G. S. Byng, R. L. Gherna, and R. A. Jensen. 1981a. Comparative allostery of 3-deoxy-*D-arabino*-heptulosonate 7-phosphate synthetase as an indicator of taxonomic relatedness in pseudomonad genera. J. Bacteriol. 145:752–59.

Whitaker, R. J., G. S. Byng, R. L. Gherna, and R. A. Jensen. 1981b. Diverse enzymological patterns of phenylalanine biosynthesis in pseudomonads are conserved in parallel with deoxyrebonucleic acid homology groupings. J. Bacteriol. 147:526–34.

White, F. F., G. Ghidossi, M. P. Gordon, and E. W. Nester. 1982. Tumor induction by *Agrobacterium rhizogenes* involves transfer of plasmid DNA to the plant genome. Proc. Nat. Acad. Sci. USA 79:3193–97.

Worsey, M. J., and P. A. Williams. 1975. Metabolism of toluene and xylenes by *Pseudomonas putida* (*arvilla*) mt-2: evidence for a new function of the TOL plasmid. J. Bacteriol. 124:7–13.

H. HESLOT

Genetic Engineering in Yeasts

20

ABSTRACT

Transformation of yeast has been achieved with both integrative and autonomously replicating plasmids. The origins of replication of the latter are
derived either from the 2 um DNA of yeast or from yeast chromosomes.
Many yeast genes have been cloned on these vectors allowing detailed
study of their structure and regulatory regions. Using appropriate yeast
promoters, foreign genes (from bacteria, man, etc.) have been expressed at
a high level in yeast.

YEAST CELL STRUCTURE-CELL CYCLE

Yeasts are unicellular eucaryotes, possessing a nucleus with chromosomes, mitochondria, vacuoles, and so on. They have for long been commercially important, and the best-known species *Saccharomyces cerevisiae*
is involved in alcoholic fermentation (wine, beer) and bread manufacture.
Haploid strains possess n = 17 chromosomes and belong to two mating
types, a and α, fusing under appropriate conditions and giving rise to diploid cells (2n = 34) that are able to sporulate. Baker's yeast is well known
genetically: more than 250 genes have been identified. However, more than
500 other species have been described, and for most of them genetic knowledge is nil or scanty.

GENETIC INFORMATION IN YEAST

Apart from the genetic information stored in the DNA of chromosomes
(3000 K base pairs), there are circular molecules of DNA (75 K bases) inside
mitochondria. During the last twenty years, mitochondrial genetics proper

has been considerably developed. In France this specific study was initiated by Ephrussi and is being carried over by groups led by Slonimski, Fukuhara, and many others. A fundamental fact that made mitochondrial genetics of baker's yeast more easily open to investigation is that cells can live even when they have lost completely their mitochondrial DNA. They then draw their energy from the glycolytic pathway. It has been shown recently that mitochondrial genes are often discontinuous (e.g., the region coding cytochrome b).

Most baker yeast strains also contain 50–100 copies per cell of a so-called 2 um plasmid (Guerineau, 1979). These are circular molecules of 2 microns length (approximately 6 K b). This plasmid has two interconvertible forms, A and B, coexisting in a 1:1 ratio inside each cell. The plasmid contains two invert repeats of 600 base pairs. It has been completely sequenced (Hartley and Donelson, 1980), and three open reading frames called Able, Baker, and Charlie have been found: two of them are responsible for the autonomous replication of the plasmid; the third is involved in the inter-conversion of A and B forms (Broach and Atkins, 1979; Broach and Hicks, 1980).

Livingstone and Hahne (1979), Nelson and Fangman (1979), and Seligy and colleagues (1980) have shown conclusively that native 2 um plasmids have a sedimentation constant of 75 S, about three times that of protein-free 2 um DNA. Moreover the 2 um and chromosomal DNA digestion patterns are very similar, indicating that both types of DNA are condensed into nucleosomes. This has been confirmed by electron microscopy.

Livingstone and Kupfer (1977) have also demonstrated that 2 um DNA replication is controlled by the same genes that control the initiation and completion of nuclear DNA replication.

Kielland-Brandt and colleagues (1980), using the karl mutation which prevents nuclear fusion, made crosses between 2 um $^+$ and 2 um° parents. They found that all progeny inheriting the nucleus from the 2 um$^+$ parent also had the plasmid, whereas only about 25–50% of the progeny with the other nucleus had the plasmid. They concluded that 2 um DNA had a nuclear location.

TRANSFORMATION

Yeast transformation with external DNA was first achieved by Hinnen and colleagues (1978). They chose as receptor cells leu2 auxotrophs having two mutations in the gene, so that the frequency of spontaneous reversion was very low. Cell walls were digested with lytic enzymes, giving rise to protoplasts, subsequently put in contact with a bacterial plasmid Col E1 containing the LEU2$^+$ yeast gene. Stable prototrophic transformants were obtained at low frequency (10^{-7} per viable cell). Analysis of transformants indicated three possible modes of integration of transforming DNA in the chromosomes: it may be additive or substitutive and integrate either in the leu2 region or in several other locations.

Then Beggs (1978) constructed chimeric plasmids containing the 2 um DNA of yeast, linked to a bacterial plasmid pMB9, as well as the LEU2$^+$ gene of yeast. Such plasmids were shown to be able to transform to prototrophy appropriate *Escherichia coli* or *S. cerevisiae* leu2$^-$ cells. Those plasmids were found to behave as autonomous replicating units, not integrated in the chromosomes.

Gerbaud and colleagues (1979) built chimeric plasmids containing the yeast URA3 gene and 2 um yeast DNA linked to a bacterial plasmid and observed high frequency of yeast transformation (10^{-4} per viable cell). Plasmids recovered from the yeast transformants were used to transform *E. coli*. Their analysis with restriction enzymes showed that in many cases the vector had recombined with the endogenous 2 um DNA of the recipient strain. The specific activity of the enzyme coded by URA3 in yeast transformants was up to 40-fold higher than the wild type. Plasmids in the transformed cells are frequently lost during vegetative multiplication giving rise to ura3 auxotrophs.

In a subsequent investigation, Blanc and colleagues (1979) found that stability of the URA3$^+$ character could be considerably increased by transforming ura3 cells devoid of any 2 um sequence.

Toh-e and colleagues (1980) constructed a plasmid pSLe 1 containing only yeast sequences, i.e., the LEU 2 gene and part of the 2 um sequence (only one of the inverted repeats). Such plasmid is stably maintained in yeast without selective pressure and does not displace normal 2 um DNA.

Toh-e and Wickner (1981) and Dobson and colleagues (1980) also showed that the 2 um DNA plasmid was often eliminated from yeast cells when they are transformed with the composite plasmid pJDB 219. Since pJDB 219 is subsequently lost with high frequency, derivatives lacking all 2 um DNA can be prepared from any strain.

Kingsman and colleagues (1979) and Struhl and colleagues (1979) found that a plasmid carrying a 1.4 Kb Eco R1 fragment from yeast TRP 1 region was capable of transforming yeast trp 1 mutants to Trp$^+$ at high frequency (10^3–10^4 transformants/ug DNA) and was capable of autonomous replication. The Trp$^+$ transformants are highly unstable, segregating Trp$^-$ cells in media containing tryptophan. The yeast DNA sequence associated with

TRP 1 that behaves as a chromosomal replicator, ars 1 (autonomously replicating sequence), has been isolated and characterized by Stinchcomb and colleagues (1979) and Tschumper and Carbon (1980).

Hsiao and Carbon (1979) have shown also that a second origin of replication (ars 2) is associated with the ARG 4 gene of yeast, but can be separated from it.

A general method for isolating DNA segments capable of autonomous replication in yeast from *S. cerevisiae* chromosomal DNA has been presented by Chan and Tye (1980) and by Beach and colleagues (1980). This technique has been extended by Stinchcomb and colleagues (1980) to isolate segments of DNA from *Neurospora crassa, Dictyostelium discoideum, Ceanhorabditis elegans, Drosophila melanogaster*, and *Zea mays* that behave as origins of replication in yeast. There is as yet no direct proof that these sequences are indeed origins of replication in the organisms from which they were extracted.

Zakian (1981) found that a segment of mitochondrial DNA of *Xenopus laevis* behaved also as an origin of replication in yeast.

Isolation of a yeast centromeric region, that of chromosome III, was first reported by Clark and Carbon (1980). When a segment of centrometric DNA is introduced into a plasmid having an origin of replication, such as ars 1 or ars 2, its stability is considerably increased. A marker gene located in such a plasmid segregates $2^+:2^-$ at meiosis. The plasmid does not pair with it normal homologous chromosome and therefore can be looked at as a minichromosome. A similar behavior has been found in the other centromeres since isolated, i.e., CEN 4 and CEN 11 (Fitzgerald et al., 1981; Mann C. et al., 1981). A functional centromeric sequence has been shown to have 627 bp CEN 4 and 858 bp CEN 11. No cross-hybridization has been observed between centromeres. However, sequencing reveals 4 regions of homology, each approximately 10 to 14 bp long.

CLONING AND EXPRESSION OF YEAST GENES IN YEASTS

Apart from LEU 2 and URA 3, mentioned above, other yeast genes have also been cloned. Shell and Wilson (1979) isolated the structural gene of galactokinase (gal 1) from *S. cerevisiae*. The gene was detected in an enzymatic digest of whole yeast DNA by hybridization with purified galactokinase m-RNA, labeled with [125]I. When a plasmid containing this gene is introduced into an *E. coli* strain deleted for the galactose operon, it causes the synthesis of low levels of yeast galactokinase activity.

Williamson and colleagues (1980) managed to isolate the ADH 1 gene of

S. cerevisiae, coding one isozyme of alcohol dehydrogenase. The plasmids containing ADH 1 were selected from a pool of plasmids containing yeast DNA by virtue of their ability to complement the growth defect of yeast strains completely lacking ADH activity. Through DNA sequencing, they were shown to contain the structural gene ADH 1.

The iso-1-cytochrome c gene of yeast has been identified and cloned by Montgomery and colleagues (1978), using a synthetic oligodeoxy nucleotide as a hybridization probe. This oligonucleotide was complementary to a region near the N terminal coding region of the yeast CYC 1 gene. Sequencing of one hundred nucleotides into the coding region of iso-1-cytochrome c yielded a DNA sequence corresponding exactly to the known protein sequence.

Subsequently, Faye and colleagues (1981) studied the 5′ termini of yeast CYC 1 gene as well as several deletions in this region, thus obtaining information on the structure of the potential promotor of CYC 1.

Ng and Abelson (1980), Gallwitz and Seidel (1980), and Gallwitz and Sures (1980) have cloned the gene for actin in *S. cerevisiae*. Actin is a major protein in all eukaryotic cells. It functions in a number of ways and seems to be involved in mitosis and cytokinesis. The actin gene has been sequenced, revealing an intervening sequence of 304 bp in the coding sequence at the 5′ end of the gene.

Hereford and colleagues (1979) have cloned sequences of yeast histone genes H2A and H2B. There are only two copies of this pair of genes within the haploid yeast genome.

Hitzman and colleagues (1980) have used an immunological screening technique to detect an *E. coli* colony containing a yeast DNA insert in plasmid Col E1, that produces antigen able to combine with antibody directed against purified 3-phosphoglycerate kinase. The hybrid plasmid was shown to contain the structural gene (PGK) of this enzyme. A functional enzyme is synthesized from the cloned gene in yeast, but not in *E. coli*.

M. J. and J. P. Holland (1979) have isolated a glyceraldehyde-3-phosphate dehydrogenase gene (GPD) from a collection of *E. coli* transformants containing segments of yeast genomic DNA. Subsequently (Holland, J. P., and M. J. Holland, 1979) they worked out the complete nucleotide sequence of the coding, as well as the flanking noncoding regions of this gene. A second GPD gene from yeast was isolated later (Holland, J. P., and M. J. Holland, 1980).

Kawasaki (1981) has isolated yeast glycolytic genes by complementation in mutants defective in growth on glucose. Transformants carrying the phosphoglucane isomeres (PGI 1), triosephosphate isomerase (TPI 1), phosphoglycerate kinase (PGK 1), phosphoglycerate mutase (PGM) or pyruvate kinase (PYK 1) gene on a multicopy plasmid have four- to tenfold higher levels than wild-type strains of their respective enzymes.

Fasiolo and colleagues (1981) have selected a plasmid containing the yeast methionyl-t-RNA synthetase gene. This plasmid transforms to prototrophy a methiony-t-RNA synthetase-impaired mutant requiring methionine. Transformants have a 30-fold increased activity of the enzyme.

Guerry-Kopecko and Wickner (1980) have introduced in a plasmid a segment of *S. cerevisiae* DNA that complements both the yeast ura 1 and *E. coli* phr D mutations in dehydro-orotate dehydrogenase.

With a view of better understanding the molecular mechanism involved in the regulation of gene expression in yeast, Struhl and David (1980) have cloned the *his*-3 gene of *S. cerevisiae* coding imidazole glycerol phosphate dehydratase.

Nasmyth and Reed (1980) have cloned the cdc 28 gene, one of several required for cell division by *S. cerevisiae*. The appropriate recombinant plasmid was detected by its ability to complement a cdc 28 ts strain and selection for clones capable of growth at the restrictive temperature.

During their work to clone the centromere of chromosome III, Clarke and Carbon (1980) have also isolated the cell cycle cdc 10, which is centromere linked.

Broach and colleagues (1979) cloned the gene encoding the arginine permease, CAN 1, which is able to complement a permease-less can 1 mutant.

Carlson and colleagues (1980) have cloned a SUC 2 gene that codes the enzyme invertase and allows yeast cells to use sucrose as carbon source. A secreted glycosylated form of invertase is responsible for sucrose utilization and its synthesis is repressed by glucose. An intracellular non-glycosylated enzyme is made at low level and appears to be constitutive. Cloning of SUC 2 has allowed characterization of the SUC RNA. In glucose-derepressed cells, which produce both forms of invertase, there are two SUC RNAS (1.8 Kb and 1.9 Kb).

CLONING AND EXPRESSION OF FOREIGN GENES IN YEAST

Panthier and colleagues (1980) have introduced the lac Z gene of *E. coli* into a plasmid able to replicate in yeast. Lac Z codes β-galactosidase and wild-type *S. cerevisiae* is completely lacking this enzymatic activity. However, yeast cells harboring a chimeric plasmid carrying lac Z have a low level of β-galactosidase, indicating that the bacterial gene is expressed in yeast. Moreover, on the basis of size and immunological criteria, no difference was detected between the coli-in-yeast β-galactosidase and the *E. coli* enzyme.

Chevallier and Aigle (1979) have shown that chimeric plasmids, able to replicate in yeast and possessing a segment of *E. coli* pBR 322 plasmid containing a DNA sequence coding β-lactamase, are able to express this enzymatic activity in yeast.

Roggenkamp and Hollenberg (1981) have purified the β-lactamase made in yeast. It is indistinguishable from pure *E. coli* β-lactamase with regard to molecular weight. In *E. coli* β-lactamase is synthesized as a preprotein comprising a leader sequence of 23 amino acids cleaved off during secretion. This suggests that *S. cerevisiae* is able to recognize and process the preprotein. The relevant protease activity appears to be membrane-bound.

Cohen and colleagues (1980) have likewise shown that the *E. coli* R-factor-derived chloramphenicol resistance (can^R) gene is functionally expressed in *S. cerevisiae*. Yeast cells harboring this gene acquire resistance to chloramphenicol, and cell-free extract prepared from these cells contains chloramphenicol acetyltransferase, the enzyme specified by can^R in *E. coli*.

Dickson (1980) transformed into *S. cerevisiae* recombinant DNA vectors carrying the B-galactosidase structural gene LAC 4, from the yeast *Kluyveromyces lactis*. All transformants-expressed β-galactosidase activity of LAC 4.β-galactosidase synthesized in either *S. cerevisae* or in *K. lactis* is the same, as judged by two-dimensional polyacrylamide gel electrophoresis. However, *S. cerevisiae* transformed with LAC 4 cannot grow on lactose, probably because a specific permease is lacking.

Henikoff and colleagues (1981) have introduced into yeast a *Drosophila* DNA sequence that complements a *S. cerevisiae* adenine-8 mutation. A poly-A-containing RNA is transcribed from the cloned *Drosophila* segment in transformed yeast cells and can account for the functional expression of the gene.

Mercereau-Puijalon and colleagues (1980) have fused *in vitro* to the beginning of the *E. coli* lactose operon a c-DNA sequence coding chicken ovalbumine. This segment was introduced into a plasmid able to replicate in both *E. coli* and *S. cerevisiae*. An ovalbumine-like protein is made in both organisms transformed with this plasmid, probably because the *E. coli* lac regulatory region has some promotor activity in yeast.

Hitzeman and colleagues (1981) have obtained the expression in yeast of the human gene coding leukocyte inferferon D. A DNA fragment, containing the 5'-flanking sequence of the gene for yeast alcohol dehydrogenase 1

(ADH 1) was trimmed to remove the ATG-translational start of ADH 1. This altered fragment was put upstream of the structural c-DNA gene for human leukocyte interferon D, containing an ATG-translational start at the beginning of the sequence for mature interferon. The combined fragments were introduced into a multicopy plasmid. Active interferon D was made, but it was neither glycosylated nor excreted.

McNeil and Friesen (1981) have constructed yeast plasmids that carry the *Herpes simplex* virus type 1 (HSV-1) thymidine kinase gene (tk). This gene is functionally expressed in yeast. Wild-type yeast is completely lacking in this enzymatic activity. It appears that the expression of tk is probably due to transcriptional read-through from a yeast promotor that is carried on a yeast DNA fragment adjacent to the HSV-1 DNA.

Jimenez and Davies (1980) have built a plasmid containing LEU 2, trp 1 replicator and the bacterial transposon Tn 601 that confers resistance to the aminoglycoside antibiotic G 418. Since this antibiotic is active on any kind of cell, microbial, vegetal, and animal, this system may supply the basis of a general cloning system.

GENE FUSIONS

Guarente and Ptashne (1981) have constructed hybrid genes between the *E. coli* lac Z gene and iso-l-cytochrome c (CYC 1) gene of *S. cerevisiae* by recombination *in vitro*. Each of these hybrid genes encodes a chimeric protein with a cytochrome c moiety at the amino terminus and an active B-galactosidase moiety at the carboxy terminus. When these hybrids are introduced into *S. cerevisiae* plasmid vectors, the B-galactosidase levels have the pattern of regulation normally seen for cytochrome c, i.e., a reduction of synthesis in cells grown in glucose. The regulatory DNA sequence of CYC 1 is the subject of intensive investigation.

Rose and colleagues (1981) constructed a plasmid that allowed selection *in vivo* of gene fusions between the *E. coli* β-galactosidase and *S. cerevisiae* URA3 genes. The URA3 gene was placed upstream of an amino-terminally deleted form of the lac Z gene. Selection for Lac$^+$ in *E. coli* yielded deletions that fused the lac Z gene to the URA 3 gene. Some of these fusion plasmids produced β-galactosidase activity when introduced into yeast. In one of the fusions to URA 3, the β-galactosidase activity had the pattern of regulation of URA 3 itself.

Martinez-Arias and Casadaban (1981) fused *in vitro* the yeast LEU 2 promotor region to the *E. coli* β-galactosidase structural gene. In yeast the synthesis of the fused β-galactosidase is regulated in the same way as the original LEU 2 enzyme: it is repressed by leucine and threonine. The β-galactosidase activity can be detected in colonies growing on agar plates containing chromogenic substrates. This allows the selection of yeast cells expressing a higher level of the enzyme.

OTHER YEAST SPECIES

Toh-e and colleagues (1981) have detected 2 um DNA-like plasmids in the osmophilic haploid yeast *Saccharomyces rouxii*. Restriction analyses of these plasmids have indicated that they differ significantly from that of 2 um DNA. There are two isomeric forms of these plasmids, suggesting the presence of inverted repeats. No homology was detected between these plasmids and 2 um DNA. A chimeric plasmid built by adding pBR 322 and the HIS 3 gene of yeast could replicate autonomously in *S. cerevisiae*.

Beach and Nurse (1981) have succeeded in transforming a leu 1 strain of the yeast *Schizosaccharomyces pombe* with hybrid plasmids containing the *S. cerevisiae* LEU 2^+ gene, a bacterial plasmid, and either the *S. cerevisiae* 2 um plasmid or autonomously replicating sequences such as ars 1.

Gaillardin and colleagues (1981) have investigated the efficiency of various replicators in the transformation of *S. pombe* and the fate of the transforming plasmids. It was found that plasmids containing part or entire 2 um DNA were subjected to spontaneous rearrangements. On the contrary, plasmids containing the ars 1 or ars 2 replicators were not modified. It was shown also that a 3 um plasmid extracted from *S. pombe* and cloned into a bacterial plasmid was capable of autonomous replication in *S. pombe*. This plasmid is presumably an extrachromosomal copy of r-DNA.

Gunge and colleagues (1981) have isolated two linear DNA plasmids (8 Kb and 12 Kb) in some strains of the yeast *Kluyveromyces lactis*. The presence of these plasmids is associated with the capacity to kill strains of *K. lactis* devoid of such plasmids or strains of *S. cerevisiae*.

In further experiments, Gunge and colleagues (1981) have transferred the two plasmids from a *K. lactis* killer strain into a non-killer strain of *S. cerevisiae* by the use of a protoplast fusion technique. Both plasmids replicated autonomously and stably in the new host cells of *S. cerevisiae* and could coexist with the resident 2 um DNA plasmid. It remains to be seen if the *K. lactis* plasmids can be used as vectors.

PROSPECTS

Yeast transformation clearly affords a powerful tool to study the structure of genes and the molecular mechanisms of their regulation.

In the applied field, it is to be expected that the cloning methodology will lead to results of industrial significance: production of metabolites, enzymes, and yeast strains with new physiological properties.

REFERENCES

Beach, D., and P. Nurse. 1981. High frequency transformation of the fission yeast *Schizosaccharomyces pombe*. Nature 290:140–42.

Beach, D., M. Piper, and S. Shall. 1980. Isolation of chromosomal origins of replication in yeast. Nature 284:185–87.

Beggs, J. D. 1978. Transformation of yeast by a replicating hybrid plasmid. Nature 275:104–9.

Blanc, H., C. Gerbaud, P. Slonimski, and M. Guerineau. 1979. Stable yeast transformation with chimeric plasmids using a 2 um circular DNA-less strain as recipient. Molec. Gen. Genet. 176:335–42.

Broach, J. R., J. F. Atkins, C. McGill, and L. Chow. 1979. Identification and mapping of the transcriptional and translational products of the yeast plasmid 2 um circle. Cell 16:827–39.

Broach, J. R., and J. B. Hicks. 1980. Replication and recombination functions associated with the yeast plasmid 2 um circle. Cell 21:501–8.

Broach, J. R., J. N. Strathern, and J. B. Hicks. 1979. Transformation in yeast: development of a hybrid cloning vector and isolation of the *CAN1* gene. Gene 8:121–33.

Carlson, M., and D. Botstein. 1982. Two differentially regulated mRNAs with different 5' ends encode secreted and intracellular forms of yeast invertase. Cell 28:145–54.

Carlson, M., B. C. Osmond, and D. Botstein. 1980. SUC genes of yeast: a dispersed gene family. Cold Spring Harbor Symp. 45:799–803.

Chan, S. M., and B. K. Tye. 1980. Autonomously replicating sequences in *Saccharomyces cerevisiae*. Proc. Natl. Acad. Sci. USA 77:6329–33.

Chevallier, M. R., and M. Aigle. 1979. Qualitative detection of penicillinase produced by yeast strains carrying chimeric yeast-*coli* plasmids. FEBS Lett. 108:179–80.

Clarke, L., and J. Carbon. 1980. Isolation of the centromere linked CDC 10 gene by complementation in yeast. Proc. Natl. Acad. Sci. USA 77:2173–77.

Clarke, L., and J. Carbon. 1980. Isolation of a yeast centromere and construction of functional small circular chromosomes. Nature 287:504–9.

Cohen, J. D., T. R. Eccleshall, R. B. Needleman, H. Federoff, B. A. Buchferer, and J. Marmur. 1980. Functional expression in yeast of the *Escherichia coli* plasmid gene coding for chloramphenicol acetyltransferase. Proc. Natl. Acad. Sci. USA 77:1078–82.

Dickson, R. C. 1980. Expression of a foreign eukaryotic gene in *Saccharomyces cerevisiae*: B-galactosidase from *Kluyveromyces lactis*. Gene 10:347–56.

Dobson, M. J., A. B. Futcher, and B. S. Cox. 1980. Control of recombination within and between DNA plasmids of *Saccharomyces cerevisiae*. Curr. Genet. 2:201–5.

Fasiolo, F., J. Bonnet, and F. Lacroute. 1981. Cloning of the yeast methinyl-tRNA synthetase gene. J. Biol. Chem. 56:2324–28.

Faye, G., D. W. Lering, K. Tatchell, B. D. Hall, and M. Smith. 1981. Deletion mapping of sequences essential for *in vivo* transcription of the iso-l-cytochrome c. gene. Proc. Natl. Acad. Sci. USA 78:2258–62.

Fitzgerald-Hayes, M., J. M. Buhler, T. G. Cooper, and J. Carbon. 1982. Isolation and sub-cloning analysis of functional centromere DNA (CEN 11) from *Saccharomyces cerevisiae* chromosome XI. Mol. Cell Biol. 2:82–87.

Gaillardin, C., P. Fournier, F. Budar, B. Kudla, C. Berbaud, and H. Heslot. 1983. Replication and recombination of 2 um DNA in *Schizosaccharomyces pombe*. Curr. Genet. 7:245–53.

Gallwitz, D., F. Perrin, and R. Seidel. 1981. The actin gene in yeast *Saccharomyces cerevisiae*: 5′ and 3′ end mapping, flanking, and putative regulatory sequences. Nucl. Acid Res. 9:6339–49.

Gerbaud, C., P. Fournier, H. Blanc, M. Aigle, H. Heslot, and M. Guerineau. 1979. High frequency of yeast transformation by plasmids carrying part or entire 2 um yeast plasmid. Gene 5:233–53.

Guarente, L., and M. Ptashne. 1981. Fusion of *Escherichia coli lacZ* to the cytochrome c gene of *Saccharomyces cerevisiae*. Proc. Natl. Acad. Sci. USA 78:2199–2203.

Guerineau, M. 1979. Viruses and plasmids in fungi. M. Dekker, New York.

Guerry-Kopecko, P., and R. B. Wickner. 1980. Cloning of the URAI gene of *Saccharomyces cerevisiae*. J. Bacteriol. 143:1530–33.

Gunge, N., and K. Sakaguchi. 1981. Intergeneric transfer of DNA-like plasmids pGK11 and pGK12 from *Kluyveromyces lactis*. J. Bacteriol. 147:155–60.

Gunge, N., A. Tamaru, F. Ozawa, and K. Sakaguchi. 1981. Isolation and characterization of linear deoxyribonucleic acid plasmids from *Kluyveromyces lactis* and the plasmid asso-ciated killer character. J. Bacteriol. 145:382–90.

Hartley, J. L., and J. E. Donelson. 1980. Nucleotide sequence of the yeast plasmid. Nature 286:860–65.

Henikoff, S., R. Tatchell, B. D. Hall, and K. A. Nasmyth. 1981. Isolation of a gene from *Drosophila* by complementation in yeast. Nature 239:33–37.

Hereford, L., K. Fahrner, J. Woolford, M. Rosbach, and D. B. Kaback. 1979. Isolation of yeast histone genes H2A and H2B. Cell 18:1261–71.

Hinnen, A., J. B. Hicks, and G. R. Fink. 1978. Transformation of yeast. Proc. Natl. Acad. Sci. USA 75:1929–33.

Hitzemann, R. A., L. Clarke, and J. Carbon. 1980. Isolation and characterization of the yeast 3-phosphoglycerokinase gene (PGK) by an immunological screening technique. J. Biol. Chem. 255:1273–80.

Hitzemann, R. A., F. E. Hagie, H. L. Levine, D. V. Goeddel, G. Ammerer, and B. D. Hall. 1981. Expression of a human gene for interferon in yeast. Nature 293:717–22.

Holland, J. P., and M. J. Holland. 1979. The primary structure of glyceraldehyde-3-phosphate dehydrogenase gene from *S. cerevisae*. J. Biol. Chem. 254:9839–45.

Holland, J. P., and M. J. Holland. 1980. Structural comparison of two non-tandemly re-peated yeast glyceraldehyde-3-phosphate dehydrogenase genes. J. Biol. Chem. 255:2596–2605.

Holland, M. J., and J. P. Holland. 1979. Isolation and characterization of a gene coding for glyceraldehyde-3-phosphate dehydrogenases from *S. cerevisiae*. J. Biol. Chem. 254:5466–74.

Hsia, C. L., and J. Carbon. 1981. Direct selection procedure for the isolation of functional centromeric DNA. Proc. Natl. Acad. Sci. USA 78:3760–64.

Jimenez, A., and J. Davies. 1980. Expression of a transposable antibiotic resistance element in *Saccharomyces cerevisiae*: a potential selection for eukaryotic cloning vectors. Nature 287:869–87.

Kawaski, G. 1981. Isolation of glycolytic genes in *S. cerevisiae*. (Personal communication.)

Kielland-Brandt, M. D., B. Wilken, S. Holmberg, and J. G. Litske Petersen. 1980. Genetic evidence for nuclear location of 2 um DNA in yeast. Carlsberg Res. Commun. 45:119–24.

Kingsman, A. M., L. Clarke, R. K. Mortimer, and J. Carbon. 1979. Replication in *Saccharomyces cerevisiae* of plasmid pBR313 carrying DNA from the yeast trpI region. Gene 7:141–52.

Livingston, D. M., and S. Hahne. 1979. Isolation of a condensed, intracellular form of the 2 um DNA plasmid of *Saccharomyces cerevisiae*. Proc. Natl. Acad. Sci. USA 76:3727–31.

Livingston, D. M., and D. M. Kupfer. 1977. Control of *Saccharomyces cerevisiae* 2 um DNA replication and cell cycle genes that control nuclear DNA replication. J. Mol. Biol. 116:249–60.

Martinez-Arias, A., and M.J. Casadaban. 1981. Regulated expression of yeast leu2 *E. coli* B-galactosidase gene fusion in yeast. (Personal communication.)

McNeil, J. B., and J. D. Friesen. 1981. Expression of the *Herpes simplex* virus thymidine kinase gene in *Saccharomyces cerevisiae*. Mol. Gen. Genet. 184:386–95.

Mercereau-Puijalong, O., F. Lacroute, and P. Kourilsky. 1980. Synthesis of a chicken ovalbumin like protein in the yeast *Saccharomyces cerevisiae*. Gene 11:163–67.

Montgomery, D. L., B. D. Hall, S. Gillam, and M. Smith. 1978. Identification and isolation of the yeast cytochrome c gene. Cell 14:673–80.

Nasmyth, K. A., and S. I. Reed. 1980. Isolation of genes by complementation in yeast: molecular cloning of a cell-cycle gene. Proc. Nat. Acad. Sci. USA 77:2119–23.

Nelson, R. G., and W. L. Fangman. 1979. Nucleosome organization of the yeast 2 um plasmid: a eucaryotic minichromosome. Proc. Natl. Acad. Sci. USA 76:6515–19.

Ng, R., and J. Abelson. 1980. Isolation and sequence of the gene for actin in *Saccharomyces cerevisiae*. Proc. Natl. Acad. Sci. USA 77:3912–16.

Panthier, J. J., P. Fournier, H. Heslot, and A. Rambach. 1980. Cloned β-galactosidase gene of *Escherichia coli* is expressed in the yeast *Saccharomyces cerevisiae*. Curr. Genet. 2:109–33.

Roggenkamp, R., B. Kustermann-Kuhn, and C. P. Hollenberg. 1981. Expression and processing of bacterial β-lactamase in the yeast. *Saccharomyces cerevisiae*. Proc. Natl. Acad. Sci. USA 78:4466–70.

Rose, M., M. J. Casadaban, and D. Botstein. 1981. Yeast genes fused to β-galactosidase in *Escherichia coli* can be expressed normally in yeast. Proc. Natl. Acad. Sci. USA 78:2460–64.

Schell, M. A., and D. B. Wilson. 1979. Cloning and expression of the yeast galactokinase gene in an *E. coli* plasmid. Gene 5:291–303.

Seligy, V. L., D. Y. Thomas, and B. L. A. Miki. 1980. *Saccharomyces cerevisiae* plasmid, Scp or 2 um: intracellular distribution, stability, and nucleosome-like packaging. Nucl. Acid Res. 8:3371–91.

Stinchcomb, D. T., C. Mann, and R. W. Davis. 1982. Centromeric DNA from *Saccharomyces cerevisiae*. J. Mol. Biol. 158:157–59.

Stinchcomb, D. T., K. Struhl, and R. W. Davis. 1979. Isolation and characterization of a yeast chromosomal replicator. Nature 282:39–43.

Stinchcomb, D. T., M. Thomas, J. Kelly, E. Selker, and R. W. Davis. 1980. Eukaryotic DNA segments. Proc. Natl. Acad. Sci. USA 77:4559–63.

Struhl, K., and R. W. Davis. 1980. Production of a functional eukaryotic enzyme in *Escherichia coli*: cloning and expression of the yeast structural gene for imidazole glycerophosphate dehydratase (his3). Proc. Natl. Acad. Sci. USA 74:5255–59.

Struhl, K., D. T. Stinchcomb, S. Scherer, and R. W. Davis. 1979. High frequency transformation of yeast: autonomous replication of hybrid DNA molecules. Proc. Natl. Acad. Sci. USA 76:1035–39.

Toh-e, A., P. Guerry-Kopecki, and R. B. Wickner. 1980. A stable plasmid carrying the yeast LEU2 gene and containing only yeast DNA. J. Bacteriol. 141:413–16.

Toh-e, A., S. Tada, and Y. Oshima. 1982. 2 um DNA-like plasmids in the osmophilic haploid yeast *Saccharomyces rouxii*. J. Bacteriol. 151:1380–90.

Toh-e, A., and R. B. Wickner. 1981. Curing of the 2 um DNA plasmid from *Saccharomyces cerevisiae*, J. Bacteriol. 145:1421–24.

Tschumper, G., and J. Carbon. 1980. Sequence of a yeast DNA fragment containing a chromosomal replicator and the TRP1 gene. Gene 10:157–66.

Williamson, V. M., J. Bennetzen, I. Young, K. Nasmyth, and B. D. Hall. 1980. Isolation of the structural gene for alcohol dehydrogenase by genetic complementation in yeast. Nature 283:214–16.

Zakian, V. A. 1981 Origin of replication from *Xenopus laevis* mitochondrial DNA promotes high frequency transformation of yeast. Proc. Natl. Acad. Sci. USA 78:3128–32.

J. L. AZEVEDO, C. L. MESSIAS, AND
W. D. SILVEIRA

Genetics and Breeding of the Entomopathogenic Fungus *Metarhizium anisopliae*: Parasexuality and Protoplast Fusion

21

GENETIC IMPROVEMENT OF MICROORGANISMS AND
BIOLOGICAL CONTROL

Methods of Microbial Breeding

Microorganisms are largely used in such industrial processes as, among others, antibiotic, vitamin, amino acids, and enzyme production or in the biological control of pests and diseases, biological nitrogen fixation, production of foods, and degradation of pollutants (Azevedo, 1983; OTA, 1981; NAS, 1979). In the same way as useful plants and domestic animals are constantly submitted to breeding programs, microbes can be manipulated genetically toward their improvement. For this it is necessary to use the variability that occurs in all microbial species. Besides the natural variability, the microbial geneticist utilizes mutagenic agents. Examples of induced mutants of industrial application are numerous regarding antibiotic production, organic acids, and vitamins among other products (Azevedo and Bonatelli, 1982; Burnett, 1975; Sermonti, 1969). Besides mutation, recombination systems can combine favorable characteristics in the same cell to produce new genotypes. Classical systems of recombination are known in bacteria (transformation, conjugation, and transduction) and in fungi (sexual cycle). Non-orthodox systems also were described such as the parasexual cycle in fungi. More recently, new recombination systems that involve recombinant DNA and protoplast fusion techniques, among others, have been described (Hopwood, 1979).

In a tropical country such as Brazil with a vast area where distinct soil types and climates exist and with different species of cultivated plants, the problem of loss of production in agriculture by pests and diseases is of great importance. The use of pesticides and fungicides is in most cases not economical, dangerous, and of questionable efficiency (Pascoal, 1983). In tropical agriculture, one must then utilize specific technologies, and among others biological control of agriculture pests by entomopathogenic fungi is a useful alternative.

Biological Control of Insects

It was Agostino Bassi (Steinhaus, 1949) who for the first time demonstrated unequivocally that diseases of insects can be caused by microorganisms. He discovered that a fungus now known as *Beauveria bassiana* was pathogenic to silkworms. Because of the economic value of silkworms and bees, the research was based first on diseases of these insects. However, following this first phase, the research on insect pathology has extended to other insects. Metchnikoff (1879), a Russian scientist, presented this first experimental work conducted with an agricultural pest, *Anisophliae austriaca*, using the fungus *Metarhizium anisopliae*. He showed the possibility of microorganisms being used for the biological control of pests.

Other groups of microorganisms have been used since then on biological control of pests, including viruses, bacteria, and protozoa. However, because of their mode of action and being disseminated easily, the fungi have been largely used in Brazil for biological control of agricultural pests, such as spittlebugs of pasture and sugar cane. These are insects of the Homoptera order, Cercopidae family, as for instance *Mahaarva posticata* (spittlebugs of sugar cane), *Zulia entreriana, Deois schach, and D. flavopicta* (spittlebugs of pastures).

The fungus *M. anisopliae* has been shown to control efficiently these insects both in laboratory and field conditions. *M. anisopliae* has been used since 1970 in sugar cane fields for the control of *M. posticata* populations in the northeast of Brazil. It has also been used to control pests of pastures in several areas of this country.

METARHIZIUM ANISOPLIAE AND BIOLOGICAL CONTROL

Biological Aspects

Metarhizium anisopliae occurs as two varieties, *minor* and *major*, which differ from each other by their conidial size (Tullock, 1976). It is an imperfect fungus (Deuteromycete) having no sexual stage of reproduction. It was described for the first time by Metchnikoff (1879) in insects that attack stored grains. This fungus has green conidia conferring to the affected insect a characteristic aspect known as green muscardine. The fungus can use several different carbon sources and can grow on a wide range of pH's, the optimum being around pH 6.9 (Veen, 1981).

Strains isolated from different Brazilian regions are all from the *minor* variety (*M. anisopliae*). Such strains as have been cultured at 28°C and have rice grains as substrate produce abundant conidiation 7–10 days after inoculation.

Insect Infection

The main pathway of insect fungus infection is via the host cuticule. The colonization is a process that begins by the germination and penetration of conidia in the surface of the body of the host insect. According to the environmental conditions (Ferron, 1977), the fungus infection can change. Abiotic factors, mainly temperature and humidity, are important in the microorganism's development. The most frequent path for fungus infection is the conidia, as in the specific case of *M. anisopliae*. Blastospores are also able to produce infection but because of the survival characteristics, conidia infect more frequently. These germinate and penetrate the insect epicuticule (Samsinakova and Misikova, 1973; Zacharuch, 1970). Oral infection occurs rarely. Penetration is performed with the help of enzymes produced by the fungus, such as chitinases, proteases, and lipases (Samsinakova et al., 1971). Also important are toxins produced by the pathogenic fungus that cause death of the host (Roberts, 1966a,b). After penetration, mycelium is produced in the interior of the host insect and all the spaces occupied by the host hemolymph are filled, which nutritionally deprives the host. In an advanced infection stage, all the host organs are attacked by the fungus. Time of penetration varies with conidia germination and enzyme production. Speed of death depends on toxin production (Roberts, 1981). For instance, in *D. flavopicta* third instar insects in 85% humidity conditions and temperature of 28°C, conidia of *M. anisopliae* penetrate the host after 72 hr and death occurs in 144 hr after contact with the pathogen (Silveira and Messias, 1982). Death time varies also according to the host. In *D. flavopicta* LT 50 (doses producing 50% death) is eleven days in adults. After this time in 85% humidity and 28°C, conidia are produced on the surface of the dead insect and can serve as inoculum to other healthy insects. Susceptibility of the hosts is variable, but until now development of host resistance to fungi has not been detected. However, it is known in insects that a resistance mechanism can be induced. Seryczynska and Basan (1975) detected in the Colorado beetle (*Leptinotarsa decemlineata*) an increase in hemocyte number in the hemolymph when *Neoplectana earpocapsae* was associated to the fungi *Beauveria bassiana* and *Paecilomyces farinosus*.

Production of Infective Units and Field Application

Production of infective units of *M. anisopliae* is being carried out in Brazil in culture medium having grains of rice as a substrate adding water (maximum amount of 50%). This is done in polypropylene bags, which are resistant to autoclave treatment. The inoculum is prepared in Roux flasks containing potato-dextrose-agar media. Conidia are collected, and a suspension is prepared for the inoculation of the polypropylene bags, which have been autoclaved previously with rice grains and water. After inoculation the bags are incubated 10–15 days. Conidia are harvested and, depending on the method used, the final material can be composed either of pure conidia or a mixture of conidia plus inert material.

Field application is carried out in the form of dry powder or aqueous suspension; generally, 10^{13} conidia are applied per hectare. This quantity of applied material depends on the purity and germination of conidia and also on the pest population and environmental conditions. The infectivity of the application, at least on sugar cane, has been calculated by the decrease of sucrose amounts. In treated areas this can be reduced by about 8% (Marques et al., 1981).

FORMAL GENETICS OF *METARHIZIUM ANISOPLIAE*

Natural Variability

Several isolates of *M. anisopliae* obtained from various regions of Brazil and other countries are now available for use in biological control. From the beginning of the studies in Brazil, results from several sources have indicated that strains can differ regarding morphology of the colonies (Aquino and Vital, 1979) and other characteristics such as growth habits, electrophoretic patterns, gas chromatography curves, conidial size, virulence, production of exogenous enzymes, resistance to inhibitors including ultraviolet light, heavy metal salts, fungicides, insecticides, and so on (Al Aidroos and Roberts, 1978; Alves, 1982; Azevedo and Messias, 1981; De

Conti et al., 1980; Haddad and Messias, 1979; Matos, 1983; Messias et al., 1978, 1983; Rosante Gomes, 1983; Rosato et al., 1981; Santos and Azevedo, 1982; Sosa-Gomes, 1983). Evaluation of this high variability is important and necessary for the application of a genetic breeding program in the species. Also, the characterization of different strains by means of distinct parameters can help to distinguish strains after a field application in areas where the fungus is already present. The system using electrophoretic patterns has proved to be especially useful. In table 1 we present several parameters that reveal the great variability found in *M. anisopliae*.

Induced Variability

As for other fungi, the isolation of induced mutants in *M. anisopliae* is based on the use mainly of ultraviolet light and alkylating agents (Okino et al., 1978; Silveira, 1983). Thanks to the use of these mutagenic agents, several types of mutants were isolated such as auxotrophic and morphological including conidial color mutants (Al Aidroos, 1980; Messias, 1981; Messias and Azevedo, 1980; Silveira, 1983; Tinline and Noviello, 1971), mutants resistant to inhibitor agents (Matos, 1983; Santos, 1978; Santos and Azevedo, 1982), and mutants with modified enzyme production (Al Aidroos and Seifert, 1980; Azevedo et al., 1981; Penalva da Silva and Azevedo, 1983). Recently a new enrichment technique to isolate auxotrophic mutants of *M. anisopliae* was developed (Silveira, 1983; Silveira and Azevedo, 1982). In this case plastic filters with six muslin layers were used. After several filtrations and measuring the time between each filtration, it was possible to get an increase of 61.3 times of auxotrophic mutants in comparison with the total isolation technique which does not include previous selection. It was after the isolation of mutants, especially the auxotrophic and conidial color mutants, that it was possible to detect the parasexual cycle in *M. anisopliae*.

Parasexuality in M. anisopliae

Parasexuality was demonstrated in several fungi (Azevedo, 1976). Especially, in the imperfect fungi, this is the only process of recombination. In this way, the discovery of the parasexual cycle in imperfect fungi, as is the case of *M. anisopliae*, is of extreme importance not only for knowledge of their genetics but also for genetic improvement of the species. The availability of auxotrophic and color mutant markers induced in a wild type strain of *M. anisopliae* (strain E9) permitted the discovery of the parasex-

TABLE I

VARIABILITY FOR DIFFERENT CHARACTERISTICS IN
SIX STRAINS OF *METARHIZIUM ANISOPLIAE*

STRAIN	CONIDIAL VOLUME (μ^3)[a]	COLONY GROWTH (0 PER CM)[b]	PRODUCTION OF SPORES IN DIFFERENT VISIBLE LIGHT[c]			ENZYME PRODUCTION[d]				SURVIVAL (%)[e]	ESTERASE PATTERN. NUMBER OF BANDS[f]
			Dark	Blue Light	Yellow Light	Amylase	Lypase	Chitinase	Protease		
A₄	21.55	3.1	44	114	19	0.66	0.66	0.57	0.81	—	4
C	—	2.8	11	18	9	1.0	0.70	0.70	0.63	—	5
E₉	12.0	2.8	183	253	30	0.76	0.75	0.69	0.84	30	5
K	—	2.8	1	7	5	1.0	0.81	0.75	0.51	3	4
M	12.0	2.8	263	223	130	—	—	—	—	18	5
P	—	2.8	80	77	72	—	—	—	—	23	—

[a] J. L. Azevedo and C. L. Messias.
[b] T. L. Vale, S. M. Frigo, C. L. Messias, and J. L. Azevedo.
[c] T. L. Vale, S. M. Frigo, C. L. Messias, and J. L. Azevedo. Production × 10⁸ in absence of light (dark), monochromatic blue and yellow lights.
[d] Y. B. Rosato, C. L. Messias, and J. L. Azevedo. Measurement in 0 colony 0 halo.
[e] J. L. Azevedo, N. M. Chacel, and P. Timo. Survival after same times of UV light treatment.
[f] E. Conti, C. L. Messias, H. M. L. Souza, and J. L. Azevedo. Same number of bands in different strains can be represented by distinct mobilities.

ual cycle in this species (Messias and Azevedo, 1980). At first, heterokaryons were forced by inoculation of roughly equal numbers of conidia from distinct auxotrophic mutants. Heterokaryotic growth was observed on a defined medium where the parent strains could not grow. Heterokaryons were at that time, described in *M. anisopliae* (Tinline and Noviello, 1971) but not diploids. To isolate diploids, conidia from typical heterokaryotic regions were suspended and about 10^6 plated onto a defined medium where the parental strains could not grow. Prototrophic colonies were recovered from this medium with a frequency of $1-2$ in 10^6 conidia plated. These colonies were shown to be diploid by several methods of such as DNA content, nuclear size, and segregation of original parental and recombinant types mainly on medium where a haploidizing agent called chloroneb (Azevedo et al., 1979) was added. By these results, we showed unambiguously the existence of a kind of parasexuality in *M. anisopliae* starting with a heterokaryon formed between two parental strains, passing through a diploid stage that was quite stable except in the presence of some haploidizing agents, which induced the formation of recombinants (table 2). This opened new possibilities for the combination of favorable markers regarding biological control in genetically improved strains. At about the same time, the existence of parasexuality in *M. anisopliae* was confirmed using a different strain of fungus (Al Aidroos, 1980). Parental strains, diploids, and recombinants were used in an attempt to find an improved strain for biological control. Although some diploids have a slightly improved virulence in relation to the parental strains, no improved recombinants were found (Riba et al., 1985). This was expected since all strains crossed were derived from the same original E9 strain. Attempts to cross strains from different regions were mainly unsuccessful probably because of incompatibility problems between them. A new approach was then used, that is, protoplast fusion.

PROTOPLAST FUSION

Among the main approaches for recombination in microorganisms developed in the seventies (Azevedo, 1979, 1983; Hopwood, 1979), a very promising one is the technology of protoplast fusion. The development of such a technique was the result of several areas of work starting with ways to isolate protoplasts, regeneration of protoplasts, and ways to facilitate fusion (Anne and Peberdy, 1975; Ferenczy, 1976). Although protoplasts of fungi have been known from many years, fusion between two protoplasts could not be increased and consequently was observed only occasionally.

TABLE 2

STRAINS OF HETEROZYGOUS DIPLOIDS
(STRAINS E$_g$) OF *METARHIZIUM ANISOPLIAE*

DIPLOID	FREQUENCY ($\times 10^6$ CONIDIA)	SECTORS DERIVED FROM DIPLOIDS	
		Type	Number
ylo pyr + + ————————— + + *pvi pab*	1.8	*ylo pyr* *pvi pab* wild type	1 15 9
ylo pyr + + ————————— + + *whi thi*	1.4	*whi thi* *ylo pyr* wild type	5 10 10

After C. L. Messias and J. L. Azevedo (modified). *ylo, pvi,* and *whi* are yellow, palevinaceous, and white color of conidia, respectively; *pab, pyr,* and *thi* are requirements, respectively, for p-aminobenzoic acid, pyridoxin, and thiamine.

Today all the previous steps to fusion are quite well known: protoplasts are produced in great quantities by the use of lytic enzymes (Ferenczy, 1981a; Peberdy, 1979). Stabilizers maintain the protoplasts, and regeneration of them is accomplished in solid medium (Ferenczy, 1981a; Peberdy, 1979). It is between production and regeneration of protoplasts that fusion can occur induced by the so-called fusogenic substances such as polyethylene glycol in the presence of Ca^{++} ions. Polyethylene glycol was used first for the fusion of plant cells, but after this discovery the same was shown to be valid for fungi as well (Anne and Peberdy, 1975). Selection of fusion products can be carried out in the same way as shown for heterokaryons and diploids (see "Parasexuality in *M. anisopliae*," above) if genetically marked strains are available. Techniques based on natural markers when crosses are done between divergent strains or distinct species (Carrau et al., 1982) or dead donors (Ferenczy, 1981b) or even liposomes (Makins et al., 1981) can also be used. Anyway, the technique of protoplast fusion opened the possibility for crosses between incompatible strains or even distinct species. Diploids or heterokaryons between divergent crosses are unstable and instability increases with the increase of diversity between the parents used in these fusions.

Several strains of *M. anisopliae* are incompatible, and consequently they cannot be crossed through conventional parasexuality. To overcome such incompatibility, the process of protoplast fusion was used in *M. anisopliae* (Silveira, 1983).

Protoplast Fusion in M. anisopliae

It was possible to obtain protoplasts from *M. anisopliae* when the osmotic stabilizer necessary for the maintenance of the protoplast was known and after identifying an appropriate lytic enzyme. Although there is variation according to the strain used, KCl 0.7M in phosphate buffer, pH 5,8, was the osmoticum that yielded the best results in several strains. After tests with different lytic enzymes such as the ones derived from *Trichoderma*, enzymes produced in our laboratory, and snail extract, it was found that a mixture of enzymes (CP plus Novozym 234) in equal amounts, which was used for several other fungi (Hamlyn et al., 1981), presented the best results, producing a highly concentrated suspension of protoplasts from *M. anisopliae*. Appropriate time of mycelium treatment with this mixture of enzymes varies between 2 to 3 hr according to the strain used (Silveira, 1983). Fusion was carried out using polyethylene glycol treatment of protoplasts derived from strains that were not able to be crossed in a normal way because of incompatibility. That was the case, for instance, of two strains, the first one (E6) isolated from Espirito Santo State and a second one (RJ) isolated in Rio de Janeiro State. When protoplasts from these strains carrying at least two auxotrophic markers each, besides morphological markers, were treated with polyethylene glycol and plated after treatment on a selective minimal medium, small colonies were able to grow. Such colonies transferred to other plates containing the same medium yielded a heterokaryotic growth. On complete medium these colonies produced sectors. Conidia from the sectors plated again on complete medium produced colonies that, when tested for growth requirements, produced not only parental types but recombinant types as well (table 3). Although typical diploid colonies were not found when such incompatible strains were crossed, the presence of recombinant types after protoplast fusion suggested that in the heterokaryons between such strains nuclear fusion must occur. The transient diploid is probably very unstable and produces mitotic recombinants. This process, forming very unstable diploids followed by spontaneous haploidization in such a way that recombinants are produced directly from a fusion product such as a protoplast or heterokaryon, was described in *Cephalosporium acremonium* (Ball and Hamlyn, 1982). In *M. anisopliae* although parameiosis does not occur in crosses involving mutant strains derived from a common ancestor, it occurs when crosses between divergent strains are employed, as in the case of the E6 and RJ strains. At any rate, protoplast fusion has been shown to be an alternative technique for crosses between incompatible strains resulting in the production of recombinants in *M. anisopliae*.

TABLE 3

SEGREGANTS FROM COLONIES DERIVED FROM PROTOPLAST FUSION
BETWEEN THE SAME AND DIVERGENT STRAINS OF *METARHIZIUM ANISOPLIAE*

CROSS	FREQUENCY OF FUSION ($\times 10^5$)	SEGREGANTS	
		Type	Number
E_6 *pvi;met;bio* $\times$ E_6 *ylo;pyr;lys*	44.1	*ylo pyr*	4
		met	2
		bio	2
		lys	5
		bio lys	5
		bio lys pyr	13
		met lys pyr	1
		parental types	13
E_6 *ylo;ade;cyc* $\times$ RJ *pvi;lys;nic*	0.88	*pvi bio*	2
		pvi ylo met	1
		ylo ade cys lys	1
		pvi lys	1
		parental types	23

After W. D. Silviera. *ade; bio; cys; lys; met; nic; pyr* are, respectively, requirements for adenine; biotin; cysteine; lysine; methionine; nicotinic acid, and pyridoxin. *pvy* and *ylo* are, respectively, palevinaceous and yellow color of conida.

CONCLUSIONS

The entomopathogenic fungus *M. anisopliae* has been shown in some insects and in some Brazilian regions to be an alternative way to control pests such as spittlebugs of pastures and pests of sugar cane. Recently, its efficiency against insects that are vectors of diseases such as Chagas' disease among others also was demonstrated (Messias, unpublished). Because it is a fungus that produces a great amount of conidia (Naves, 1980; Sena et al., 1979) and is inocuous for animals (El-Kadi et al., 1983; Pereira and Ventura, 1979; Ventura, 1981), the possibility of using it on a large scale to control insects is great. The genetic variability found among strains of *M. anisopliae* indicate that genetics can play an important role in the production of more virulent strains or strains with high specificity against particular insects. Also genetic improvement can be carried out aiming at the production of strains better adapted to certain climatic conditions or strains with rapid and higher conidia production, which are important factors on an industrial scale. In Brazil, because of the diversity found in different regions of the country, it is necessary to think in terms of the use of distinct improved strains according to pest, culture, climatic conditions, and so on. The introduction of new characteristics in strains already being used can be done by a classical parasexual way or by protoplast fusion. Possibly, protoplast fusion technique plus recombinant DNA technology will enable generation of new entomopathogenic fungi strains useful for biological control.

REFERENCES

Al Aidroos, K. 1980. Demonstration of parasexual cycle in the entomopathogenic fungus *Metarhizium anisopliae*. Can. J. Genet. Cytol. 22:309–14.

Al Aidroos, K., and D. W. Roberts. 1978. Mutants of *Metarhizium anisopliae* with increased virulence toward mosquito larvae. Can. J. Genet. Cytol. 20:211–19.

Al Aidroos, K., and A. M. Seifert. 1980. Polysaccharide and protein degradation, germination, and virulence against mosquitoes in the entomopathogenic fungus *Metarhizium anisopliae*. J. Inv. Pathol. 36:29–34.

Alves, S. B. 1982. Caracterizacao, padronizacao, e producao de *Metarhizium anisopliae* (Metsch.) Sorok. Livre Docencia, Thesis—ESALG—Univ. of S. Paulo 110 pp.

Anne, J. L., and J. F. Peberdy. 1975. Induced fusion of fungal protoplasts following treatment with polyethlene glycol. Arch. Microbiol. 105:201–5.

Aquino, M. L. N., and A. F. Vital. 1979. Caracteres culturais de alguns isolados de *Metarhizium anisopliae* (Metschnkoff) Sorokin. Pesquisa Agrop. Pernambuco. 3:21–28.

Azevedo, J. L. 1976. Variabilidade em fungos fitopatogenicos. Summa Phytopathol. 2:2–15.

Azevedo, J. L. 1979. Novos processos de recombinacao em microrganismos *I⁰ Simp. Enol. Vit. Inst. Biotec.* Caxias do Sull, RS. pp. 6–18.

Azevedo, J. L. 1983. *Genetica de Microrganismos no Brasil.* FEALQ, Piracicaba, SP. 400 pp. (in press).

Azevedo, J. L., and R. Bonatelli, Jr. 1982. Genetics of the overproduction of organic acids. In V. Krumphanzl, B. Sikita, and Z. Vanec (eds.), Overproduction of microbial products, pp. 432–50. Academic Press, New York.

Azevedo, J. L., and C. L. Messias. 1981. Tamanho de conidios em diferentes isolates de *Metarhizium anisopliae*. Bol. Grupo Pesq. Biol. Contr. 2:5–8.

Azevedo, J. L., E. P. Santana, and R. Bonatelli, Jr. 1979. Resistance and mitotic instability to chloroneb and 1,4-oxathlin in *Aspergillus nidulans*. Mut. Res. 48:163–72.

Azevedo, J. L., N. M. Chacel, and P. Timo. 1981. Resistencia a luz ultravioleta e mutantes de *Metarhizium anisopliae*. VIII Reuniao Gen. Microorg. Belo Horizonte, 8:1 (abstract).

Ball, C., and P. F. Hamlyn. 1982. Genetic recombination studies with *Cephalosporium acremonium* related to the production of the industrially important antibiotic cephalosporium C. Rev. Bras. Genet. 5:1–13.

Bonatelli, R. Jr., J. L. Azevedo, and G. V. Valent. 1983. Parasexuality in a citric acid producing strain of *Aspergillus niger*. Rev. Bras. Genet. 6:399–405.

Burnett, J. H. 1975. Mycogenetics. John Wiley and Sons, New York.

Carrau, J. L., J. L. Azevedo, P. Sudbery, and D. Campbell. 1982. Methods for recovering fusion products among oenological strains of *Saccharomyces cerevisiae* and *Schizosaccharomyces pombe*. Rev. Bras. Genet. 5:221–26.

Congress, United States, Office of Technology Assessment. 1981. Impacts of applied genetics: microrganisms, plants, and animals. Washington D.C. 331 pp.

De Conti, E., C. I. Messias, H. M. L. Souza, and J. L. Azevedo. 1980. Eletrophoretic variation in esterases and phosphatases in eleven wild-type strains of *Metarhizium anisopliae*. Experientia. 36:293–94.

El-Kadi, K. K., L. S. Xara, P. F. Matos, J. V. N. Rocha, and D. P. Oliveira. 1983. Effects of the entomopathogen *Metarhizium anisopliae* on guinea pigs and mice. Environ. Entomol. 12:37–42.

Ferenczy, L. 1976. Protoplast fusion in fungi. *In* D. Dudits, P. Farkas, and P. Maliga (eds.), Cell genetics in higher plants, pp. 171–82. Akademia Kiado, Budapest.

Ferenczy, L. 1981a. Microbial protoplast fusion. In D. Glover and D. A. Hopwood (eds.), Genetics as a tool in microbiology, pp. 1–34. Society for General Microbiology, Cambridge.

Ferenczy, L. 1981b. Influence of relative humidity on the development of fungal infection caused by *Beauveria bassiana* (Fungi-Imperfecti, Moniliales). Imagines of Alanthoscelides obtectus (Col.: Bruchidael). *Entomophaga* 22:393–96.

Haddad, C. R. B., and C. L. Messias. 1979. Influencia de espalhantes adesivos e de inseticidas na germinacao e desenvolvimento de *Metarhizium anisopliae*. Bol. Grup. Pesq. Biol. Contr. 1:6.

Hamlyn, P. F., R. E. Bradshaw, F. M. Mellon, C. M. Santiago, J. M. Wilson, and J. F. Peberdy. 1981. Efficient protoplast isolation from fungi using commercial enzymes. Enzymes and Microbial Techn. 3:321–25.

Hopwood, D. A. 1979. The many faces of recombination. *In* O. K. Sebek and A. I. Laskin (eds.), Proceedings Third Int. Symp. Genet. Microorg., pp. 1–9. American Society for Microbiology, Washington, D.C.

Makins, J. F., R. Rodicio, K. F. Chater, and D. A. Hopwood. 1981. Liposome protoplast fusion in the biochemical and genetic manipulation of microorganisms. FEMS Symp. Overpr. Microbial. Prod. p. 236. Hradec-Kralove.

Marques, E. J., A. M. Vilasboas, and C. E. F. Pereira. 1981. Orientacoes tecnicas para a producao de fungo entomogeno *Metarhizium anisopliae* (Metschn.) em laboratorios setoriais. Bol. Tec. Planalsucar, Piracicaba. 3:5–23.

Matos, A. J. A. 1983. Crescimento e mutantes de *Metarhizium anisopliae*. M.S. thesis, UNICAMP, Campinas, S.P. 78 pp.

Messias, C. L. 1979. Parassexualidade em *Metarhizium anisopliae* (Metsch.) Sorokin. Ph.D. thesis, ESALQ, Univ. de Sao Paulo, 73 pp.

Messias, C. L., and J. L. Azevedo. 1980. Parasexuality in the Deuteromycete *Metarhizium anisopliae*. Trans. Brit. Mycol. Soc. 75:473–77.

Messias, C. L., J. L. Azevedo, E. De Conti, and H. M. L. Souza. 1978. Aspectos biologicos e inducao de mutantes em *Metarhizium anisopliae*. 3d Congres. Latino-Americano Entomol. Soc. Bras. Entomol. p. 69. Itabuna, Bahia (abstract).

Messias, C. L., D. W. Roberts, and A. T. Griff. 1983. Pyrolysis-gas chromatography of the fungus *Metarhizium anisopliae*: an aid to strain identification. J. Inv. Pathol. 42:393–97.

Metchnikoff, E. 1879. "Diseases of the larva of the grain weevil. Insects harmful to agriculture". Issue III. The grain weevil. Commission attached to the Odessa Zemstro Office for the Investigation of the Problem of Insects Harmful to Agriculture, Odessa (in Russian).

National Academy of Sciences. 1979. Microbial processes: Promising technologies for developing countries. Washington, D.C. 195 pp.

Naves, M. A. 1980. As cigarrinhas das pastagens e sugestoes para sue controle. Circular Tecnica no. 3. CPAC-Embrapa, Brasilia, D.F. 27 pp.

Okino, L. A., J. C. Silva, A. L. L. Santos, C. L. Messias, and J. L. Azevedo. 1978. Determinacao da sobrevivencia de *Metarhizium anisopliae* e duas especies de *Apergillus* a radiacao gama. O. Solo. 70:32–36.

Pascoal, A. D. 1983. O onus do medelo da agricultura industrial. Rev. Bras. Tecnol. 14:17–27.

Peberdy, J. F. 1979. Fungal protoplasts: isolation, reversion, and fusion. Ann. Rev. Microbiol. 33:21–39.

Penalva Da Silva, F. P., and J. L. Azevedo. 1983. Esterase pattern of a morphological mutant of the Deuteromycete *Metarhizium anisopliae*. Trans. Brit. Mycol. Soc. 81:161–63.

Pereira, F. F. L., and J. A. Ventura. 1979. Observacoes sobre a inoculacao de *Metarhizium anisopliae* (Metsch.) Sorokin no camundongo albino. I. Efeito de imunosupressores. Bol. Grupo Pesq. Biol. Control 1:5.

Riba, G., J. L. Azevedo, C. L. Messias, and W. D. Silvera. 1985. Studies on the inheritance of virulence in the entomopathogenic fungus *Metarhizium aniospliae*. Jour. Inv. Pathol. 46:20–25.

Roberts, D. A. 1966a. Toxins from the entomogenous fungus *Metarhizium anisopliae*. I. Production in submerged and surface cultures, and in inorganic and organic nitrogen media. J. Inv. Pathol. 8:212–21.

Roberts, D. A. 1966b. Toxins from the entomogenous fungus *Metarhizium anisopliae*. II. Symptons and detection in moribund hosts. J. Inv. Pathol. 8:222–27.

Roberts, D. A. 1981. *Microbial control of pests and diseases. 1970–1980*. Academic Press, Inc. London. p. 442–464.

Rosante Gomes, O. C. 1983. Resistencia a metais pesadoes e defensivos agricolas em *Metarhizium anisopliae* (Metsch) Sorokin. M.S. thesis, ESALQ, Univ. S. Paulo. 90 pp.

Rosato, Y. B., C. L. Messias, and J. L. Azevedo. 1981. Production of extracellular enzymes from isolates of *metarhizium anisopliae*. J. Inv. Pathol. 38:1–3.

Samsinakova, A., and S. Misikova. 1973. Enzyme activities in certain entomophagous representatives of Deuteromycetes (Moniliales) in relationship to their virulence. Ceska Mycol. 27:55–60.

Samisinakova, A., S. Misikova, and J. Leopold. 1971. Action of enzymatic systems of *Beauveria bassiana* on the cuticle of the greater max mottle larvae (*Galleria mellonella*). J. Inv. Pathol. 18:322–30.

Santos, A. L. L., and J. L. Azevedo. 1982. Resistencia do fungo entompatogenico *Metarhizium anisopliae* em relacao a tres fungicidas. Rev. Microbiol. 13:272–78.

Sena, R. C., V. L. B. Cavalcanti, and M. L. N. Aquino. 1979. Producao em larga escala de *Metarhizium anisopliae* (Metsch.) Sorokin, na CODECAP. Anais Simp. Ciencias Brasicas Aplicadas. Publ. 19. ACIESP, S. Paulo. Pp. 176–77.

Sermonti, G. 1969. Genetics of antibiotic producing microorganisms. Wiley Interscience, London.

Seryczynska, H., and C. Basan. 1975. Defensive reactions of L^3L^4 larvae of the Colorado beetle to the insecticidal fungi *Paecilomyces fainous* (Dicks) Brown and Smith, *Paecilomyces fumoso-roseus* (Wize), *Beauveria bassiana* (Bolo/ Unill) (Fungi Imperfecti: Moniliales). Boll. l'Academ. Polon. Sci. Ser. Biol. 23:227–94.

Silveira G. A. R., and C. L. Messias. 1982. Infeccao de *Deois* sp. por *Metarhizium anisopliae*: tempo de penetracao e colonizacao. Cienca e cultura 34:760 (abstract).

Silveira, W. D. 1983. Obtencao e fusao de protoplastos em *Metarhizium anisopliae* (Metsch) Sorokin. M.S. thesis, ESALQ, Univ. S. Paulo. 153 pp.

Silveira, W. D., and J. L. Azevedo. 1982. Obtencao de mutantes pela tecnica de enriquecimento em *Metarhizium anisopliae*. IXA Reuniao Annual Genet. Microorg. 9:12.

Sosa-Gomes, D. R. 1983. Caracterizacao e padronizacao de ll isolados de *Metarhizium anisopliae* (Metsch) Sorokin, 1883, provenientes de diferentes regioes do Brasil. M.S. Thesis, ESALQ, Univ. S. Paulo, 117 pp.

Steinhaus, E. A. 1949. *Principles of insect pathology*. McGraw-Hill, New York. 757 pp.

Tinline, R. D., and C. Noviello. 1971. Heterocaryons in the entomogenous fungus *Metarhizium anisopliae*. Mycologia 63:701–12.

Tullock, M. 1976. The genus *Metarhizium*. Trans. Brit. Mycol. Soc. 66:407–11.

Vale, T. L., S. M. Grigo, C. L. Messias, and J. L. Azevedo. 1981. Variabilidade em linghagens haploides de *Metarhizium anisopliae*. VII Congr. Bras. Entomol. Campinas, S.P. 347 pp.

Veen, K. 1981. Recherches sur la maladie due à *Metarhizium anisopliae* chez le criquet pelerin. Landbouwhogeschol Wageningen Netherland 68:1–77.

Ventura, J. A. 1981. Testes de seguranca na utilizacao do *Metarhizium anisopliae*. Bol. Grupo Pesq. Contr. Biol. 2:11

Zacharuch, R. Y. 1970. Fine structure of the fungus *Metarhizium anisopliae* infecting three species of larval elateridae (Coleoptera). I. Dormant and germinating conidia. J. Inv. Pathol. 15:63–80.

E. CHARTONE-SOUZA

Bacterial Behavior

22

ABSTRACT

The study of bacterial behavior and of chemotaxy in particular has advanced greatly after the introduction of new methodologies including the techniques of gene manipulation and cloning, which combine the analytical powers of genetics and biochemistry. Considerable progress has been made at the molecular level in terms of learning about the structure and function of many of the elements that participate in the bacterial sensory system, i.e., the receptors, messengers, processing center, and the motor response provided by the flagellum.

INTRODUCTION

One of the common types of behavior among living beings is tactism, which is motion in favor of or against a known stimulus. In addition to responding to chemotaxy, where the stimulus is a chemical agent, bacteria can respond to light, temperature, and magnetism, including the gravitational field of the earth. It is fascinating how these minuscule cells, which at the same time are real organisms, respond to external stimuli and adapt to the environment (Adler, 1976). Today we know that bacteria have receptors on their surfaces, as higher organisms have eyes and a nose, and also have a primitive system of information processing and a motor response.

A system is more interesting when it is also relevant to other species. Although it would be an exaggeration to say that "if you saw one cell you saw them all" (Koshland, 1980), biochemistry has shown that nature repeats certain patterns and thus the most primitive cells share many characteristics with the most complex organisms.

Chemotaxy, which is the basis of bacterial behavior and which will be discussed hereafter, was discovered at the end of the past century (Engelmann, 1881; Pfeffer, 1888). The discussion of this subject becomes pertinent when we consider that the study of behavior at the unicellular and molecular level is founded on bacteria, *Escherichia coli* in particular, the main tool and "star" of genetical engineering, since it represents the simplest behavioral system and doubtlessly the most extensively studied in its biochemical and genetical aspects. In addition, the great advances in understanding of the intimate mechanisms of bacterial chemotaxy are due to the application of recombinant DNA and cloning techniques that have permitted the definition of practically all gene products involved in information processing (Silverman and Simon, 1977; DeFranco and Koshland, 1981).

SOME STUDY METHODS

The study of bacterial chemotaxy generated many papers soon after its discovery (Engelmann, 1881; Pfeffer, 1888), but interest in this area of research started to lag in the twenties. The subject began to be intensely explored again after 1966, and research has continued until today thanks to the formulation of defined media (Adler, 1966, 1973; Adler and Templeton, 1967) and to the elaboration of quantitative and semiquantitative methods (Adler, 1969; Dahlquist et al, 1972; Macnab and Koshland, 1972; Tso and Adler, 1974; Armitage et al., 1977; Chartone-Souza, 1979).

The "capillary method" (Pfeffer, 1888) consists of the introduction of a capillary tube containing the solution under study into bacterial suspension (fig. 1), thus creating a gradient, and of observation at the microscope. If the solution under study is an attractant, bacteria will accumulate around the opening of the tube and inside it (positive chemotaxy); if it is a repellent, the opposite will occur, i.e., the bacteria will move away (negative chemotaxy). This method has been successively modified and made quantitative when convenient dilutions of the bacterial suspension in the capillary tube were seeded onto solid culture medium and colonies were later counted (Adler, 1969).

A new strategy was then introduced into the method for the study of negative chemotoxy, i.e., the repellent was placed in the bacterial suspension and the bacteria that took shelter in the capillary were counted (Tso and Adler, 1974) (fig. 1).

Another method—the "plate method" (Adler, 1966)—widely used in positive chemotaxy studies consists of inoculating a bacterial suspension into the center of a Petri dish containing soft agar (0.2 to 0.3%), some salts,

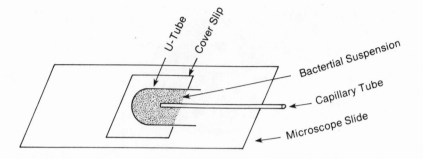

Fig. 1. Apparatus used in the capillary method.

and the metabolizable attractant. The bacteria themselves, by consuming the local supply of attractant, create the gradient, which is detected and followed by them with the consequent formation of a halo, visible to the unaided eye, whose radius grows around the original inoculum as shown in figure 2. For the study of negative chemotaxy (Tso and Adler, 1974), the bacterial suspension is seeded onto the medium described earlier at sufficient concentrations to cause turbidity. Agar blocks or disks soaked with the probable repellent are distributed. The bacteria soon start moving away, forming empty haloes around the repellent source (fig. 3). Study of the areas covered by wild bacteria in the dish during formation of the halo permits the derivation of mutants both for positive and negative chemotaxy (figs. 2–3).

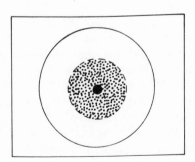

Fig. 2. Plate containing an attractant.

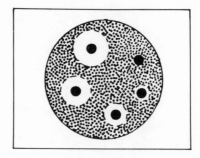

Fig. 3. Discs containing different concentrations of repellents (Tso and Adler, 1974).

The defined gradient method (Dahlquist et al., 1972) can be used to study both attractants and repellents. The authors invented a "populational migration apparatus" that measures the dislocation of a bacterial population in a space gradient by light scattering. In this case, the changes in agent concentration affecting the bacteria are generated by their motion through the gradient.

For the study of the "time gradient," i.e., of "bacterial memory," a small device has been invented whereby the bacteria are submitted to abrupt changes (0.2 second) in attractant or repellent concentration (Macnab and Koshland, 1972). Thus, a spatially uniform environment in terms of a chemical agent is abruptly transformed into another environment that is also spatially uniform. The bacterial response, i.e., the path of the bacteria in liquid medium after mixing and therefore the creation of a time gradient, is photographed in a dark field using strobe lights. The presence of a gradient, whether a space or a time one, is essential for chemotaxy to occur.

The measurement of the time taken by the bacteria to adapt, i.e., to return to a normal motility pattern after mixing, provides an idea of the intensity and nature of the chemical agent analyzed. This, therefore, is a good method for the evaluation of mechanisms and for the rapid screening of attractants and repellents. In this respect, the method may be simplified (Tsang et al., 1973) by placing the bacterial suspension between a glass slide and coverslip and adding small amounts of attractant or repellent with the aid of a micropipette. After diffusion of the chemical agent, the change in the pattern of motility can be followed easily at the microscope.

The study of bacterial "memory," which is done using the time gradient, became quantitative when the bacterial pathways were photographed and when the frequency of the bacterial "tumbles" were measured as a function of stimulus and time (Spudich and Koshland, 1975).

The "Coulter Counter" method (Armitage et al., 1977) consists of the measurements of the frequency of suspended bacteria that pass through a polycarbonate membrane into a buffer containing the stimulating agent at a given time and under negative or positive chemotactic influence.

The "conjugation method" (Chartone-Souza, 1979) was developed as an attempt to detect chemotactic behavior between strains having opposite sex reactions, i.e., strains that are donors and receptors of conjugant plasmids. The method is based on the differential frequency of "specific pari formation" in crosses between bacterial strains that are donors and receptors of conjugant R plasmid (derepressed, if possible), with (+) and without (−) motility, according to the combinations listed below and subdivided into two series:

Series 1

Donor (motility)[1]		Receptor (motility)[2]
E. coli R +	×	*E. coli* K12 −
E. coli R +	×	*E. coli* K12 +
E. coli R −	×	*E. coli* K12 −
E. coli R −	×	*E. coli* K12 +

Series 2

K12 Donor (motility)[3]		Receptor (motility)[4]
E. coli R +	×	*E. coli* K12 −
E. coli R +	×	*E. coli* K12 +
E. coli R −	×	*E. coli* K12 −
E. coli R −	×	*E. coli* K12 +

1. The mutant without motility was obtained from the wild *E. coli* strain and R plasmid donor.

2. Mobile *E. coli* K12 strain, R plasmid receptor, with chromosome markers for drug resistance (e.g., streptomycin, nalidixic acid, rifamycin) not found in the R plasmid of the donor strain. The mobile mutant was obtained from this strain of *E. coli* K12.

3. Transconjugant *E. coli* 12 strains both mobile and immobile, and therefore R plasmid donors obtained by conjugation of the donor *E. coli* from item 1 with *E. coli* K12 from item 2.

4. The same *E. coli* K12 strains as described in 2.

The frequency of transconjugant formation is considered to be proportional to the formation of specific pairs between R-plasmid donor and receptor cells, according to the approach already used by researchers studying the kinetics of conjugation (Nelson, 1951; Hayes, 1957; Curtiss, 1969), who, however, consider the frequency of chromosomal gene recombination to be an indication of the formation of specific pairs. It has been suggested (Nelson, 1951; Hayes, 1957; Curtiss, 1969) that in these cases, whenever possible, markers should be used that do not require integration with the chromosome of the receptor cell to manifest themselves, according to the procedure adopted in the present method. On the other hand, the presence or absence of motility has been explored already in studies on the conjugation and aggregation of donor and receptor cells (Collins and Broda, 1975).

Mixed cultures must contain variable cell concentration. At lower concentrations, a fall in random cell encounters obviously occurs that favors the objectives of the method. The use of media with different viscosity (nutrient broth or nutrient broth with 0.1% agar added) is also useful.

After 60 minutes of incubation, aliquots of appropriate dilutions of mixed cultures are seeded onto selective media to obtain the titer and the transconjugant.

POSITIVE AND NEGATIVE CHEMOTAXY

A large body of data has been collected with the use of appropriate methodology, and about thirty bacterial genera have been studied already, with greater emphasis on *Escherichia* and *Salmonella*. *Escherichia coli*, for example, has been shown clearly to be able to respond chemotactically to many attractants, most of them nutrients such as sugars, vitamins, and amino acids. Let me add here that several amino acids such as leucine, isoleucine, and valine are repellents, although they are metabolizable. The great majority of repellents, however, are toxic substances or products of bacterial excretion such as acids and alcohols. Therefore, correlation between the metabolism ability of a substance and its ability to attract or repel bacteria is not a constant occurrence (Koshland, 1980; Weibull, 1960). Some substances such as glycerol, pyruvate, and others that are metabolized are attractants, among them some sugar analogues such as D-glucose, a galactose analogue (Adler, 1973), and some amino acids such as x-methyl-aspartate (Mesibov and Adler, 1972). In addition, a mutant that loses the ability to metabolize an attractant such as galactose continues to be attracted by it. Another interesting fact is the case of phenol, which is an attractant for *E. coli* but a repellent for *Salmonella typhimurium* (Lederberg, 1956).

Attractant amino acids can be detected by two receptors, those for serine and aspartate (Mesibov and Adler, 1972), whereas repellent amino acids can be detected only by the leucine receptor. Eleven receptors have been reported thus far for sugars (Adler, 1975; Koshland, 1980). Other receptors have also been described, as discussed below. A total of 30 receptors have been described for *E. coli* and *S. typhimurium*, 20 of attractants and 10 of repellents (Koshland, 1980). This variability in the behavioral responses obtained in the laboratory, many of which are apparently incoherent, has raised difficulties in speculation on the evolutionary reasons that led to this behavioral picture. Things begin to gain coherence only when we consider that, in the environment, bacteria are usually exposed to a pool of molecules, such as, for example, the 20 amino acids produced by protein degradation in most living beings. Thus, in terms of parsimony, only the two receptors for serine and aspartate are needed for bacteria to find all other amino acids. On the other hand, repellent amino acids have side chains similar to small molecules that are toxic. Tyrosine and tryptophan, for example, contain indole and phenol side chains. What is much more difficult to explain is that phenol is an attractant for *E. coli* and a repellent for *S. typhimurium* even though the two organisms are genetically very close (Koshland, 1980).

CHEMOTAXY IN INTERACTIONS

Chemotaxy seems to play an important role in microbial communities, which depend on a wide chemical diversity maintained by a complete series of interactions (Chet and Mitchell, 1976). It was discovered, for example, that some preying bacteria are attracted by their own prey. This occurs with *Pseudomonas*, which is a compulsory or optional parasite of yeasts (Straley and Conti, 1974). This capacity must offer some selective advantage to bacteria in a natural habitat where prey is scarce. Other notable examples of chemotaxy refer to some phytopathogenic (Chet et al., 1973) and symbiotic bacteria such as *Rhizobium* in relation to leguminous plants, which seem to attract this bacterium by means of nutrients or other factors in the rhizosphere (Chet and Mitchell, 1976).

J. L. Azevedo and I studied the interaction of donor and receptor strains of the conjugant R plasmid of *E. coli* (Chartone-Souza, 1979). The plate method (Adler, 1966), modified, and conjugation method (Chartone-Souza, 1979) described above were used in these studies. When mobile *E. coil* cultures such as those described above were inoculated at two sites in a plate containing semi-solid medium, the figures formed after the development and joining of the bacterial haloes were of two types. If the two inocula were from a strain of the same mating type, either donor or receptor, the figures were symmetrical. In this case, the haloes were circular and did not intermingle when touching. On the other hand, if one of the inocula was from a donor strain and the other from a receptor strain, the figures formed were asymmetrical because of a deformation of the internal growth halo of the donor strain that projected toward the receptor strain through a chemotactic band, as shown in figure 4.

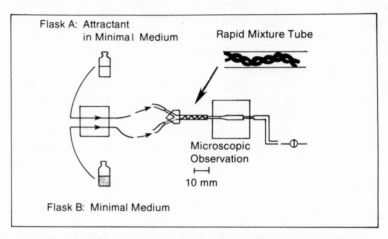

Fig. 4. Symmetrical interactions between strains of the same mating type (a - donor/donor; b - receptor/receptor), and asymmetrical interaction between strains of opposite mating types, with the presence of a chemotactic band (c - donor/receptor - on the right).

In another experiment, also designed to investigate possible chemotaxy between strains of opposite mating types, the "conjugation method" (Chartone-Souza, 1979) described in item 2 was used. The donor strain initially used was a strain of *E. coli* BH 100, carrier of conjugant R plasmid, with five markers for resistance, and the receptor was a strain of *E. coli* K12 without plasmids and with initial chromosome resistance to streptomycin followed by resistance to rifamycin also, after mutation. In addition to these differences, *E. coli* Bh 100 is lac+ and *E. coli* K 12 is lac−, a fact that facilitated the study in terms of obtaining transconjugants and the titers of mixed cultures.

In the two series of crosses described above, similar results were obtained, i.e., transconjugants originated at highest frequency from mixed cultures in which the donor strain was mobile. Figures 5 and 6 contain the results of one of the experiments.

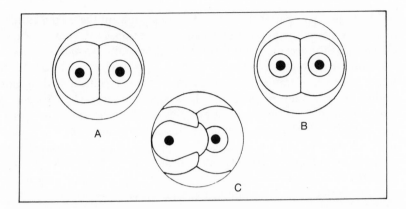

Fig. 5. Number of transconjugants originating from the mixed cultures shown below ($\pm 5 \times 10^{-7}$/ml) between *E. coli* BH 100 and *E. coli* K12 with (+) and without (−) motility.

BH100 Mot.$^+$ × K12 Mot.$^-$
BH100 Mot.$^+$ × K12 Mot.$^+$
BH100 Mot.$^-$ × K12 Mot.$^+$
BH100 Mot.$^-$ × K12 Mot.$^-$

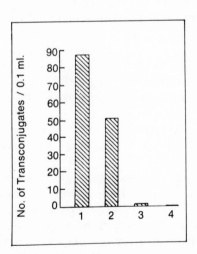
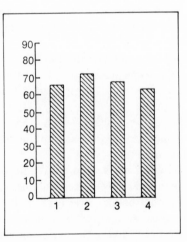

Fig. 6. Number of transconjugants originating from the mixed cultures shown below ($\pm 2.5 \times 10^{8}$/ml) between *E. coli* BH 100 and *E. coli* K12 with (+) and without (−) motility:

BH100 Mot.$^+$ × K12 Mot.$^-$
BH100 Mot.$^+$ × K12 Mot.$^+$
BH100 Mot.$^-$ × K12 Mot.$^+$
BH100 Mot.$^-$ × K12 Mot.$^-$

The results are that at the lowest concentrations in the mixed cultures (5×10^7ml), a significantly higher frequency of transconjugants was associated with the presence of motility in the donor strain. No such association was observed at higher concentrations (2×10^8ml). This may be due to the high frequency of random cell encounters. These results, added to those obtained by the "plate method" (fig. 4), permit us to suggest the existence of a possible attractant substance excreted by the receptor strain and specifically detected by the donor strain. In this case, the donor cell may move through the gradient of the attractant until it meets the receptor cell, thus starting conjugation. Actually, other authors working with *S. typhimurium* (Bezdek and Soska, 1972) also suggest the possible existence of chemotaxy between strains of opposite mating types.

MECHANISMS OF CHEMOTAXY

Chemotactic mechanisms have been elucidated in detail and biochemical methods including gene manipulation and other special techniques.

Mutants to different levels of the sensory system have been obtained both in *E. coli* and *S. typhimurium* (Koshland, 1980; DeFranco and Koshland, 1981; Armstrong et al, 1967; Hazelbauer et al, 1969):

1. at the receptor level where they eliminate in the response to specific chemical agents (specific chemotactic mutants);

2. at the level of signaling or messenger membrane components (*tar, tsr, trg* proteins) whose mutants (*tar-, tsr-, trg-*) eliminate the response to two or more receptors at the same time, thus demonstrating the focusing role of these messenger components in relation to the signals received by the receptors;

3. at the central processing level, where they eliminate the response to all chemotactic signals simultaneously (generalized *che-* or chemotactic mutants).

On the basis of present knowledge, the bacterial sensory system is organized along conceptual lines very similar to those for the organization of higher beings. Briefly (Koshland, 1980), about thirty receptors are already known, which are located at the periphery of the cell (i.e., on the membrane and in the periplasmic space) and receive the signals of attractants and repellents from the external environment. These signals are processed by a sensory "apparatus" that includes twelve proteins, nine of which are located in the cytoplasm and three on the membrane. The last step in this processing is a signal given to the flagellum, which can suddenly change its direction of rotation, thus producing a tumble that causes the bacterium to change direction, or can suppress the tumbles and continue to dislocate the bacterium in an almost straight trajectory. Thus a motor response occurs at the end of this processing system.

RECEPTORS

The receptors in bacterial chemotaxy are of a protein nature, similar to the receptors of all other species. Almost all of them are specific and have great affinity for one or two chemical agents, with lower or little affinity for a limited series of other molecules. This is actually a general peculiarity of protein specificity (Koshland, 1980). The first receptor to be purified and studied quantitatively was galactose-binding protein (Zukin et al, 1977). Galactose and glucose bind to this receptor with great affinity, with dissociation constants of 2×10^{-7} and 1×10^{-7} respectively. These two molecules have small differences in configuration, whereas other molecules with greater differences, such as arabinose, lactose, and fructose, have much lower affinity, with dissociation constants of 4×10^{-5}, 6×10^{-4}, and 6×10^{-3}, respectively. As a rule, it can be said that some agents strongly interact with the receptors, whereas others only interact weakly.

As previously mentioned (Koshland, 1980), a total of 20 receptors have been identified thus far that are attractants and repellents. The attractant receptors are: glucytol, N-acetyl glucosamine, D-fructose, D-galactose, D-glucose, D-mannose, maltose, mannitol, ribose, D-sorbitol, trealose, aspartate, serine, Mg^{2+}, citrate, nitrate, 0^{2-}, fumarate, and arabinose. Repellant receptors are: acetate, isopropanol, leucine, indole, benzoate, Ni^{2+}, CO^{2+}, and S^{2-}. Phenol, H^+ and OH^- can be both attractants and repellents.

The sensory power of a bacterium can therefore be defined as a product of the specificity of each individual receptor multiplied by the combination of receptors located at its periphery.

Many receptors are inducible, such as those of ribose, galactose, maltose, and other sugars, whereas a few are constitutive, as is the case for serine and aspartate (Koshland, 1980). The first three are located in the periplasmic space, whereas others have been found on the membrane.

An interesting fact is related to the multiple function of chemoreceptors. Thus galactose- and ribose-binding proteins also play a role in the transport system (Oxender, 1972); the glucose receptor also participates in the phosphotransferase system (Adler and Epstein, 1974); and the oxygen, nitrate, and fumarate receptors are identified as enzymes involved in electron transport (Taylor et al., 1979). Although the transfer and chemotactic systems initially share the binding proteins, they are entirely independent thereafter, each having its own mutants and following its own path (Ordal and Adler, 1974).

TRANSMISSION OF INFORMATION BY THE RECEPTORS TO MESSENGER ELEMENTS

Receptors must transmit the variations detected in attractant and repellent concentrations to their flagella. Three hypotheses hae been advanced in an attempt to solve this problem (Engelmann, 1981): diffusion of a chemical messenger to the base of the flagelum, direct protein-protein interaction, and depolarization of the cytoplasmic membrane.

For periplasmic receptors, the first step in this process appears to be direct physical interaction between receptor molecules and messenger proteins (*trg, tsr, tar*) located on the membrane (Koshland, 1980). For this to occur, the receptor protein should be induced to change its conformation by the stimulating agent (chemoeffector), thus becoming able to interact with the messenger protein. This hypothesis was shown to be acceptable by tests such as transfer of fluorescent energy, optical rotary scattering, and others (Koshland, 1980). Through this interaction, the message would be transferred to the interior of the cell and from there to the "motor" of the flagellum after passing through the processing center. Several lines of evidence suggest that this is the mechanism of periplasmic receptors (Strange and Koshland, 1976). Clear evidence is provided by the competition existing between various molecules for the few messenger elements. There are mutants, for example, that have lost tactism to ribose and galactose simultaneously (Ordal and Adler, 1977). However, they are mutants not of the receptor proteins but rather of the gene that codes protein *trg*, located on the membrane to which galactose and ribose receptors seem to bind (Koshland, 1980). Therefore these two receptors can use the same messenger protein in a funneling or converging process. Other messenger proteins have been characterized as well: *tsr* and *tar*. The receptors of serine, benzoate, indole, leucine, H^+, and aminoisobutyrate converge toward *tsr*; and the receptors of aspartate, maltose, and others converge toward *tar* (Silvermann and Simon, 1977). The three proteins coded by genes *trg, tsr*, and *tar* are acceptors of the methyl group and have molecular weights of approximately 60,000 daltons. However, other receptors use other messenger elements, such as the receptors that operate through the phosphotransferase and electron acceptor systems (Koshland, 1980).

CENTRAL PROCESSING SYSTEM AND FLAGELLAR RESPONSE

The signals intercepted by the receptors and transferred to the messengers must be processed for the stimuli to be correctly interpreted and translated, finally, into a motor action by the flagellum.

The genetics and structure of flagella have been reviewed recently (Silverman and Simon, 1977). In wild *E. coli*, the flagellum consists of a long left-handed helical filament of flagellin (thousands of sub-fibrils) that fits inside a curved connection (hook) also consisting of a single protein. This part fits, in turn, in the basal body, which contains at least nine proteins arranged in a series of disks. Disk L is located in the internal membrane, where energy is available.

A total of 25 genes code the flagellum, although only one gene (*hag*) is needed to code flagellin. The energy that causes the flagellum to move comes from a protomotive force and not from ATP (Manson et al., 1977).

Most mutants produce flagellum-free (Fla⁻) phenotypes, and others have apparently normal but nonfunctioning flagella (Mot⁻).

A bacterium such as wild *E. coli* moves at random in three dimensions in a medium without gradient, whether consisting of an attractant or a repellent (Berg and Brown, 1972): a short, almost straight run (about 1.0 sec.) is followed by a tumble (about 0.1 sec.) that changes the direction of dislocation, and so on (Adler, 1976).

On the other hand, bacteria suffer fewer tumbles and therefore move along a straight line for a longer period of time when submitted to increasing concentrations of attractants (Macnab and Koshland, 1972; Berg and Brown, 1972) and suffer more tumbles when submitted to decreasing concentrations of attractants (Macnab and Koshland, 1972). The opposite occurs with repellents (Tsang et al., 1973). Under any kind of conditions, the presence of gradient is needed for chemotaxy to occur. Bacteria require a longer time to recover from stimuli that suppress tumbles (attractants) than from stimuli that increase them (repellents) (Macnab and Koshland, 1972; Berg and Brown, 1972; Tsang et al., 1973), which might mean that more than one type of adapting mechanism is present (Parkinson, 1977). It also has been discovered that *E. coli* and other bacteria are able to detect both space gradients (a higher concentration on the right than on the left, for example) and time gradients (a higher concentration now than before, for example). This time gradient "memory" exists both for attractants (see the time gradient apparatus described above) and repellents (Tsang et al., 1973).

More recently it has been shown that the basis of bacterial motion and chemotaxy is the clockwise and counterclockwise rotation of the flagellum (Silverman and Simon, 1974). It also has been observed that the attractant induces counterclockwise rotation, whereas the repellent causes clockwise rotation (Larsen et al., 1974). Since it was already known that the addition of an attractant inhibits the tumbling frequency and therefore increases motion along a straight line, whereas the repellent acts in the opposite manner, it became clear that in bacterial motion the trajectory along a straight line is caused by the inversion of this rotation. The effect of flagellum rotation on tumbling has been confirmed in experiments with generalized chemotactic mutants (*Che⁻*) that never tumble and whose flagella always rotate counterclockwise. In contrast, in other *Che⁻* mutants that always tumble their flagella rotate only clockwise. Thus, the critical role of tumbling in bacterial chemotaxy was definitely established. The flagellar filaments, when rotating counterclockwise and impelling the bacterium toward an attractant, organize in bundles (Macnab and Koshland, 1974), each flagellar filament maintaining its left-handed helical shape (Koshland, 1980). Conversely, the bundles separate during the clockwise inversion of flagellar rotation. The tumbling phenomenon seems to arise from this asynchrony of the flagella rotating in different directions. In this case, the bacterium is pulled by the flagella and can move away from the repellent (Adler, 1976; Koshland, 1980). However, the concept of two flagellar shapes, a left-handed helix of 2.1 A° per turn and a right-handed helix of 1.05 A° per turn, added to the study of several mutants with inverted behavior, suggested a new model for bacterial tumbling (Koshland, 1980; Khan et al, 1978).

Some indication of how the tumbling frequency is controlled came from the discovery that chemotaxy requires methionine. Without methionine, wild bacteria swim along an almost straight line without tumbling. Thus methionine is necessary to invert flagellar rotation clockwise (Adler, 1976; Koshland, 1980). Today we know that genes *R* and *B* of the processing center are involved in the methylating process of messenger proteins *tar*, *tsr*, and *trg*: the *che R* gene, by coding the enzyme methyl-transferase (Springer et al., 1977), and the *che B* gene by coding methyl-esterase. Seven other genes of the processing center are known: *A, C, S, V, W, Y*, and *Z*, as well as almost all their products. Gene handling and cloning techniques were of great value during this phase (Koshland, 1980; Silverman and Simon, 1977; DeFranco and Koshland, 1981). Not many more genes than these nine should exist, since this number remains unchanged despite intensive investigation (Koshland, 1980). Genes *che C* and *che V* act between the central processing system and the flagellar motor. Both code proteins essential in chemotaxy and also seem to be involved in flagellar structure. All indications are to the effect that these proteins interact with the response regulators by converting information into flagellar motion.

Several interactions of the products of these genes have been established by complementation tests (Koshland, 1980). It is also interesting that *che* genes are controlled by the operons of the flagellar genes. This probably occurs as an economy measure.

Another interesting fact is the ability of bacteria to integrate multiple stimuli from attractants and repellents, giving in some cases exact additive responses (Adler, 1976; Koshland, 1980).

Finally it should be mentioned that after the response to a known chemotactic stimulus, bacteria adjust again to the new environment. Adaptation, one of the most important properties of biological systems, is currently a subject of intensive investigation.

In conclusion, we can state that much is known today about the sensory system of bacteria from the moment when the receptors are stimulated up to the motor response and later adaptation, although little is known about the intimate details of these mechanisms.

REFERENCES

Adler, J. 1966. Chemotaxis in bacteria. Science 153:708–16.
Adler, J. 1969. Chemoreceptors in bacteria. Science 166:1588–97.
Adler, J. 1976a. Chemotaxis in bacteria. Ann. Rev. Biochem. 44:341–56.
Adler, J. 1976b. The sensing of chemicals by bacteria. Sci. Amer. 234:40–47.
Adler, J. 1973. A method for measuring chemotaxis and use of the method to determine optimum conditions for chemotaxis by *Escherichia coli*. J. Gen. Microbiol. 74:77–91.
Adler, J., and B. Templeton. 1967. The effect of environmental conditions on the motility of *Escherichia coli*. J. Gen. Microbiol. 46:175–84.
Adler, J., and W. Epstein. 1974. Phosphotransferase-system enzymes as chemoreceptors for certain sugars in *Escherichia coli* chemotaxis. Proc. Natl. Acad. Sci. USA 71:2895–99.
Armitage, J. P., D. P. Josey, and D. G. Smith. 1977. Simple quantitative method for measuring chemotaxis and motility in bacteria. J. Gen. Microbiol. 102:199–202.
Armstrong, J. B., J. Adler, and M. M. Dahl. 1967. Non chemotactic mutants of *Escherichia coli*. J. Bacteriol. 93:390–98.
Berg, H. C., and D. A. Brown. 1972. Chemotaxis in *Escherichia coli* analyzed by three-dimensional tracking. Nature 239:500–504.
Bezedek, M., and J. Soska. 1972. Sex-determined chemotaxis in *Salmonella typhimurium*. Folia Microbiol. 17:366–69.
Chartone-Souza, E. 1979. Quimiotaxia entre linhagens doadoras e receptoras de plasmidio R em *Escherichia coli*. Tese de doutorado. Instituto de Genetica, USP, Piracicaba, Sao Paulo.

Chet, I., S. Fogel, and R. Mitchell. 1971. Chemical detection of microbial prey by bacterial predators. J. Bacteriol. 106:863–67.

Chet, I., Y. Henis, and R. Mitchell. 1973. Effect of biogenic amine cannabinoids on bacterial chemotaxis. J. Bacteriol. 115:1215–18.

Chet, I., and R. Mitchell. 1976. Ecological aspects of microbial chemotactic behavior. Ann. Rev. Microbiol. 30:221–39.

Collins, J. F., and P. Broda. 1975. Motility, diffusion, and cell concentration affect pair formation in *Escherichia coli.* Nature 258:722–23.

Curtiss III, R. 1969. Bacterial conjugation. Ann. Rev. Microbiol. 23:69–136.

Dahlquist, F. W., P. Lovely, and D. E. Koshland, Jr. 1972. Quantitative analysis of bacterial migration in chemotaxis. Nature New Biol. 236:120–23.

DeFranco, A. L., and D. E. Koshland, Jr. 1981. Molecular cloning of chemotaxis genes and overproduction of gene products in the bacterial sensing system. J. Bacteriol. 147:390–400.

Engelmann, T. W. 1881. Neue Methode zur Untersuchung der Sauerstoffausscheidung pflazicler und thiericher Organismen. Pflugers Arch. Ges. Pysiol. 25:285–92.

Hayes, W. 1957. The kinetics of the mating process of *Escherichia coli* J. Genet. Microbiol. 16:97–119.

Hazelbauer, G. L., R. E. Mesibov, and J. Adler. 1969. *E. coli* mutants defective in chemotaxis toward specific chemicals. Proc. Natl. Acad. Sci. USA 64:1300–1307.

Khan, S., R. M. Macnab, A. L. DeFranco, and D. E. Koshland, Jr. 1978. Inversion of a behavioral response in bacterial chemotaxis: explanation at the molecular level. Proc. Natl. Acad. Sci. USA 75:4150–54.

Koshland, D. E., Jr. 1980. Bacterial chemotaxis as a model behavioral system. Raven Press, New York.

Larsen, S. H., R. W. Reader, F. N. Kort, W. W. Tso, and J. Adler, 1974. Change in direction of flagellar rotation is the basis of the chemotactic response in *Escherichia coli.* Nature 249:74–77.

Lederberg, J. 1956. Conjugal pairing in *E. coli.* J. Bacteriol. 71:497–98.

Macnab, R. M., and D. E. Koshland, Jr. 1972. The gradient-sensing mechanism in bacterial chemotaxis. Proc. Natl. Acad. Sci. USA 69:2509–12.

Macnab, R. M., and D. E. Koshland, Jr. 1974. Bacterial motility and chemotaxis: light-induced tumbling response and visualization of individual flagella. J. Mol. Biol. 84:399–406.

Manson, M. D., P. Tedesco, H. C. Berg, F. M. Harold, and C. Van der Drift. 1977. A protonmotive force drives bacterial flagella. Proc. Natl. Acad. Sci. USA 74:3060–64.

Mesibov, R., and J. Adler. 1972. Chemotaxis toward amino acids in *Escherichia coli.* J. Bacteriol. 112:315–26.

Nelson, T. C. 1951. Kinetics of recombination in *Escherichia coli.* Genetics 36:162–75.

Ordal, G. W., and J. Adler. 1974. Properties of mutants in galactose taxis and transport. J. Bacteriol. 117:517–26.

Oxender, D. L. 1972. Membrane transport. Ann. Rev. Biochem. 47:777–814.

Parkinson, J. E. 1977. Behavioral genetics in bacteria. Ann. Rev. Genet. 11:397–414.

Pfeffer, W. 1888. Uber chemotaktische Bewegungen von Bakterien. Flagellation und Volvocineen. Unters. Botan. Inst. Tubingen 2:582–663.

Silverman, M., and M. Simon. 1974. Flagellar rotation and the mechanism of bacterial mobility. Nature 249:73–74.

Silverman, M., and M. Simon. 1977. Bacterial flagella. Ann. Rev. Microbiol. 31:397–419.

Springer, M. S., M. F. Goy, and J. Adler. 1977. Sensory transduction in *Escherichia coli*: a requirement for methionine in sensory adaptation. Proc. Natl. Acad. Sci. USA 74:183–87.

Spudich, J. L., and D. E. Koshland, Jr. 1975. Quantitation of the sensory response in bacterial chemotaxis. Proc. Natl. Acad. Sci. USA 72:710–13.

Stock, J. B., and D. E. Koshland, Jr. 1978. A protein methylesterase involved in bacterial sensing. Proc. Natl. Acad. Sci. USA 75:3659–63.

Straley, S. C., and S. F. Conti. 1974. Chemotaxis in Bdellovibrio bacteriovorus. J. Bacteriol. 120:549–51.

Strange, P. G., and D. E. Koshland, Jr. 1976. Receptor interactions in a signalling system: competition between ribose receptor and galactose receptor in the chemotaxis response. Proc. Natl. Acad. Sci. USA 73:762–66.

Taylor, B. L., J. B. Miller, H. M. Warrick, and D. E. Koshland, Jr. 1979. Electron acceptor taxis and blue light effect in bacterial chemotaxis. J. Bacteriol. 140:567–73.

Tsang, N., R. Macnab, and D. E. Koshland, Jr. 1973. Common mechanism for repellents and attractants in bacterial chemotaxis. Science 181:60–63.

Tso, W. W., and J. Adler. 1974. Negative chemotaxis in *Escherichia coli*. J. Bacteriol. 118:560–76.

Warrick, H. M., B. L. Taylor, and D. E. Koshland, Jr. 1977. The chemotactic mechanism of *Salmonella typhimurium*: mapping and characterization of mutants. J. Bacteriol. 130:223–31.

Weibull, C. 1960. Movement. *In* I. C. Gunsalus and R. Y. Stanier (eds.), The bacteria, pp. 153–205. Academic Press, New York.

Zukin, R. S., P. G. Strange, L. R. Heavy, and D. E. Koshland, Jr. 1977. Properties of the glactose bindng protein of *Salmonella typhimurium* and *Escherichia coli*. Biochemistry 16:381–86.

ISAAC ROITMAN AND MONICA A. C. CARREIRA

Eukaryotic Protists as a Model for Genetic Engineering

23

Specific genetic alterations in eukaryotic cells have been made difficult by the scarcity of available vectors. In genetic engineering of plants, the utilization of *Agrobacterium tumefasciens* plasmids is promising (Thomashon et al., 1980). The utilization of recombinants of SV-40 virus is also important in the development of genetic manipulation in eukaryotic cells (Hamer et al., 1980; Muzyczka, 1979).

The genetic engineering of eukaryotic cells is still in a primitive stage in comparison with protist cells. In the study of yeast, the utilization of transformation made possible the insertion of yeast and bacteria genes in the yeast chromosome. The fungus protoplast fusion technique is another method that allows a great development in genetic recombination (Peberdy, 1979).

The protozoa from the Trypanosomatidae family contain a granular structure, the kinetoplast. This structure contains mitochondrial DNA that represents about 20% of the total DNA (McGhee and Cosgrove, 1980). The kinetoplast has a DNA-dependent RNA polymerase and a protein synthesis system (Zaitseva et al., 1977) and consequently is an excellent model for studies of the mitochondrial DNA expression.

Several insect trypanosomatids harbor an endosymbiont with characteristics of a prokaryotic cell. The study of such association is important to understand intracellular parasitism, and it can be an evolutionary tool concerned with the theory of the endosymbiont origin of organelles in eukaryotic cells.

One of those insect trypanosomatids that contains symbionts is *Crithidia deanei*, isolated from the hemipteron predator *Zelus leucogrammus*. In contrast to other trypanosomatids that have a nutritional requirement of

10 amino acids, 7 vitamins, hemin, and purine, *C. deanei* can be cultivated in the presence of 3 amino acids, 4 vitamins, and in the absence of hemin and purine. Probably the growth factors required by other trypanosomatids but not by *C. deanei* are being synthesized by the endosymbiont (Mundim et al., 1974). It has been shown that the endosymbiont can synthesize the purine ring by a *de novo* pathway (Ceron et al., 1979). It has also been shown that the endosymbiont plays a role in the surface sugars composition of *C. deanei* (Esteves et al., 1982). In addition, *C. deanei* cannot grow at 34°C (Mundim et al., 1974).

After chloramphenicol treatment, it was possible to obtain the aposymbiotic strain of *C. deanei*. This strain has nutritional requirements similar to other insect trypanosomatids that do not harbor symbionts (Mundim and Roitman, 1977).

Attempts have been made to transfer the symbiont from *C. deanei* to another trypanosomatid that locks a symbiont. The possibility of a phenotypic alteration in the receptor cell does not involve manipulation of recombinant-DNA or the utilization of vectors.

Herpetomonas samuelpessoai has been used as the receptor cell. This organism was also isolated from *Z. leucogrammus* and can be cultivated in a defined medium at 37°C. It has the classical nutritional requirements of insect trypanosomatids, including the requirements of hemin and purine (Roitman et al., 1972). Attempts to transfer the symbiont from *C. deanei* to *H. samuelpesoai* were made with cell fusion by the use of polyethylene glycol (PEG).

The cells of both organisms were obtained after cultivation in a defined medium (Roitman et al., 1972) for 36 hr at 28°C. Each batch of cells was washed 2 times with saline (NaCl 0.85%). The pellets were reunited and washed with saline + $CaCl_2$ 0.01 M. The pellet was resuspended in 2 ml of PEG (6,000) 35% solution and incubated for various intervals in the range of 2–40 min at 28°C. By microscopic examination we showed that even after 40 min incubation, the cells remained motile and morphologically normal. The cells, after being treated with PEG, were inoculated in selective conditions: a defined medium (Roitman et al., 1972) without hemin or purine and incubation carried out at 37°C for an attempt on the isolation of a receptor cell (*H. samuelpessoai*) with symbionts. In a series of experiments, no growth was observed in those selective conditions. By electron microscopic studies of the pellet after PEG treatment, we found only cell aglutination without cytoplasmic bridge. Those results explain the failure of the transfer of the symbiont from *C. deanei* to *H. samuelpessoai*.

In order to obtain a real cell fusion, several compounds have been tried, mainly those that have a membrane-denaturing action: urea, formamide, triton X-100, NP-14, NP-40, NP-X, DMSO, DOC, trypsin, pepsin, EDTA, and concanavalin A. Triton X-100 was selected as the one that in combination with PEG produces a large number of clumps; however, this association was not suitable for the transfer of the symbiont.

Physical contact, in the presence of PEG, by plating, centrifugation, and millipore filtration plus the liberation of the symbiont by freezing of *C. deanei* plus subsequent contact with *H. samuelpessoai* also yielded negative results.

REFERENCES

Ceron, C. R., R. A. Caldas, C. R. Felix, M. H. Mundim, and I. Roitman. 1979. Purine metabolism in trypanosomatids. J. Protozool. 26:479–83.

Esteves, M. J. G., M. H. Mundim, A. F. B. Andrade, J. Angluster, W. Souza, I. Roitman, and M. E. A. Periera. 1982. Cell surface carbohydrates in *Crithidia deanei*: influence of the endosymbionte. Eur. J. Cell Biol. 26:244–48.

Hamer, D. H., K. D. Smith, S. H. Boyer, and P. Leder. 1979. SV40 recombinants carrying rabbit B-globin gene coding sequences. Cell 17:725–36.

Hamer, D. H., M. Kaehler, and P. Leder. 1980. A mouse globin gene promoter is functional in SV40. Cell 21:697–708.

McGhee, R. B., and W. B. Cosgrove. 1980. Biology and physiology of the lower trypanosomatids. Microbiol. Rev. 44:140–73.

Mundim, M. H., I. Roitman, M. A. Hermans, and E. W. Kitajima. 1974. Simple nutrition of *Crithidia deanei*, a reduviid trypanosomatid with an endosymbiont. J. Protozool. 21:518–21.

Mundim, M. H., and I. Roitman. 1977. Extra nutritional requirements of artificially aposymbiotic *Crithidia deanei*. J. Protozool. 24:329–31.

Peberdy, J. F. 1979. Fungal protoplasts: isolation, reversion, and fusion. Ann. Rev. Microbiol. 33:21–39.

Roitman, C., I. Roitman, and H. P. Azebedo. 1972. Grown of an insect trypanosomatid at 37°C in defined medium. J. Protozool. 19:346–49.

Thomashow, M. F., R. Nulter, M. D. Chilton, F. R. Blattner, A. Powell, M. P. Gordon, and E. W. Nester. 1980. Recombination between higher plant DNA and the Ti plasmid of *Agrobacterium tumefasciens*. Prot. Natl. Acad. Sci. USA 77:6448–52.

Zaitseva, G. N., A. A. Kolesnikov, and A. T. Shirshov. 1977. The genetic system of kinetoplast in trypanosomatides. Mol. Cell Biochem. 14:47–54.

Characteristics of Nucleotide Sequences of Rho Mutants Derived from the COB Region in *Saccharomyces cerevisae* D273-10B

24

ABSTRACT

The junction points of 12 petite mutants that span the COB region in yeast mitochondrial DNA have been examined for sequence features that might be related to the excision event. It appears that these ethidium bromide induced rho mutants can arise from almost any location in the mitochondrial genome, although some preferential sites exist.

INTRODUCTION

The petite (rho) mutation in *Saccharomyces cerevisiae* turned out to be extremely important for the genetic analysis of mitochondrial DNA in association with antibiotic resistance, mit and syn mutants (Locker et al., 1979; Slonimski and Tzagoloff, 1976; Coruzzi et al., 1979). To further understand the organization and expression of the organelle DNA, a necessary step was the detailed analysis of the molecule through nucleotide sequence documentation of selected areas. The usual approach is the cloning of fragments of mitochondrial DNA in a vector plasmid (Martin et al., 1979) with subsequent screening of the clones with suitable radioactive RNA or DNA probes. The alternative way proved to be simpler and quicker, relying on the characteristics of the petite mutant caused by a deletion that eliminates most of the mitochondrial DNA. The remaining fragment usually is reiterated in tandem many times, generating a large molecule similar in size to the original genome (Locker et al., 1979). The mitochondrion is able to propagate with high fidelity these molecules that represent an amplified source of genic or subgenic fragments that are easy to characterize genetically with suitable markers and can be purified readily for restriction, hybridization, or sequence analysis (Macino and Tzagoloff, 1979; Nobrega and Tzagoloff, 1980ab). This natural cloning technique so far seems to be restricted to the yeast mitochondrial genome, and the investigations carried out by different laboratories unraveled a fascinating story that encompasses the human and yeast organelle genomes and their particular features as overall organization, split genes, tRNA complement, condon usage, and genetic code (Hall, 1979; Dujon, 1979; Borst and Grivell, 1981).

In this communication, the nucleotide sequences that flank the excision points (or junction sites) of several rho mutants derived from the COB region of *S. cerevisiae* D273-10B have been examined.

RESULTS AND DISCUSSION

During studies that led to the complete nucleotide sequence of the cytochrome b split gene in mitochondrial DNA, a number of ethidium bromide induced rho mutants were isolated and used to sequence the region and narrow down the location of several mit markers. The sequence of 6200 bp encompassing the gene and the intergene region downstream has been published (Nobrega and Tzagoloff, 1980a). Also the 750 bp near the glutayl-tRNA gene has been sequenced (Nobrega and Tzagoloff, 1980c). The remaining sequences (~600 bp) where documented, ascertained later (Bonitz et al., 1982) and fill the gap that includes most of the leader sequence ahead of the cytochrome b gene. The parent rho (DS400/A12) was isolated from the wild-type strain (D273-10B/A21) after ethidium bromide mutagenesis. This mutant was the starting point for further dissection and amplification of subgenic fragments by induction of secondary or tertiary rho mutants with ethidium bromide (Nobrega and Tzagoloff, 1980b). Eleven such petite mutants have been sequenced and precisely localized over the main sequence. In figure 1 a map of DS400/A12 is shown along with the location of the sequences represented in other derived rho. It became apparent that certain locations are more prone to serve as the start or end point of a rho. Seven such regions were located (fig. 1). In five petites, the end points are within 0 to 7 bases apart. In the other two, the separation is 90 and 35 base pairs, these regions being occupied by high G + C clusters characteristic of yeast mitochondrial DNA (Tzagoloff et al., 1980). Only one location (−104 to −109) sits in a 100% A + T rich region at least for 50 bases upstream or downstream from the site, although there is one GC pair at position −104.

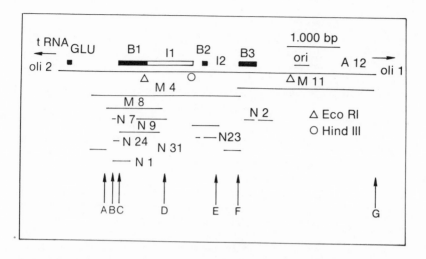

Fig. 1. Relative position of DS400/A12 and derived petites. The location of tRNA GLU gene and cytochrome b exons (B_1, B_2 and B_3) is shown. Ori: location of the ori-type sequence; A: n 31 (−390) 90 bp G + C-rich cluster, m 8 (+1137); E: n 23 (+2446), n 28 (+2451); F: m 11 (+2765) 35 bp G + C-rich cluster, m 4 (+2800); G: A12 and m 11 (+6125).

At the other locations, even if the particular junction site is 100% A + T, one can locate within 0–21 bases in its vicinity a moderately or high G + C sequence.

The nucleotide sequences at the junction points for twelve rho are presented in figure 2. Only two rho are spontaneous mutants. The general features discussed above hold for most junctions. An interesting feature appears very clearly in m 4 and less sharply in A 12 and m 11: a 12 bp G + C-rich region that precedes the junction point ends the unit length of m 4. This G + C-rich cluster has a nucleotide sequence common to numerous such clusters found in yeast mitochondrial DNA (Tzagoloff et al., 1980).

This feature of m 4 is similar to the 23 base long sequences located in the same relative position in spontaneous petites studied by Bernardi and co-workers (Baldacci et al., 1980). These authors consistently found these regularities of end sequences in rho excised from A + T-rich and G + C-rich regions. Comparing their results with our collection of induced mutants, it seems that m 4 type petites are less frequent, the ethidium bromide treatment generating rho almost anywhere although preferences are still evident. Many of their rho also contain the ori (Zamaroczy et al., 1981) or rep

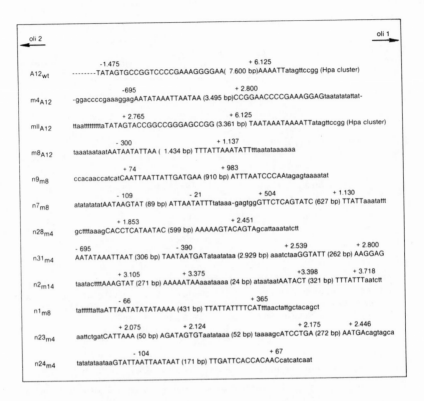

Fig. 2. Sequence around the end points of DNA regions retained in different petites induced with ethidium bromide (except n 2 and n 1, which are spontaneous secondary rho). The parent petite is indicated under the mutant designation. The heavy line shows the rho⁻ repeat unit.

(Blanc and Dujon, 1980) sequence that encompasses about 265 bp and seems to impart suppressivity to the particular petite that carries them. Only two of our mutants carry such a sequence: A 12 and m 11. Most rho are tandem arrays of the retained sequence derived from a single deletion event. Three (n 7, n 2, and n 23) contain internal deletions as well.

It becomes clear that inside the yeast mitochondria there is a remarkable DNA replicating machinery, able to propagate and amplify almost any fragment of its sequence. The petite mutation has proved to be extremely convenient and efficient as a natural cloning device for the isolation and study of mitochondrial genes. Its potential has not been exhausted.

REFERENCES

Baldacci, G., M. Zamaroczy, and G. Bernardi. 1980. Excision sites in the GC clusters of the mitochondrial genome of yeast. FEBS Lett. 144:234–36.

Blanc, H., and B. Dujon. 1980. Replicator origins of the yeast mitochondrial DNA responsible for suppressiveness. Proc Natl. Acad. Sci. USA 77:3942–46.

Bonitz, S. G., G. Homison, B. E. Thalenfeld, A. Tzagoloff, and F. G. Nobrega. 1982. Assembly of the mitochondrial membrane system: processing of the apocytochrome by precursor RNA's in *Saccharomyces cerevisiae* D273-10B. J. Biol. Chem. 257:6268–74.

Borst, P., and L. A. Grivell. 1981. Small is beautiful: portrait of a mitochondrial genome. Nature 290:443–44.

Coruzzi, G., M. K. Trembath, and A. Tzagoloff. 1979. The isolation of mitochondrial and nuclear mutants of *Saccharomyces cerevisiae* with specific defects in mitochondrial functions. Methods in Enzymology 56:95–106.

Dujon, B. 1979. Mutants in a mosaic gene reveal functions for introns. Nature 282:777–78.

Hall, B. D. 1979. Mitochondria spring surprises. Nature 282:129–30.

Locker, J., A. Lewin, and M. Rabinowitz. 1979. The structure and organization of mitochondrial DNA from petite yeast. Plasmid 2:155–81.

Macino, G., and A. Tzagoloff. 1979. Assembly of the mitochondrial membrane system: partial sequence of a mitochondrial ATPase gene in *Saccharomyces cerevisiae*. Proc. Natl. Acad. Sci. USA 76:131–35.

Martin, N. C., D. L. Miller, and J. E. Donelson. 1979. Cloning of yeast mitochondrial DNA in the *Escherichia coli* plasmid pBR 322: identification of tRNA genes. J. Biol. Chem. 254:11729–34.

Nobrega, F. G., and A. Tzagoloff. 1980a. Assembly of the mitochondrial membrane system: complete restriction map of the cytochrome b region of mitochondrial DNA in *Saccharomyces cerevisiae* D273-10B. J. Biol. Chem. 255:9821–27.

Nobrega, F. G., and A. Tzagoloff. 1980b. Assembly of the mitochondrial membrane system: DNA sequence and organization of the cytochrome b gene in *Saccharomyces cerevisiae* D273-10B. J. Biol. Chem. 255:9828–37.

Nobrega, F. G., and A. Tzagoloff. 1980. Assembly of the mitochondrial membrane system: structure and location of the mitochondrial glutamic tRNA gene in *S. cerevisiae*. FEBS Lett. 113:52–54.

Slonimski, P. P., and A. Tzagoloff. 1976. Localization on mitochondrial DNA of mutations expressed in a deficiency of cytochrome oxidase and/or coenzyme QH_2-cytochrome c reductase. Eur. J. Biochem. 61:27–41.

Tzagoloff, A., M. P. Nobrega, A. Akai, and G. Macino. 1980. Assembly of the mitochondrial membrane system: organization of yeast mitochondrial DNA in the Oli region. Curr. Genet. 2:149–57.

Zamaroczy, M., R. Marotta, G. Fauzeron-Fonty, R. Goursot, M. Mangin, G. Baldacci, and G. Bernardi. 1981. The origins of the mitochondrial genome of yeast and the phenomenon of suppressivity. Nature 292:75–78.

F. J. LARA, D. M. GLOVER, A. ZAHA,
A. J. STOCKER, R. V. SANTELLI, M. T. PUEYO,
S. M. DE TOLEDO, AND J. F. GARCIA

Application of Recombinant DNA Techniques to the Study of Gene Amplification in *Rhynochosciara*

25

In the salivary gland chromosomes of larvae of flies of the family Sciaridae (Diptera), one observes formation of the so-called DNA puffs. These puffs are characterized by the fact that, in the sites that give rise to them, either previous to or simultaneously with transcription there is amplification of the DNA bands involved in puff formation (Breuer and Pvaan, 1955; Ficq and Pavan, 1957; Rudkin and Corlette, 1957). Thus, these puffs differ significantly from the RNA puffs that have been studied, particularly in *Drosophila* and *Chironomus*. For reviews of the work on RNA puffs, see Beermann (1969), Edstrom (1974), Daneholt (1974), and Ashburner and Bonner (1979).

The DNA puffs are a special feature of the salivary gland in the larvae of Sciaridae, and all occur during the last cycle of DNA replication, which takes place in these organs at the end of larval life (Machado-Santelli and Basile, 1975). The DNA puffs constitute the first case in which gene amplification was detected in somatic cells (Breuer and Pavan, 1955; Ficq and Pavan, 1957; Rudkin and Corlette, 1957).

The DNA puffs seem interesting on several counts. One would like to know: (1) what is their role in the biology of the organism in which they are found; (2) what is the amount of amplification that takes place; (3) how is the DNA contained in the band that forms DNA puffs organized in the chromosomes, both previous to and after the amplification phenomenon; and last but not least, (4) what are the mechanisms that bring about amplification of specific regions of the genome at specific times in the course of development.

In our laboratory, we have selected larvae of the Brazilian fly *Rhynchos-ciara americana* as the object for a study in depth of the phenomenon of gene amplification in *Sciaridae* for the following reasons: (1) the larvae of this organism are among the largest of the *Sciaridae* (Dreyfus et al., 1951); (2) the females of this species lay about 1,000 eggs, and the larvae that originate from them keep body contact from the time they hatch until their emergence and development with a great degree of synchrony; (3) the animals can be raised under laboratory conditions (Lara et al., 1965); (4) in the salivary glands of their larvae there are about 10 DNA puffs, where DNA amplification starts about 6 to 8 days before pupation (Berendes and Lara, 1975). Our initial efforts with this organism were centered in elucidating the biological functioning of the "DNA puffs." Our first significant progress in this direction stems from the finding that there is a correlation between the degree of expansion of the B-2 DNA puff in the proximal section of the salivary gland and the appearance in the cells of this region of a 16S poly(A)-RNA (Okretic et al., 1977). Subsequently we found that this RNA hybridized preferentially to the DNA from the B-2 DNA puff (Bonaldo et al., 1979); this indicates strongly that this RNA is transcribed therein. More recently, we identified an 18S RNA, which hybridizes "*in situ*" to the C-3 puff (de Toledo, 1978).

Winter and colleagues (1977) detected five new peptides that are synthesized in the gland only at the time in which the DNA puffs are formed. They also found that two of these peptides could be assigned as being codified by specific DNA-puffs, one to the B-2 and the other to the C-3 puff. Now we have succeeded in transcribing *in vitro* both the 16S poly(A)-RNA and 18S poly(A), and the main translation products found correspond to the peptides that had been previously correlated with the functioning of the B-2 and C-3 DNA puffs respectively. In all, the evidence just presented indicates strongly that the "DNA puffs" are involved in messenger RNA production.

The proteins transcribed from the DNA puff messengers are important components of the salivary gland secretion at the time in which the DNA puffs appear (Winter et al., 1977). It also was found (Winter et al., 1977) that the proteins formed at this time are equivalent to the increase in protein content of the communal cocoon in which the larvae pupate.

We have taken advantage of the fact that the DNA puffs are involved in messenger RNA formation to clone the DNA sequences of the chromosomal bands that give rise to them. To this end, we prepared cDNA from messengers obtained from the salivary gland at the time in which the DNA puffs are present in the gland, with the use of reverse transcriptase. The single strand cDNA library obtained was made double-stranded, and tailed with about 50 dG residues and finally annealed with Eco RI-digested pAT153 plasmid (Twigg and Sherratt, 1980) that had been previously tailed with dC residues. After annealing and ligation, the recombinant plasmids so obtained were used to transform *Escherichia coli* HB 101 to tetracycline resistance that is determined by a gene in pAT153 in which there are no Eco RI sites. The transformed colonies so obtained were screened by hybridization to nick-translated p32-labeled cDNA made from messenger obtained from glands at the time in which the DNA puffs are present (period 5), (see Terra et al. [1973] for a definition of these periods) as well as to a similar cDNA that was made, however, from messengers obtained from glands at the time in which the DNA puffs are absent (period 3). The colonies giving positive hybridization with the first of these cDNA's were identified as containing information from developmentally regulated genes and were selected for further study. These clones were considered as strong candidates for recombinants containing information from the DNA puff sequences. Our studies have been concentrated on two recombinants, designated pcRa 3.65 and pcRa 3.81.

We found that RNA transcribed from these plasmids hybridized *in situ* to the C-3 and C-8 DNA puffs respectively. We have used the DNA from these plasmids to measure the magnitudes of the amplification that takes place at the sites to which they have complementary information. To this end, we carried out quantitative Southern hybridization experiments (Southern, 1975; Lis et al., 1978) in which labeled DNA from pcRa 3.65 and pcRa 3.81 were hybridized to salivary gland DNA from animals in period 3 of development (before the DNA puffs are formed) and to DNA from glands of animals in period 6 of development (after the DNA puffs are formed). These DNAs are designated DNA3 and DNA6, respectively. We found that with both plasmids, hybridization against DNA6 was 14–16 fold stronger than against DNA3.

Due to the purity of the probes that can be obtained by recombinant DNA technology, the above-mentioned results constitute a solid proof that DNA amplification really does take place in the cells of the salivary gland of this organism. Furthermore, these measurements must be considered as the most accurate measurements obtained to date of the amplification phenomenon in *Rhynchosciara*.

We have also prepared a library of *Rhynchosciara* genomic DNA. This was done by inserting a selected size population of EcoRI fragments of *Rhynchosciara* salivary DNA into phage gt B. A total of 8×10^6 recombinant phages was obtained. The size of this gene bank is sufficiently great to represent a genome 400 times greater than that of *Rhynchosciara*. It is, therefore, almost a certainty that the sequences of the DNA puffs are contained in this bank. We have now started to identify the recombinant phages that contain such sequences. To this end, we have used DNA from pcRa 3.65 and pcRa 3.81 as probes. We have just identified the first of these phages from a screen that involved examining 10^5 recombinants, and we are now using such phages to walk along the chromosomes, both upstream and downstream from the sequences corresponding to the pcRa 3.65 and pcRa 3.81. The results that we hope to obtain from this study will help us understand how the sequences involved in DNA puff formation are arranged both before and after the amplification phenomenon.

We have also used pcRa 3.65 and pcRa 3.81 to identify the messengers in the salivary gland that hybridize to the *Rhynchosciara* DNA sequences they contain. pcRa 3.65 hybridizes to a 1.95 kb RNA. Also, we are studying the kinetics of the appearance of these RNAs in the salivary gland. Our results indicate that the RNAs complementary to the pcRa 3.65 and pcRa 3.81 clones have different kinetics in the course of development and that the onset of transcription, at least in the case of pcRa 3.65, appears to be coincident with the onset of amplification as inferred from DNA-DNA hybridization data.

The results mentioned above indicate that the application of recombinant technology to the approach to this problem permits us to obtain data with a depth that would have been impossible before the advent of these techniques.

ACKNOWLEDGMENT

This work was mainly supported by a grant from the "Stiftung Volkswagenwerk" from Hanover, F.R.G. D.M.G. is from the Department of Biochemistry, Imperial College of Science and Technology, London, England, and was a visiting scientist under the Brazilian National Research Council—Royal Society exchange Program. A.J.S. was a visiting scientist supported by FAPESP. A.Z. is a post-doctoral fellow from CNPq. J.F.G. is a graduate student under the United Nations Program for Development, and S.M. de T. is from the Chemistry Institute of the University of Campinas, S.P., Brazil.

REFERENCES

Ashburner, M., and J. J. Bonner. 1979. The induction of gene activity in *Drosophila* by heat shock. Cell 17:417–62.

Beerman, W. 1969. Differentiation at the level of the chromosomes. *In* W. Beerman (ed.), Cell differentiation and morphogenesis, pp. 24–54. North Holland Publishing Company, Amsterdam.

Berendes, H. D., and F. J. S. Lara. 1975. RNA synthesis a requirement for hormone induced DNA amplication in *Rhynchosciara americana*. Chromosoma 50:259–74.

Bonaldo, M. F., R. V. Santelli, and F. J. S. Lara. 1979. The transcript from DNA puff of *Rhynchosciara*, and its migration to the cytoplasm. Cell 17:832–37.

Breuer M. R. and C. Pavan. 1955. Behavior of polytene chromosomes of *Rhynchosciara angelae* at different stages of larval development. Chromosoma 7:371–86.

Daneholt, B. 1974. Transfer of genetic information in polytene cells. Int. Rev. Cytol. 39:417–62.

de Toledo, S. M. 1978. Pufes de DNA e a sintese de RNA Mensageiro em *Rhynchosciara americana*. Ph.D. Thesis, University of Sao Paulo, Brazil.

de Toledo, S. M., and F. J. S. Lara. 1978. Translation of messeges transcribed from the DNA puffs of *Rhynchosciara*. Biochem. Biophys. Res. Commun. 85:160–66.

Dreyfus, A., E. Nonato, M. E. Breuer, and C. Pavan. 1951. Cromossomos politenicos em varios orgaos de *Rhynchosciara angelae*. Rev. Brazil. 11:439–50.

Edstrom. J. E. 1974. Polytene chromosomes in studies of gene expression. *In* H. Busch (ed.), The cell nucleus, pp. 293–332. Academic Press, New York.

Ficq, A., and C. Pavan. 1957. Autoradiography polytene chromosomes of *Rhynchosciara angelae* at different stages of larval development. Nature 180:983–84.

F.J.S. Lara, H. Tamaki, and C. Pavan. 1965. Laboratory culture of *Rhynchosciara angelae*. Amer. Nat. 99:189–91.

Lis, J., L. Prestidge, and D. S. Hogness. 1978. A novel arrangement of tandemly repeated genes at major heat shock site in *Drosophila melanogaster*. Cell 14:901–19.

Machado-Stantelli, G. M. and R. Basile. 1975. DNA replication and DNA puffs in salivary chromosomes of *Rhynchosciara*. Ciencia e Cultura. 27:167–74.

Okretic, M. C., J. S. Penoni, and F. J. S. Lara. 1977. Messenger like RNA synthesis and DNA chromosomal puffs in salivary glands of *Rhynchosciara americana*. Arch. Biochem. Biophys. 178:158–65.

Rudkin, G. T., and S. L. Corlette. 1957. Disproportionate synthesis of DNA in a polytene chromosome region. Proc. Natl. Acad. Sci. USA 43:964–68.

Southern, E. M. 1975. Detection of specific sequences among DNA fragments separated by gel electrophoresis. J. Mol. Biol. 98:503–17.

Terra, W. R., A. G. de Bianchi, A. G. Gambarini, and F. J. S. Lara. 1973. Haemolynph amino-acids and related compounds during cocoon production by the larvae of fly *Rhynchosciara americana*. J. Insect Physiol. 19:2092–2106.

Twigg, A. T., and D. Sherrat. 1980. Trans-complementable copy number mutants of plasmid ColE1. Nature (London) 283:216–18.

Winter, C. E., A. G. de Bianchi, W. R. Terra, and F. J. S. Lara. 1977. A relationship between newly synthesized proteins and DNA puff pattern in salivary glands of *Rhynchosciara americana*. Chromosoma 61:193–206.

Winter, C. E., A. G. de Bianchi, W. R. Terra, and F. J. S. Lara. 1977. The giant DNA puffs of *Rhynchosciara americana* code for polypeptides of the salivary gland secretion. J. Insect Physiol. 23:1455–59.

D. M. GLOVER, C. F. ADDISON, D. CICCO,
A. McCLELLAND, J. ROSS MILLER,
H. RHOIHA, AND D. F. SMITH

The Application of DNA Cloning Techniques to Analyze Two Sets of Genes from *Drosophila melanogaster*

26

ABSTRACT

We describe two sets of genes from *Drosophila melanogaster*. The first set consists of three developmentally regulated genes that encode the major protein found in the hemolymph of third instar larvae. The second set is the genes coding for ribosomal RNA. These genes are associated with two types of insertion elements. The major type of insertion (type I) has many structural features that suggest it is capable of transposition. We discuss these features together with some recent observations we have made on the amplification of rDNA and its associated insertion sequences in the phenomenon of rDNA magnification.

INTRODUCTION

The genetics of *Drosophila melanogaster* together with the cytological advantages of its chromosomes make the organism attractive for the study of gene expression using cloned DNAs. Its genome consists of four pairs of chromosomes containing a diploid DNA content of $2 \times 165{,}000$Kb. In certain tissues, chromosomal replication occurs without concomitant cell division, and the resulting sister chromatids become laterally aggregated and form the polytene chromosomes. Classical cytogenetics has localized many genes to particular bands of the polytene chromosomes. It is also possible to localize nucleotide sequences on the polytene chromosomes by hybridization *in situ*. Nucleic acids are labeled with ^{3}H and then hybridized to chromosome squashes. The site of hybridization can be detected autoradiographically. This technique has been invaluable in localizing the origins of cloned segments of *Drosophila* DNA. Once a segment of DNA from a given locus has been cloned, then it is possible to use sequences from the end of the segment to screen libraries of cloned random segments for homologous DNA. In this way, one can obtain overlapping segments of DNA and so "walk" from one chromosomal region to another. The trail can be followed by hybridization *in situ*, and one can used cloned DNA from rearranged chromosomes to give "short cuts."

Perhaps one of the most significant contributions of recombinant DNA technology is the study of several types of moderately repetitive DNA sequences in the *Drosophila* genome. It is clear that many of these are transposable elements. They hybridize *in situ* to many polytene chromosome sites, but these sites differ from one strain of fly to another. There are several families of such sequences; but many are flanked by repetitive elements, and like prokaryotic transposons cause a duplication of the sequence at the site of their insertion (Dunsmuir et al., 1981). Already molecular evidence has been obtained that a variety of unstable mutations at the white locus is due to transposable elements (Rubin, personal communication). The chromosomal DNA containing an eye color gene, known to all biology students, was cloned using the knowledge of its association with a transposable element "copia" in the white apricot (wa) mutation (Gehring and Paro, 1980). Copia sequences are rare in *D. similans*, and also clones of copia sequences from flies carrying wa were hybridized *in situ* to *D. simulans* chromosomes. In this way, it was possible to identify among all the possible chromosomal segments having copia those that hybridized to the 3C region (Bingham et al., 1981). The possibility of introducing certain classes of transposable elements into virtually any gene by the phenomenon of hybrid dysgenesis should enable this general approach to be applicable to the isolation of many *Drosophila* genes.

In this article, we will describe work from our laboratory to characterize two sets of *Drosophila* genes at the molecular level. The first set encodes the major polypeptides synthesized at the end of the third larval instar of the organism's development. The second set is the ribosmal genes, a complex gene family associated with a group of enigmatic isertion sequences.

The larval serum proteins (NSPs) are the most abundant polypeptides synthesized exclusively in the fat body and represent a set of coordinately expressed and developmentally regulated genes. How they function has not been exactly defined, but it has been suggested that the homologous proteins in blowflies are involved in the formation of the adult cuticle (see review [Roberts and Brock, 1981]). Consistent with this interpretation is the observation that larvae with a variant form of one of the proteins (NSP2) develop into flies with defective abdominal cuticles. We have characterized the genes for the major larval serum protein (NSP1), which consists of random hexameric associates of three polypeptide chains (α, β, and γ) of 78–83 kilodaltons. The alleles coding electrophoretic variants of the a and b polypeptides have been localized using standard genetic techniques. Chromosome deficiencies in the vicinity of the map position have been made heterozygous with the variant chromosomes in order to localize the cytogenetic interval that would uncover the α and β genes. In this way a has been mapped to 11A7-11B9 on the X chromosomes and B to 21D2-22A1 on the second chromosome. Similarly a y null allele has been mapped between 61A1 and 61A6 on the left arm of the third chromosome (Roberts and Evans-Roberts, 1979). We cloned these three genes by screening a library of *D. melanogaster* chromosomal DNA (Maniatis et al, 1978) for recombinant phage that would hybridize strongly with 32p cDNA against third instar larval fat body RNA, but that would not hybridize with 32p cDNA against embryonic RNAs. DNAs from strongly hybridizing phage were tested for their respective abilities to select mRNAs that could be translated *in vitro* and give the NSP1 polypeptides. The phages containing the α, β, and γ NSP1 genes hybridize *in situ* to the polytene regions 11A, 21D, and 61A, respectively, in excellent agreement with the genetic studies (fig. 1). The three genes have extensive homologies within their nucleotide sequences. We have investigated the extent of this homology by measurement of the lengths of heteroduplexes seen in the electron microscope and by thermal close evolutionary relationships between the α and β genes. The β gene also has considerable homology to the γ gene, with about 11% of the nucleotides having mismatch. The α and γ genes are the most distantly related with a shorter stable duplex and 11–15% mismatch in this region. An evolutionary model consistent with these data would be duplication from an ancestral gene giving rise to β and γ gene lines, followed more recently by the appearance of the a gene by duplication of the β gene. During this evolutionary period, the three genes have been dispersed onto each of the three major chromosomes.

We have analyzed the fine structure of the three genes by the S1 mapping technique of Berk and Sharp (1977) and by DNA sequencing (McClelland et al., 1981). The organizational features of the genes are highly conserved.

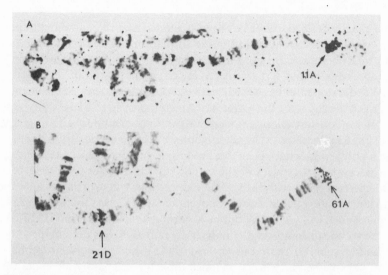

Fig. 1. Insitu hybridization of cloned chromosomal segments containing the LSP1, α, β, and γ genes (panels A, B, and C respectively). The arrows point to silver grains deposited over the indicated polytene regions.

Each has a short intervening sequence about 400 nucleotides from its 5′ terminus (fig. 2). This is a common feature of several insect genes that have been examined at this level of resolution. The nucleotide sequences at the junction of the intervening sequences with the coding sequences have strong homology to other such junctions in a wide variety of eukaryotes, reflecting a common mechanism of removal of the intervening sequence from the primary transcript. We are now beginning to analyze the nucleotide sequence of null alleles of these genes in collaboration with Roberts's laboratory. The nulls that we have so far examined have the gene sequence within their chromosomes and yet it is not expressed into RNA or protein. We anticipate that these mutations will fall into two main categories: mutations in sequences common to the 5′ ends of all eukaryotic genes that are thought to be involved in binding polymerase and promoting transcription and secondly, mutations in sequences required in the developmental regulation of gene activity.

RIBOSOMAL GENES

Our interest in the ribosomal genes centers around two types of sequences that are found inserted into the 28S gene (fig. 3). These were in fact the first eukaryotic genes described as containing intervening sequences,

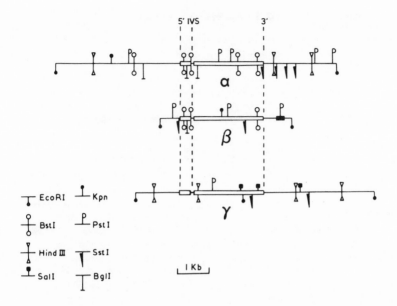

Fig. 2. Physical maps of the three LSP1 genes and the solid block, the gene for a small nuclear polyA-RNA.

but we know that the *Drosophila* rDNA units that contain insertions are unusual and are rarely transcribed. It is the rDNA units that do not contain insertions that serve as templates in rDNA synthesis (Long and Dawid, 1979; Kidd and Glover, 1981). There are two nonhomologous types of insertions, which are located at different sites in the gene for the 1.95Kb subunit of the 28S rRNA separated by 51−73 nucleotide pairs (Roiha and Glover, 1980; 1981). The type I insertions occur in about 60% of the rDNA units of the X chromosomes and not in the rDNA of the Y chromosomes (Tartof and Dawid, 1976; Wellauer et al, 1978). They fall into three main size classes with mean lengths of roughly 5Kb, 1Kb, and 0.5Kb.

The DNA sequence at the right-hand junctions with rDNA of each of these three classes is identical, in contrast to the left-hand junctions. The most common type I insertion is the 5Kb element, to the left of which there is a 9 nucleotide deletion of rDNA sequences (Roiha and Glover, 1980). A high proportion of the type I sequences also occurs outside the nucleolus in the chromocentral heterochromatin, frequently in tandem arrays (Kidd and Glover, 1980). In one stretch of 5 such tandemly arranged type I elements, we have shown that each unit is flanked on the left by a 20 nucleotide

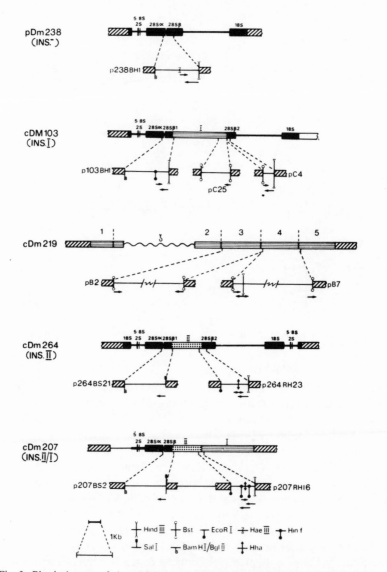

Fig. 3. Physical maps of cloned DNAs containing ribosomal genes and their sequence insertions. The expanded maps show subcloned segments from which the junction sequences were determined. The arrows indicate the extent of nucleotide sequence determined from single end-labeling experiments. 28S rRNA or Drosophila consists of four polynucleotide chains (2S, 5.8S, 28Sα, and 28Sβ) held together by hydrogen bonding. The insertions occur in the 28S gene. cDm103 contains a type I insertion, cDm264 a type II insertion, and cDm207, both types. cDm219 originates from heterochromatin outside the nucleolus. It contains 5 tandemly arranged type I elements.

segment (B″ [fig. 4]) identical to that flanking the 5Kb insertion in rDNA and on the right by a 13 nucleotide segment of rDNA (Roiha and Glover, 1980). This strongly suggests that these sequences were originally located within the rDNA and have subsequently undergone transposition to the heterochromatic sites. The shorter type I insertions represent a coterminal subset of sequences from the right-hand side of the 5Kb insertion, suggesting that they are derived from the longer element. They are flanked by a duplicated segment of rDNA, present only once in uninterrupted units. The length of the duplicated sequence varies from 7 to 15 nucleotides (fig. 5; duplicated sequence indicated by asterisks) (Roiha and Glover, 1981). Such sequence duplications have been found to flank a variety of transposable elements in both prokaryotes and eukaryotes. Since the *Drosophila* rDNA insertions are always found at the same site, it is likely that their transposition would occur by a process of site-specific recombination analogous to that of phage genome integration into, and excision from, *Escherichia coli* chromosome. The duplicated sequence would in this case correspond to the duplicated att site sequence found on either side of the bacteriophage genome in lysogens. There is abundant evidence that many sequence rearrangements have occurred in this region of 28S. The inference of a mechanism for site-specific recombination must, however, remain somewhat speculative because we have not yet directly demonstrated ongoing transposition.

The low levels of transcription of rDNA units containing insertions are intriguing. The deletion of rDNA sequences adjacent to the 5Kb type I insertions would preclude these rDNA units from acting as functional rRNA templates. One might still have expected abortive transcription of these units, especially since they have sequences identical to uninterrupted units in the putative promotor region at the 5' end of the transcription unit (Long et al., 1981). Long and Dawid (1979) could, however, only detect on the order of 6–10 molecules of high molecular weight transcripts containing type I sequences per nucleus and about 50 IKb RNA molecules in the cytoplasm of embryonic cells. This should be compared with about 1,300 molecules per nucleus of the 8Kb primary transcript from uninterrupted units. These results are consistent with the observation by electron microscopy of long silent regions interspersed with active transcription units in nucleolar chromatin from embryonic cells (McKnight and Miller, 1976). Transcription of the insertions does not even occur at a significant level in fly strains with the bobbed phenotype (Long et al., 1981). These flies have shorter bristles, slow development, and other pleiotropic effects as a consequence of deletions of more than 50% of their rDNA, and must be under

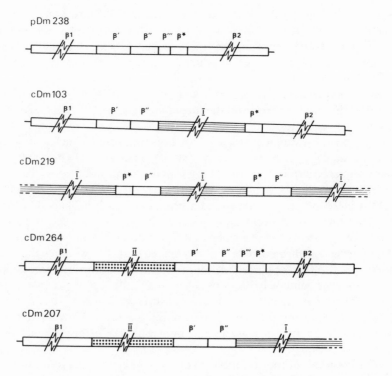

Fig. 4. A schematic comparison of the segments of the 28S gene flanking the insertions. The lengths of the segments are as follows: β'', 20 nucleotides; β''', 9 nucleotides; $\beta*$ 13 nucleotides.

strong pressure to use all their available rDNA. A similar situation probably pertains with *D. hydei*, where the severity of the bobbed phenotype has been correlated with the number of uninterrupted rDNA units and not the total number of rDNA units (Franz and Kunz, 1981).

The type II insertions that occur in about 15% of the rDNA on both X and Y chromosomes of *D. melanogaster* are transcribed at comparably low levels. We find, however, that the type II transcripts are only in the nucleus, the most abundant corresponding in length to the most common type II insertion (3.4Kb) (Kidd and Glover, 1981). There is, in addition, a set of higher molecular weight transcripts linked mainly to rDNA upstream in the rDNA transcription unit from the insertion. The different patterns of transcription of the two types of insertion that we have observed might

a) Insertions not flanked by duplications:

              LEFT JUNCTION   insertion 0.747kb   RIGHT JUNCTION

RI9

         GAAAATTTGGTATAACTGGACTGGATGCATATCTT------<u>ATCCGAAAAGCATACAT</u>TGTCCCTATCTACTACTATCTAGCAA

Dm103   insertion approximately 5kb

GAATGGATTAACGACGACGTGTTTTCGTTGCGCT-------------------<u>ATCCGAAAAGCATACAT</u>TGTCCCTATCTACTACTATCTAGCAA

b) Insertions flanked by duplications:

         LEFT JUNCTION   insertion 0.525(0.530)kb   RIGHT JUNCTION

RI10

GAATGGATTAACGAGATTCCTACTGTCCCTATC<u>TACTACTACTACCACAGT</u>TCCGCTG------ATCCGAAAAGCATACAT<u>TGTCCCTATCTATCTACTATCTAGCAA</u>
                      *******   *****************

MB27   insertion 1.006(1.007)kb

GAATGGATTAACGAGATTCCTACTGTCCC<u>TGTCT</u>TAGCTGGGAGCAG----------<u>ATCCGAAAAGCATACAT</u>TGTCCCTATCTACTACTATCTAGCAA
                             *******   *******

RI9   insertion 0.047kb

GAATGGATTAACGAGATTCCTACTGTCCCTATCT<u>ACTT</u>GGAACACGCACGTAAAT--<u>ATCCGAAAAGCATACAT</u>TGTCCCTATCTACTACTATCTAGCAA
                            *****************   *****************

c) Uninterrupted rDNA:

        <u>GAATGGATTAACGAGATTCCTACTGTCCCTATCTACTATCTAGCGAA</u>

Fig. 5. Nucleotide sequences at the junction of several type I elements with rDNA. The cloned segment R110 contains two insertions in the 28S gene.

reflect the different structures of their junctions with the rDNA sequence. There is no duplication of rDNA sequences on either side of the insertion, but we do see a striking tract of about 20dA residues at the right-hand junction. The most abundant transcripts containing type II sequences seem to have one end point at this dA tract, suggesting either that transcription terminates here or that newly synthesized RNA is rapidly cleaved at this site. We find great variation in the concentration of insertion transcripts between strains. Flies carrying the y + Y chromosomes have three orders of magnitude more type II transcripts than many wild-type strains. This observation may go some way toward explaining the EM observations of Chooi (1979) that nurse cells of flies carrying the y + Y chromsome have a significant proportion of long transcription units.

The heterogeneity of the insertion sequences and also of the nontranscribed spacer region gives a complex, chromosome-specific restriction pattern, which has been used to follow the inheritance of rDNA. Many *Drosophila* tissues contain polytene chromosomes in which the rDNA undergoes fewer rounds of replication relative to sequences in euchromatic regions. Spear and Gall (1973) showed that the rDNA content of diploid tissues is proportional to the number of nucleolus organisers (NOs) in the genome. The rDNA content of the polytene salivary gland chromsomes is proportionally 2–8 fold lower than in diploid cells and reaches a roughly constant level irrespective of the number of NOs. The restriction patterns of rDNA from these tissues show that ribosomal genes from only one NO are replicated during polytenization, and within this NO not all genes are replicated equally (Endow and Glover, 1979; Endow, 1980). The replication of rDNA from only one of the X chromosomes in females explains the finding that rDNA levels are the same in X/X polytene cells as in X/O polytene cells, but the molecular mechanism of their differential replication is not understood.

An inherited increase in rDNA genes occurs in male flies when rDNA-deficient (bb) chromosomes are placed together. We have followed the cleavage pattern of rDNA during magnification when the Xg2 tybb chromosome is placed against the Ybb chromosome (Ritossa et al., 1971). In previous reports of this system, a stepwise increase in rDNA in successive generations when the Xg2 tybb chromosome is repeatedly placed against Ybb- has been described. Our observations on the rDNA patterns of single flies indicate an approximate fourfold increase in the rDNA content of premagnified males in the first generation. In subsequent generations, the rDNA returns to wild-type levels, which are approximately twice those on the original bobbed X chromosome. We find that all types of rDNA unit undergo this amplification. More surprisingly, there is concomitant amplification of the type I sequences outside the nucleolus. Southern hybridization analyses of X chromosomes with heterochromatic deficiencies have allowed us to localize these sequences to the blocks of heterochromatin on either side of the nucleolus organiser. The rDNA amplication in premagnified males therefore involves a large block of the X heterochromatin and not just the nucleolus organizer.

Two models have previously been put forward to explain rDNA magnification. The instability of the magnified bobbed chromosomes in early generations has been suggested by Ritossa and colleagues (1973) to be due to extrachromosomal copies of rDNA not yet integrated into the chromosome. These extrachromosomal copies are thought of as being transcriptionally inactive because the premagnified flies have an extreme bobbed phenotype. Our observations fit quite well with Ritossa's model, except that large blocks of rDNA and flanking heterochromatin would have to become extrachromosomal. It is tempting to speculate that the type I elements themselves would play a part in such a process. There is abundant evidence of recombinational events at their termini, and they could conceivably serve as endpoints in the recombinational excision of large blocks of rDNA. In any case, a fourfold increase in rDNA content could not be explained by a single mitotic unequal exchange of sister chromatids, as proposed by Tartof (1973). That is not to say an unequal exchange mechanism could not be operating in other chromosomes undergoing rDNA magnification. Tartof has described magnifying males somatically mosaic for bb/bb+, which give a higher frequency of revertants than do nonmosaic males. He also argues that bobbed lethal mutations arise reciprocally with magnified chromosomes as one would expect from an unequal exchange mechanism. There is still no compelling molecular evidence that would favor one of these models over the other, but then it is perhaps naïve to expect a single molecular mechanism in light of the varied forms of recombinational activity that occur in eukaryotic organisms. We are currently examining rDNA magnification on other chromosomes, and also the state of the amplified rDNA in premagnified flies in order to try to address this problem.

REFERENCES

Berk, A. J., and P. V. Sharp. 1977. Sizing and mapping of early adenovirus in RNA's by gel electrophoresis of SI Endonuclease-digested hybrids. Cell 12:721–32.

Bingham, P. M., R. Levis, and G. M. Rubin. 1981. Cloning of DNA sequences from the white locus of *Drosophila melanogaster*. Cell 25:693–704.

Chooi, W. Y. 1979. The occurrence of long transcription units among the X and Y ribosomal genes of *Drosophila melanogaster*: transcription of insertion sequences. Chromosoma 74:57–81.

Dunsmuir, P., W. J. Brorein, M. A. Simon, and G. M. Rubin. 1980. Insertion of the *Drosophila* transposable element copia generates a 5-Base pair duplication. Cell 21:575–90.

Endow, S. A., and D. M. Glover. 1979. Differential replication of ribosomal gene repeats in polytene nuclei of *Drosophila*. Cell 17:597–605.

Endow, S. A. 1980. On ribosomal gene compensation in *Drosophila*. Cell 22:149–56.

Franz, G., and W. Kunz. 1981. Intervening sequences in ribosomal RNA genes and bobbed phenotype in *Drosophila hydei*. Nature 292:638–40.

Gehring, W., and R. Paro. 1980. Isolation of a hybrid plasmid with homologous sequences to a transposing element of *Drosophila melanogaster*. Cell 19:897–909.

Kidd, S. J., and D. M. Glover. 1980. A DNA segment from *Drosophila melanogaster* which contains five tandemly repeating units homologous to the major rDNA insertion. *Cell* 19:103–19.

Kidd, S. J., and D. M. Glover. 1981. *Drosophila melanogaster* ribosomal DNA containing Type II insertions is variably transcribed in different strains and tissues. J. Mol. Biol. 151:645–62.

Long, E. O., and I. B. Dawid. 1979. Expression of ribosomal DNA insertions in *Drosophila melanogaster*. *Cell* 18:1185–96.

Long, E. O., M. L. Rebbert, and I. B. Dawid. 1981. Nucleotide sequence of the initiation site for ribosomal RNA transcription in *Drosophila melanogaster*: comparison of genes with and without insertions. Proc. Natl. Acad. Sci. USA 78:1513–17.

Long, E. O., M. Collins, B. I. Kieffer, and I. B. Dawid. 1981. Expression of the ribosomal DNA insertions in bobbed mutants of *Drosophila melanogaster*. Mol. Gen. Genet. 182:377–84.

Maniatis, T., R. C. Hardison, E. Lacy, J. Lauer, C. O'Connell, D. Quan, G. K. Sim, and A. Efstratiadis. 1978. The isolation of structural genes from libraries of eucaryotic DNA. Cell 15:687–97.

McClelland, A., D. F. Smith, and D. M. Glover. 1981. Short intervening sequences close to the 5′ ends of the three *Drosophila* larval serum protein 1 genes. J. Mol. Biol. 152:257–72.

McKnight, S., and O. Miller. 1976. Ultrastructural patterns of RNA synthesis during early embryogenesis of *Drosophila melanogaster*. Cell 8:305–19.

Ritossa, F., C. Malva, E. Boncinelli, F. Graziana, and L. Polito. 1971. The first steps of magnification of DNA complementary to ribosomal RNA in *Drosophila melanogaster*. Proc. Natl. Acad. Sci. USA 68:1580–84.

Ritossa, F., F. Scalenghe, N. Dituri, and A. M. Contini. 1973. Cold Spring Harbor Symp. Quant. Biol. 38:483–90.

Roberts, D. B., and H. N. Brock. 1981. The major serum proteins of Dipteran larvae. Experientia 37:103–10.

Roberts, D. B., and S. Evans-Roberts. 1979. The genetic and cytogenetic localization of the three structural genes coding for the major protein of *Drosophila* larval serum. Genetics 93:663–79.

Roiha, H., and D. M. Glover. 1980. Characterization of complete type II insertions in cloned segments of ribosomal DNA from *Drosophila melanogaster*. J. Mol. Biol. 140:341–55.

Roiha, H., and D. M. Glover. 1981. Arrangements and rearrangements of sequences flanking the two types of rDNA insertion in *Drosophila melanogaster*. Nature 290:749–53.

Smith, D. F., A. McClelland, B. N. White, C F. Addison, and D. M. Glover. 1981. The molecular cloning of a dispersed set of developmentally regulated genes which encode the major larval serum proteins of *Drosophila melanogaster*. Cell 23:441–49.

Spear, B. B., and J. G. Gall. 1973. Independent control of ribosomal gene replication in polytene chromosomes of *Drosophila melanogaster*. Proc. Natl. Acad. Sci. USA 70:1359–63.

Tartof, K. D., and I. B. Dawid. 1976. Similarities and differences in the structure of X and Y chromosome rRNA of *Drosophila*. Nature 263:27–30.

Tartof, K. D. 1973. Unequal mitotic sister chromatid exchange and disproportionate replication as mechanisms regulating ribosomal RNA gene redundancy. Cold Spring Harbor Symp. Quant. Biol. 38:491–500.

Wellauer, P. K., I. B. Dawid, and K. D. Tartof. 1978. X and Y chromosomal ribosomal DNA of *Drosophila*: comparison of spacers and insertions. Cell 14:269–78.

N. J. ALEXANDER AND R. W. DETROY

Fermentation of Cellulose/Hemicellulose to Ethanol: Exploitation Using Recombinant DNA Technology

27

ABSTRACT

The efficient transformation of lignocellulosics, hemicelluloses (xylans), lignin, and starches to fermentable sugars, chemical feedstocks, and fuels is untested territory for the application of biochemical genetics: recombinant DNA technology, DNA transformation, cell fusion. These genetic techniques can be focused on microorganisms that transform plant biomass to useful chemicals and fuels. Exploitation by molecular biology is mandated in numerous areas of liquid fuel production involving fermentations: alcohol for hemicellulose, 2, 3-butanediol production, alcohol tolerance, cellulase synthesis and regulation, and lipid/fatty acid synthesis by microbial systems. Various approaches to the application of biochemical genetics to the conversion of cellulose/hemicellulose to ethanol are discussed.

INTRODUCTION

Ethanol[1] is manufactured synthetically from petrochemicals or biologically through the process of fermentation by microorganisms. Until recently most of the ethanol in the United States was produced synthetically, primarily because of a cheap and abundant supply of oil. However, with the rapidly increasing cost of oil and increasing dependence on foreign oil suppliers, numerous institutions in many countries are looking at the biological production of ethanol.

To produce ethanol for fuel use, increased supplies of necessary substrates must be made available. The simple sugar, glucose, is the usual substrate to support fermentations involving the yeast *Saccharomyces*. Fermentation of glucose may be effected either by enzymatic or chemical degradation of biomass. Grains (such as corn, wheat, rice) serve as excellent resources for starch, which can be readily converted to glucose either biologically or chemically. However, these grains are used mostly as food sources. More abundant alternative substrates are agricultural residues such as corn stalks, wheat straw, wood chips, and fruit pulps. The carbohydrates of these materials comprise three main polymers: lignin, cellulose, and hemicellulose. Chemical/biological pretreatments of residues yield mixtures of glucose, sucrose, xylose, and lignin monomers. The complex plant polysaccharides are not readily available for fermentation without pretreatment by acids, bases, or milling steps (Detroy et al., 1980). Numerous procedures are available for enzymatic hydrolysis of lignocellulosics to produce fermentable sugars (Flickinger, 1980).

Most plant fibers contain cellulose, hemicellulose, and lignin in approximate ratios of 4:3:3 (Tsao, 1978). Cellulose is a homogeneous polymer of glucose, whereas hemicellulose molecules often are polymers of pentoses (xylose and arabinose), hexoses (mannose), and a number of sugar acids. Lignin, a polyphenolic macromolecule, is relatively higher in C and H and lower in O content than are cellulose and hemicellulose. Hydrolysis of hemicellulose to mono- and oligosaccharides can be accomplished with either acids or enzymes under moderate conditions (Browning, 1967). Unlike hemicellulose, cellulose is resistant to hydrolysis. Cellulose fibers generally consist of a highly ordered crystalline structure of cellulose surrounded by a lignin seal, which becomes a physical barrier to easy hydrolysis. The difficulty encountered in obtaining fast and complete hydrolysis of the secondary hydroxyl-linked polysaccharides is due to the inherently more resistant β-1, 4-glucan materials.

The strong crystalline structure and lignin barrier limits cellulose hydrolysis by either acids or enzymes. Acids are nonspecific catalysts; they attack cellulose as well as lignin. On the other hand, enzymes (cellulases) are specific catalysts that convert cellulose into glucose with little byproduct. Three enzymes are involved in the degradation of cellulose to glucose: (a) β-1, 4-glucano-glucanydrolase (C_x, endocellulase); (b) β-1, 4-glucan cellobiohydrolase (C_1, cellobiohydrolase); (c) β-1, 4-glucosidase (cellobiase) (Flickinger, 1980). A number of microorganisms produce these enzymes, notably *Trichoderma reesei*, but the overall rate of saccharification is slow (Ladisch et al., 1981). Significant improvements in cellulase activity must be achieved before these enzymes can be produced on a commercial scale.

Hemicellulose is one of the major components of renewable resources, comprising upward of 35% of the plant material (Dunning and Lathrop, 1945). Hydrolysis of hemicellulose yields a mixture of sugars, primarily D-xylose, as the major component. For complete utilization of biomass-derived sugars in the production of fermentation alcohol, the conversion of both the cellulosic and hemicellulosic components is required. Until recently, no yeasts had been reported that would ferment pentoses, although some were capable of aerobic metabolism (Barnett, 1976). The utilization of xylose, a 5-C sugar from hemicelluloses, is becoming more important because of the recent discovery that *Pachysolen tannophilus*, a yeast, is capable of fermenting xylose to ethanol (Schneider et al., 1981; Slininger et al., 1982). Other yeasts are able to assimilate aldopentoses but are not able to ferment them to alcohol unless exogenous isomerases are added to the growth medium (Wang et al., 1980). Thus, the utilization of *P. tannophilus* for fermentation of xylose is exceedingly attractive because no additional enzymes need to be added to the system.

Work currently is being done at NRRC to increase the yields of glucose/xylose from lignocellulosics. Recombinant DNA and cell fusion technology are being applied to the microorganisms that hydrolyze polysaccharides and ferment the sugars to ethanol in order to construct the most efficient process for converting plant biomass to fuels.

RESULTS AND DISCUSSION

Saccharification and Alcohol Fermentation of Lignocellulosics

Preliminary studies have been concerned with primary chemical or thermal modifications of wheat straw residues (WS) for subsequent enzymatic hydrolysis to glucose for ethanol fermentations by *Saccharomyces uvarum*.

A column reactor vessel as described by Detroy and colleagues (1982) was utilized for experiments with 300–500 g of straw. Various chemical reagents were pumped continuously through the temperature-controlled column with aeration to optimize delignification, as depicted in figure 1. Saccharification of delignified WS was achieved by treatment with *Trichoderma viride* cellulase (10 IU/g dry weight residue).

When 300–500 g of straw was treated with 4.0% NaOH for 6 hr, 15 to 18% of the biomass was lost. Conversion of the cellulose of WS to fermentable glucose in the various trials ranged between 30 to 50%. The crude fungal cellulase preparations used to treat the modified straw yielded also a substantial quantity of the fermentable 5-C sugar, xylose.

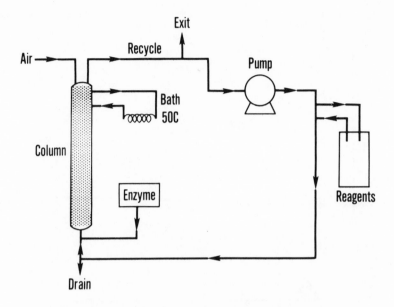

Fig. 1. Reactor column for pretreatment, saccharification, and fermentation.

Ethanol production from the sugars in the column inoculated with *S. uvarum* at 10^8 cells/ml was 30 to 42% of theoretical yield. These lower fermentation values may be due partially to the presence of endogenous substances generated in the alkali modification process, coupled to dilute sugar concentrations of 1 to 3% in the column reactor aqueous phase.

Fermentation of Xylose and Wheat Straw Hemicellulose Hydrolyzates

Pachysolen tannophilus strain NRRL 2460 is capable of an ethanol fermentation of xylose under initial aeration conditions for generation of high cell populations. Detroy and colleagues (1982) have reported batch fermentations with 70 g/L D-xylose, resulting in 0.3 g ethanol/g of pentose metabolized in 6 days. Figure 2 is of the replication cycle of *P. tannophilus* at 25°C on 7% xylose. Cell populations double every 24 hr for approximately 3 days. Ethanol concentrations of 2.0% were achieved by 6 days with complete utilization of the xylose available.

Our most recent experiments with *P. tannophilus* involve fermentation of xylose from crude WS hydrolyzates. The yeast produced 7.2 g (0.72%) ethanol from a hydrolyzate containing 43 g/L xylose (4.3%) as shown in

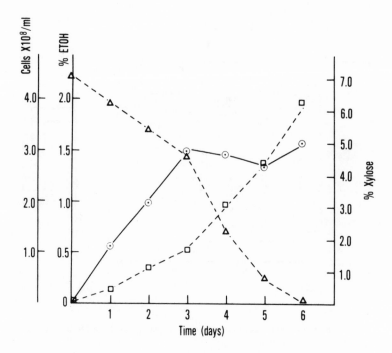

Fig. 2. Ethanol production by *Pachysolen tannophilus* NRRL 2460 on 7% xylose. (·—· cells x 10 /ml; △—△ % xylose; □—□ % ethanol).

figure 3. All of the xylose was consumed within 4 days. The suboptimal yield of ethanol is probably due to the presence of lignin by-products and degraded sugar derivatives in the concentrated WS hydrolyzates. Further optimization of both the processing of hydroyzates and the fermentation is under way in our laboratory.

Figure 4 is of an overall schematic with an initial chemical pretreatment to yield a xylose/pentosan component plus the cellulosic residue for ethanol production. From 500 g of WS, one obtains 400 g cellulosic pulp after chemical pretreatment with 4% NaOH for 6 hr. The liquor contains some 40 g of fermentable D-xylose, which, although suboptimal, supports the *P. tannophilus* 5-C fermentation. Treatment of the cellulosic pulp with cellulase (10 IU/g) for 6 hr yields 105 g of fermentable sugar, which is only 60% of the available glucose in the pulp. Addition of *S. uvarum* cells to the saccharified material yields 42 g ethanol (80% of theoretical amount) in 48 hr.

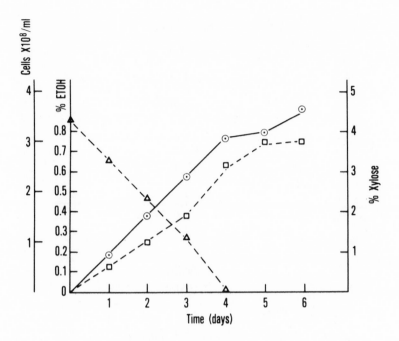

Fig. 3. Ethanol production by *Pachysolen tannophilus* NRRL 2460 on wheat straw hydrolyzates (4.3% xylose). ($\cdot$—$\cdot$ cells x 10 /ml; $\triangle$—$\triangle$ % xylose; $\square$—$\square$ % ethanol).

Application of Genetic Engineering to Microbial Fermentation Systems

Molecular genetic techniques incorporating recombinant DNA and cell fusion methodology offer new opportunities for the improvement of microorganisms involved in fermentations. Traditional methods of strain improvement (UV/chemical mutagenesis) are being supplanted by techniques involving the transfer of nuclear genes from one organism to another (often across generic barriers) by vectors or plasmids. Although there are variations on the actual techniques of genetic engineering, an overall scheme is presented in figure 5 for transformation of a gene from one organism to *Escherichia coli*, a commonly used host organism.

This scheme includes use of antibiotic resistance genes carried by the plasmid for the detection of transformed *E. coli*. A donor organism is selected that carries the gene desired, and the DNA is isolated and then restricted with an appropriate endonuclease. Restriction endonulceases may

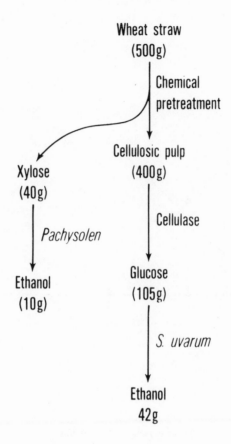

Fig. 4. Fermentation schematic for ethanol from wheat straw.

make either blunt or staggered cuts, with the latter nicking the phosphodiester bonds of the DNA duplex a few bases apart and resulting in DNA fragments with complemetary single-stranded termini. If both the isolated plasmid and the donor DNA are cut with the same endonuclease, then the single-stranded tails of DNA are homologous, which will facilitate the insertion of the foreign DNA into the plasmid. The selection of a restriction enzyme is important. An enzyme recognizing a 6-base sequence will statistically yield fragments large enough to code most proteins (Dally et al., 1981), yet the same enzyme should make only one cut in the selected plasmid, preferably in a marker gene. The enzyme selected in figure 5 fragments

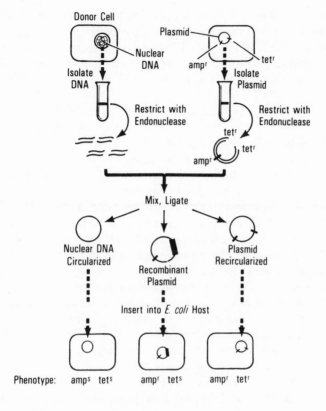

Fig. 5. Scheme illustrating transformation of *Escherichia coli*.

the donor DNA into various sizes while making only one cut in the plasmid DNA, in the gene responsible for resistance to the antibiotic tetracycline. After the DNAs have been restricted, they are mixed and the enzyme ligase is added. T4 DNA ligase, in the presence of ATP, repairs the break in the phosphodiester bonds between two adjacent bases. Three types of circular molecules may result: the nuclear DNA fragments may circularize; the plasmid may recircularize on itself; a piece of donor DNA may be inserted into the plasmid forming a recombinant molecule. To select for the recombinant plasmid, the DNA-plasmid-ligase mix is added to recipient *E. coli* cells (carrying no plasmids) and incubated to allow for the uptake of DNA; the cells then are plated onto medium containing the antibiotic ampicillin.

Cells that have not received any plasmid or that have received only circular-ized donor DNA will be killed. Those cells receiving a plasmid will be ampicillin-resistant. Of the ampicillin-resistant cells, only the ones pheno-typically tetracycline-sensitive will carry recombinant plasmids. The inser-tion of donor DNA into the tetracycline-resistance locus stops the expres-sion of this gene.

Recombinant DNA technology is now rapidly expanding into the indus-trially important yeast *S. cerevisiae* (Hinnen et al., 1978; Beggs, 1978). The organisms used as hosts are typically well characterized and genetically understood. Although *E. coli* may not be the best production organism, it certainly has been well studied and thus is suitable for studying the effects of the introduction of foreign genetic material.

Selection of an appropriate vector must take into account the hosts to be used. In order for the plasmid to increase in copy number and be main-tained in the cell, the origin of replication for the plasmid must remain intact. If the plasmid is to be maintained in both prokaryotic and eukary-otic cells, two origins must be included on the plasmid—one recognized by the prokaryotic host, the other by the eukaryotic host. Table 1 is a list of several plasmids available for use in transformations. The expression of eukaryotic genes in *E. coli* may not be seen because an increasing number of eukaryotic genes are found to be split genes. The enzymes necessary for processing of this RNA are found only in eukaryotic cells, and thus a eu-karyotic split gene transferred into a prokaryotic host will not be ex-pressed. However, *E. coli* serves as a suitable host for the cloning and am-plification of yeast genes that may then be put back into yeast and the expression assayed.

Application of recombinant DNA/cell fusion techniques is important in the investigation of hydrolytic enzymes. The enzymes listed in table 2 are important in the degradation of complex carbohydrates to simple, ferment-able sugars. Certain organisms produce useful quantities of one particular enzyme or enzyme complex but use the simple sugars produced solely for growth. Using recombinant DNA methods, it is now possible to transfer the genes responsible for carbohydrate hydrolysis into another organism, which can then ferment the simple sugars to a useful substance.

The cellulase complex produced by *T. reesei* has been well studied (Eriksson, 1981) and is likely to be the one chosen for initial recombinant DNA studies. Although the endoglucanase and exocellobiohydrolase may be coordinately controlled (Shoemaker et al., 1981), it is unlikely they are closely lined genes, and, as such, both genes will not be transferred on one

TABLE 1

VEHICLES FOR TRANSFORMATION

Plasmid	Host	Markers
pBR	*E. coli*	amp^r tet^r
YIp	*E. coli*/yeast	Yeast integrating plasmid; amp^r tet^r, yeast gene; low transformation
YEp	*E. coli*/yeast	Yeast episomal plasmid; amp^r tet^r, yeast gene, 2μ DNA; high transformation
YEp	*E. coli*/yeast	Yeast centromere; high stability
Ti	Plants	Tumor-inducing plasmid

TABLE 2

HYDROLYTIC COMPLEXES

Enzyme	Possible Donor	Possible Recipient	Remarks
Amylase	*Bacillus*/yeast	*E. coli*/yeast/fungi	Increased enzyme production; fermentation
Cellulase	*Trichoderma reesii*	*E. coli*/yeast/fungi	Increased enzyme production; fermentation of cellulose
Xylose fermentation	*Pachysolen tannophilus*	*E. coli*/yeast/fungi	Increased fermentation; increased enzyme production
Xylanases	*Trichoderma/Penicillium*	*E. coli*/yeast	Hemicellulose, xylan degradation
Glycolytic	*Escherichia coli*/yeast/fungi	*E. coli*/yeast/fungi	Increased enzyme production

piece of DNA to a plasmid. Thus, the transfer of all three genes involved in cellulose hydrolysis will require three separate transformations.

Until recently, successful fermentation of xylose by yeast required the addition of exogenous isomerases (Wang et al., 1980). However, *P. tannophilus* converts xylose and crude xylan-hydrolysates (Detroy et al., 1982) to ethanol. This yeast strain may be improved, or become more useful, by the addition of the gene coding amylase, so that the new organism will ferment both xylose and amylose. Further modifications might include genetic deregulation of the fermentation pathway or an increase in gene dosage for a rate-limiting step in the degradation.

Removal of feedback controls will be a necessary component for increasing metabolite flow in the microbes evaluated. The isolation and cloning of the glycolytic genes (Holland and Holland, 1979; Thompson et al., 1979) should allow for the genetic deregulation of this pathway and lead to an increase in the end products.

SUMMARY

The demonstration of microbial systems for direct fermentation of glucose and xylose to ethanol provides a new dimension in the conversion of biomass to liquid fuels, especially ethanol. Processes for the conversion of cellulose and hemicellulose to ethanol could be integrated into existing cellulose-conversion plants where hemicellulose-derived pentoses are underutilized. Future conversion industries would incorporate both the grain- and residue-derived sugars for optimum production of fermentation alcohol.

Application of recombinant DNA technology for genetic engineering of microorganisms to increase the efficiency of conversion of complex carbohydrates to simple sugars undoubtedly will play a major role in the fermentation industry in the future. The effect will be realized not only in the utilization of renewable resources but also in the many diverse products formed by microbial fermentations.

1. The mention of firm names or trade products does not imply that they are endorsed or recommended by the U.S. Department of Agriculture over other firms or similar products not mentioned.

REFERENCES

Barnett, J. A. 1976. The utilization of sugars by yeasts. Adv. Carbohydr. Chem. and Biochem. 32:125–234.

Beggs, J. D. 1978. Transformation of yeast by a replicating hybrid plasmid. Nature 275:104–9.

Browning, B. L. 1967. Isolation and separation of the hemicelluloses. *In* Methods of wood chemistry, vol. 2, chap. 26. John Wiley & Sons, New York.

Dally, E. L., D. E. Eveleigh, B. S. Montenecourt, H. W. Stokes, and R. L. Williams. 1981. Recombinant DNA technology: food for thought. Food Technol. 35:26–33.

Detroy, R. W., R. L. Cunningham, R. J. Bothast, M. O. Bagby, and A. Herman. 1982. Bioconversion of wheat straw cellulose/hemicellulose to ethanol by *Saccharomyces uvarum* and *Packysolen tannophilus*. Biotechnol. Bioengr. 24:1105–13.

Detroy, R. W., and L. A. Lindenfelser, G. St. Julian, Jr., and W. L. Orton. 1980. Saccharification of wheat straw cellulose by enzymatic hydrolysis following fermentative and chemical pretreatment. Biotechnol. Bioengr. Symp. 10:135–48.

Dunning, J. W., and E. C. Lathrop. (1945). Sugars from wood: the saccharification of agricultural residues. Indust. Engr. Chem. 37:24–29.

Eriksson, K.-E. 1981. Cellulases of fungi. *In* A. Hollaender (ed.), Trends in the biology of fermentations for fuels and chemicals, pp. 19–32, Plenum Press, New York.

Flickinger, M. C. 1980. Current biological research in conversion of cellulosic carbohydrates into liquid fuels: how far have we come? Biotechol. Bioengr. 22 (Suppl. 1):27–48.

Hinnen, A., J. B. Hicks, and G. R. Fink. 1978. Transformation of yeast. Proc. Natl. Acad. Sci. USA 75:1929–33.

Holland, M. J., and J. P. Holland. 1979. Isolation and characterization of a gene coding for glyceraldehyde-3-phosphate dehydrogenase from *Saccharomyces cerevisiae*. J. Biol. Chem. 254:5466–74.

Ladisch, M. R., J. Hong, M. Voloch, and G. T. Tsao. 1981. Cellulase kinetics. *In* A. Hollaender (ed.), Trends in the biology of fermentations for fuels and chemicals, pp. 55–83. Plenum Press, New York.

Schneider, H., P. Y. Wang, Y. K. Chan, and R. Maleszka. 1981. Conversion of D-xylose into ethanol by the yeast *Pachysolen tannophilus*. Biotechnol. Lett. 3:89–92.

Shoemaker, S. P., J. C. Raymond, and R. Bruner. (1981). Cellulases: diversity amongst improved *Trichoderma* strains. *In* A. Hollaender (ed.), Trends in the biology of fermentatoins for Fuels and Chemicals, pp. 89–109.

Slininger, P. J., R. J. Bothast, J. E. Van Cauwenberge, and C. P. Kurtzman. 1982. Conversion of D-Xylose to ethanol by the yeast *Pachysolen tannophilus*. Biotechnol. Bioengr. 24:371–84.

Thompson, J., P. D. Gerstenberger, D. E. Goldberg, E. Gociar, A. Orozco de Silva, and D. G. Fraenkel. 1979. Col E1 hybrid plasmids for *Escherichia coli* genes of glycolysis and the hexose monophosphate shunt. J. Bacteriol. 137:502–6.

Tsao, G. T. 1978. Cellulosic material as a renewable resource. Process Biochem. 13:12–14.

Wang, P. Y., B. F. Johnson, and H. Schneider. 1980. Fermentation of D-xylose by yeasts using glucose isomerase in the medium to convert D-xylose to D-xylulose. Biotechnol. Lett. 2:273–78.

F. C. A. TAVARES

Genetic Engineering for Ethanol Production

28

ABSTRACT

Industrial ethanol from biomass fermentation stands among the most cred-
ited renewable energy sources. Although growing steadily in Brazil, industrial yields could be additionally increased by directing efforts toward better efficiency in processing technology. Superior yeast strains adequate for
the alcohol production factories are necessary. Among other yeast properties high conversion of substrate to ethanol and tolerance to high levels of
the product in beer are desired characters which could also contribute to a
significant reduction in stillage volumes. Promising strains were obtained to
attend such requirements and some of them have been tested on an industrial scale and those results are presented here. Genetic engineering offers
attractive alternatives for the program and expected improvements could
be advanced for substrate conversion, new processing technologies and
residual effluent conversion.

INTRODUCTION

Ethanol produced by fermentation of biomass is one of the most important renewable energy sources and a promising fine chemical available to
replace part of petroleum products. Although economically produced
worldwide by catalytic hydration of the ethylene from petroleum, industrial ethanol obtained through yeast fermentation of sugar is becoming
competitive on the international market mainly as a consequence of rising
oil prices.

Brazil for many years has used ethanol as automotive fuel in mixture
with gasoline and recently as hydrate alcohol. The demand as a fuel has
increased production, which growth has been largely due to the open-

ing of new areas for sugar cane plantations and installation of new factories. This growth is evident: in 1976 crop year, production was about 654 million litres of alcohol (Anonymous, 1979); this year Sao Paulo State alone is producing 2.8 billion litres, and 1985 production was expected to reach over 10 billion litres. Although production has been growing rapidly, better yields can be obtained with no additional fixed investments by directing efforts toward increasing efficiency in processing technology. Efficient conversion of sugar to ethanol and enhancement of the product level in beer are included among the most important variables that may affect overall efficiency (Tavares, 1980). In order to accomplish such objectives, yeast improvement and processing optimization are required. Results are presented here regarding yeast improvement for ethanol production and an overview of complementing possibilities offered by genetic engineering. For the ethanol industry, a significant contribution for novel substrate utilization and the development of efficient processing technologies with limited stillage volumes is expected.

GENETIC APPROACH FOR EFFICIENCY IN ETHANOL PRODUCTION AND RESULTS OF A BRAZILIAN PROGRAM OF YEAST GENETIC IMPROVEMENT

Genetic manipulation of microorganisms stands among the most efficient methods in biotechnology. Microbial improvement has gained the development of new processes in fermentation, and new products could be obtained that represent a significant technological breakthrough in the fermentation industry (Sebek and Laskin, 1979). To obtain special strains of microorganisms, a variety of very ingenious techniques has been advanced (Elander and Chang, 1979; Hopwood, 1981). It will be emphasized that the techniques generally apply the principle of induced mutation followed by appropriate selection of desired strains (Clarke, 1976) and that mutants are even used to make feasible controlled crosses (Merrick, 1976). These methods are the foundations of strain development programs that led to the great expansion of the fermentation industry. Mutations have also been useful in elucidating the biosynthetic pathways of natural products besides the improvement of desired characters and bioconversion of industrial concern (Wang et al., 1979).

Genetic recombination of industrial strains has been valuable in programs where the crosses between parents are easily accomplished, which is the case for yeast improvement (Tavares, 1974). In parasexual crosses, although an efficient method with good practical results, sometimes limitations arise because of undesired side effects caused by the complementing mutations necessary for crossing and also to the low frequency of recovered recombinants (Hopwood et al., 1977).

Generation of recombinants in high frequencies can be obtained by applying the technique of polyethylene glycol-mediated protoplast fusion, which also expands the possibilities of transfer of genetic material and of genetic diversity (Demain, 1980). Viable and stable recombinants can be obtained after further selection, and successful interspecific protoplast fusion and recombination, for example, have been accomplished between *Penicillium chrysogenum* and *Penicillium cyaneo-fulvum*, *Aspergillum nidulans* and *Aspergillus rugulosus* (Peberdy, 1979), candida and endomycopses (Provost et al., 1978), and in various streptomycetes (Demain, 1980). Recombination permits combining the yield-increase mutations and obtaining a superior producer before carrying out further mutagenesis and also permits obtaining the synthesis of new products (Bleck, 1979). It also opens up the possibility of rational use of natural variability for the improvement of industrial microorganisms such as has been done with higher organisms (Tavares, 1974; Azevedo et al., 1978).

The advances of molecular genetics and especially of DNA sequencing, chemical DNA synthesis, DNA enzymology, and cloning—bases of recombinant DNA technology—offer novel opportunities for the fermentation industry. This way production of mammalian peptides, primary metabolites, microbial enzymes, antibiotic and other secondary metabolites, and liquid fuels and chemicals using fermentation methods is a reality or is expected to be obtained soon (Demain, 1980).

In our laboratory, special emphasis is given to recombinational methods, and we conduct programs that make use of natural variability following methods developed after adaptation of plant breeding techniques (Tavares, 1974). Although almost unknown, population and quantitative genetics for microbial breeding purposes can give results of practical application. In order to optimize the genetic gains possible to be achieved, planned experiments are conducted following statistical and biometrical principles that are elaborate. Variance analysis applied to such experiments gives information that helps the selection of superior genotypes with the advantage of avoiding undesired mutation side effects. Most characters of economically important microorganisms depend on many genes being expressed and some quantitative characters such as growth rate, velocity of fermentation, vigor, mass production, and so on, hardly can be improved efficiently by mutation-selection techniques alone. This is particularly the case in developing high-yielding yeast for ethanol production (Tavares, 1980). Some results are presented below.

Evaluation of different yeast strains under restrictive conditions of temperature and osmotic pressure, following variance analysis and statistical tests, permitted us to select a pool of superior strains in which ethanol production and temperature tolerance were consistent in further evaluations (Tavares, 1980). In figure 1, I present the relative performance of such strains in a comparative assay to an industrial strain whose value is represented by the broken line. From the 24 selected strains, ten were at least 10% superior to the industrial strain at 40°C but only six were superior at 35°C. Besides other parameters of industrial concern, ethanol production, temperature tolerance, and strain stability were the main characters studied under variable stressing conditions. Data for a group of 12 hybrids and selected strains appear in table 1, where variance analysis for ten successive fermentations and respective means and variances of production are presented.

Homeostasis of the strains could be statistically estimated, and the values showed that some hybrids present smaller fluctuations of ethanol production. One of the hybrids (M300A) was chosen to be propagated for tests in industry where comparison was made to an industrial strain taking into account the parameters of productivity, velocity of fermentation, percentage of yeast cells in the fermentation wort, and ethanol production. The general performance of the hybrids was consistent with laboratory data indicating that scale-up problems were not maximized.

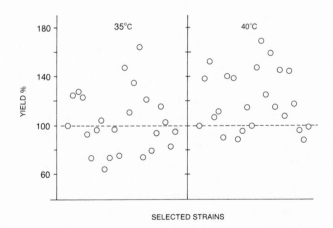

Fig. 1. Ethanol yield of selected strains relative to tester strain values in two temperatures and 25% sugars.

TABLE 1

VARIANCE ANALYSIS

Sources	DF	MS
Treatments	11	65.6126**
Error A	24	1.8299
Plots	35	
Fermentations	9	13.1557**
T × F.......................	99	5.0479
F × Replications	18	0.4808
Error B	198	0.3371
Total	359	
Treatments (T)..............	11	
Strains (S)	10	71.3113**
Groups	1	8.6258
T × F.......................	99	
S × F.......................	90	5.4122**
G × F	9	1.4046

Means and Variances of Ethanol by Volume (%)

IZ209	:	2.20	± 0.1818
IZ1215	:	5.15	± 0.7713
IZ1904	:	5.79	± 0.1699
IZ671	:	3.40	± 0.8520
FF	:	5.87	± 1.2657
M300	:	7.92	± 0.0645
IZ672	:	8.35	± 0.1909
M301	:	6.48	± 0.5687
IZ666	:	8.16	± 0.4508
IZ669	:	7.10	± 0.2989
IZ853	:	8.22	± 0.2242
M309	:	9.68	± 0.1870

The hybrid M300A was then approved for use in industrial fermentations, and produced in 1980 about 200 million litres of ethanol in four industrial plants of Sao Paulo. In 1983, fermentation with this hybrid has produced more than 600 million litres, with high efficiency for sugar conversion (about 90% general mean). Fermentations were conducted in distilleries of up to 600,000 litres of ethanol daily capacity.

For small factories (so-called mini- and microdistilleries of daily production up to 5,000 litres of ethanol), different yeast strains were developed. After transfer of flocculent ability from brewery strains to some of our strains for ethanol fermentation, high efficiency was obtained for processing small volumes. In table 2, I present an assay of different strains in stressing conditions of fermentation wort with addition of stillage molasses and high temperature that also illustrates the importance of environmental

TABLE 2

ETHANOL PRODUCTION OF FLOCCULENT YEAST IN MOLASSES AND MOLASSES PLUS
STILLAGE MOLASSES. TEMPERATURE 35° C AND DURATION OF FERMENTATION 14:00 HOURS

EXPERIMENTS	STRAINS	MOLASSES			MOLASSES + STILLAGE MOLASSES (30%)		
		X	S	CV	X	X	CV
I	Test	5.50	0.78	14.23	5.42	0.61	11.27
	A	6.12	1.19	19.50	5.68	0.41	7.26
	B	5.87	1.06	18.10	5.32	0.40	7.56
	C	5.62	1.04	18.45	5.46	0.34	6.21
	D	5.41	1.50	27.63	5.30	0.45	8.57
	E	5.93	0.59	9.97	5.24	0.40	7.61
	F	5.27	0.96	18.18	4.95	0.33	6.89
		5.67	0.99	17.57	5.34	0.44	8.33
II	Test	6.04	0.37	6.20	4.97	0.92	18.47
	G	5.69	0.91	15.95	5.32	0.52	9.73
	H	5.22	0.76	14.62	4.12	1.11	27.09
	I	6.17	0.37	6.03	5.97	0.44	7.36
	J	6.20	0.39	6.34	5.36	0.51	9.47
	K	5.31	1.44	27.04	5.19	0.44	8.58
	L	5.44	1.43	26.37	5.53	0.57	10.40
		5.72	0.92	16.13	5.21	0.83	15.86

conditions for selection. Some of the strains were tested in batch and continuous fermentation systems in pilot plant scale. The new yeasts for ethanol fermentation have shortened the time for processing, thus enhancing productivity.

As illustrated, the biometrical approach can be of significant value in strain evaluation for both programs based on induced or natural variability techniques.

The genetic approach for ethanol production for Brazilian industries has to take into account the substrate conversion and the efficiency in processing technologies. The substrate saccharose from sugar cane will predominate and much work is needed on improvement for which novel methodologies have been developed (Azevedo et al., 1978). Starch will for some time be dependent on economical processing, but this will probably be eased with appropriate microorganisms for local production of amylolitic enzymes for further fermentations or with developed yeast and bacteria able to ferment starch. In developing a program of *Aspergillus* sp. improvement for glucoamylase production, the results already obtained are presented in table 3, where original and selected strains are compared. Enzyme activity titres are more than 50% higher than economically feasible for industrial processing of starch with crude enzyme extracts. The efficiency of starch conversion is above 90%.

TABLE 3

Genetic Improvement of *Aspergillus* SP. for Glucoamylase Production. Relative Values of Enzyme Activity are Compared to an Industrial Tester Strain for Every Evaluation Testing Assay. Unity of Activity Equals the Quantity of Enzyme (G) for Conversion of a 0.5% Starch Suspension (W/V) to Obtain 10 mg of Reducing Sugar

Strains	Original Activity		1° Selection Cycle	2° Selection Cycle	Best Strains
	EA	EA%	EA%	EA%	EA%
A. awamorii	14.98	117	135	166	172
A. niger (ATCC 22343)	15.42	127	109	144	155
A. niger (ATCC 13497)	11.79	103	123	124	129

Novel substrates and microorganisms are under study in many countries for ethanol production (Guiraud et al., 1981; Leyness and Doelle, 1981; Euriquez et al., 1981; Azhar, 1981; Wilke et al., 1981). Cellulose from biomasses such as agricultural, forestry, and municipal residues represents a vast quantity of carbohydrates potentially convertible to liquid fuels. Research and development activities in the use of these cellulosic residues for ethanol production traditionally have focused in the areas of pretreatment, saccharification, and fermentation, but direct microbiological degradation of biomass by cellulotic microorganisms is an attractive alternative. Anaerobic and aerobic processes are not yet efficient, and there remain many technical and economic unknowns to be resolved (Detroy et al., 1981; Canevascino and Galten, 1981; Augerinos and Brew, 1980).

Engineered microorganisms, however, can be very useful in converting pentoses present in molasses or dextrins and amylose fractions of sugar cane juice. The conversion of starch to ethanol by yeast fermentation has also been attempted in our laboratory. The first step of the program is to obtain, through protoplast fusion, a yeast strain with glucoamylase activity and ethanol fermentation ability as well. The results already obtained after transconjugant selection of *Saccharomyces cerevisiae* and *S. diastaticus* crosses have permitted isolation of a strain whose test tube ethanol production is given in table 4. Ethanol yields are expressed by volume (%) and represent mean values of four successive fermentations in dextrinized starch plus salts medium.

A problem encountered in large-scale ethanol production is the large stillage volumes, which are about twelve times the alcohol production from molasses. In Brazil some alternatives to overcome the problem have been proposed such as stillage concentration. But the most practical way to solve a significant part of the problem is to return stillage to the crop field as a fertilizer (Rodella and Ferrari, 1977). We have tested the alternative to recycle stillage in the industrial plant to dilute molasses. Industrial fermentations with up to 30% stillage have caused no major problem for the yeast performance since mean conversion efficiency of sugar was closed to 90%. Another alternative we expect to be available, will be the aerobic conversion of stillage from molasses for simultaneous phosphate solubilization and biomass production (Cardoso, personal communication). This way the pollutant effluent will become an important substrate for further fermentation. It is expected that with the above processes and conducting fermentations with high levels of ethanol in beer the stillage problem from molasses will be minimized.

TABLE 4

ETHANOL PRODUCTION WITH DEXTRINIZED STARCH BY A
SELECTED STRAIN. MEAN VALUES IN ETHANOL BY
VOLUME (%) AND 14 HOURS OF FERMENTATION TIME

DEXTRINIZED STARCH	ETHANOL YIELD	
	30°C	35°C
15%	7.78	7.79
20%	8.04	7.60
25%	5.60	5.73

CONCLUSION

The present status of ethanol production in Brazil presents challenging problems to be solved and certainly genetic engineering complementing traditional genetic techniques can offer the most efficient answer.

Substrate conversion rates with improved microorganisms represent one of the most important economic factors. Efficient yeast strains for hexose fermentation and the possibility of introduction of pentose conversion genes and amylotic genes will certainly contribute to better utilization of the substrates. Alternative techniques were proposed for utilization of cellulosic materials and other unusual substrates in fermentation industry. However, attention is called to the problem of the large stillage volumes from molasses, possibly one of the most important pollutants today in Brazil, to be solved by improved microorganisms.

ACKNOWLEDGMENTS

I am indebted to J. Johnston, U. A. Lima and C. Panchal for providing strains of yeast; the Conselho Nacional de Desenvolvimento Cientifico e Tencnologico (CNPq); the Fundacao de Amparo a Pesquisa do Estado de Sao Paulo (FAPESP); Zanini S/A Equipamentos Pesados, Usina Santa Elisa, Usina da Pedra e Usina Vale do Rosario, Sao Paulo; and Grupo Getec, Rio de Janeiro.

REFERENCES

Amer, G. I., and S. W. Brew. 1980. Microbiology of lignin degradation. *In* G. T. Tsao (ed.), Annual reports on fermentation processes, 4:68–103. Academic Press, New York.

Anderson, E. 1979. Brazil sets lofty goals for ethanol. Chem. Eng. News 57:15–16.

Azevedo, J. L., F. C. A. Tavares, and M. R. M. Cruz. 1978. Genetic improvement of yeasts. *In* M. Bacila et al. (eds.), Biochemical genetics of yeasts, p. 563.

Azhar, A. 1981. Alcohol fermentation of sweet potato. 1. Acid hydrolysis and factors involved. Biotech. BioEng. 23:879–86.

Augerinos, G. C., and S. W. Brew. 1980. Direct microbiological conversion of cellulosic to Ethanol. Ann. Rep. Ferm. Proc. (Tsao, G. T., ed.) 4:165–91.

Canevascini, G., and C. Galten. 1981. A comparative investigation of various cellulase assay procedures. Biotech. BioEng. 23:1573–90.

Clarke, P. H. 1976. Mutant Isolation. *In* K. D. MacDonald (ed.), Genetics of industrial microorganisms, pp. 15–28. Academic Press, New York.

Crocomo, O. J. 1981. Genetic engineering for the sugarcane plant system. *In* Genetic engineering and biotechnology proceedings, pp. 29–32. Piracicaba, SP, Brazil.

Demain, A. L. 1980. The new biology: opportunities for the fermentation industry. *In* G. T. Tsao (ed.), Annual reports on fermentation processes, 4:193–208. Academic Press, New York.

Detroy, R. W., L. A. Lindenfelsen, S. Sommer, and W. L. Orton. 1981. Bioconversion of wheat straw to ethanol: chemical modification, enzymatic hydrolysis, and fermentation. Biotech. BioEng. 23:1527–35.

Elander, R. P., and L. T. Change. 1979. Microbial culture selection. *In* H. L. Peppen and D. Perhman (eds.), Microbial technology, pp. 243–302. Academic Press, New York.

Euriquez, A., R. Montalvo, and M. Canales. 1981. Variation of Bagasse crystallinity and cellulase activity during the fermentation of *Cellulomonas* bacteria. Biotech. BioEng. 23:1431–36.

Fleck. W. F. 1979. *In* O. K. Sebek and A. I. Laskin (eds.), Genetics of industrial microorganisms, pp. 117–22. American Society for Microbiology, Washington, D.C.

Guiraud, G. P., J. Daurelles, and P. Galzy. 1981. Alcohol production from Jerusalem artichokes using yeasts with inulinase activity. Biotech. BioEng. 23:1461–65.

Hopwood, A. 1981. The genetic programming of industrial microorganisms. Sci. Amer. 245:90–125.

Hopwood, D. A., H. M. Wright, M. J. Bibb, and S. N. Cohen. 1977. Genetic recombination through protoplast fusion in *Streptomyces*. Nature 268:171.

Lyness, E. D., and H. W. Doelle. 1981. Fermentation pattern of sucrose to ethanol conversions by *Zymomonas mobilis*. Biotech. Bio Eng. 23:1449–60.

Merrick, M. L. 1976. Hybridization and selection for penicilin production in *Aspergillus nidulans*: a biometrical approach to strain improvement. *In* K. D. MacDonald (ed.), Genetics of industrial microorganisms, pp. 229–42. Academic Press, New York.

Peberdy, J. F. 1979. Genetics of industrial microorganisms. (O. K. Sebek and A. I. Laskin eds.). 192–196. Amer. Soc. Microbiol.

Provost, A., C. Bourguignon, P. Fornier, A. M. Fibert, and H. Heslot. 1978. Intergenic hybridization in yeasts throughout protoplast fusion. FEMS Microbiol. Let. 3:309–12.

Rodella, A. A., and S. E. Ferrari. 1977. A Composicao da vinhaca e efeitos de sua aplicacao como fertilizante na cana-de-acucar. Brazil Acucareiro. 90:6–13.

Sebek, O. K., and A. I. Laskin (eds.). 1979. Genetics of industrial microorganisms. Amer. Soc. Microbiol. Publ.

Tavares, F. C. A. 1980. Microbial breeding in Brazil: yeast breeding for ethanol production. *In* Proceedings of the second Japan–Brazil symposium on science and technology, pp. 185–91.

Tavares, F. C. A. 1974. Principios do melhoramento em funcao de parametros biometricos aplicados a *Saccharomyces uvarum* Beijerinck. Ph.D. thesis, Piracicaba, S.P. 100 pp.

Wang, D., C. Cooney, A. Demain, P. Dunill, A. Humphrey, and M. Lilly. 1979. Bioconversions. In Fermentation and enzyme technology, pp. 34–37. John Wiley and Sons, New York.

Wilke, C. R., R. D. Yang, A. F. Sciamanna, and R. P. Freitas. 1981. Raw materials evaluation and process development studies for conversion of biomass to sugars and ethanol. Biotech. Bio Eng. 23:163–83.

WILLIAM R. STROHL AND ROBERT M. PFISTER

Automation in Modern Microbial and Tissue Fermentation Processes

29

INTRODUCTION

What Is Fermentation?

Fermentation may be defined as a process by which products are obtained from organisms cultured in contained systems. Fermentation classically consists of four components: the fermentor hardware or vessels, the sensors that monitor the system, the medium, and the organisms. In this paper we shall describe a fifth component of modern fermentation: the computer-control system.

The fermentation industry traces its roots to Egyptian times and before, with the natural production of wine from grapes carried out by anaerobic culture of the grapes with their indigenous yeasts. It was not until the late 1800s, however, that Pasteur proved that the fermentation process was due to the action of the yeasts. Once that link was established, a burgeoning fermentation industry was started for the bioproduction of acetone, butanol, and glycerol needed for World War I.

Likewise natural products from other microorganisms and from plants have been used in folk remedies as herbal cures, or in production of foods such as pickles, sauerkraut, or soy sauce, for centuries. Most of the active products of these processes have now been identified and purified, and the processes for producing them have been optimized. Over the past sixty years, thousands of useful biochemicals, antibiotics, and pharmacologically-active drugs produced by bacteria, fungi, plants, and now even animal tissue cells, many of which have had a major impact on society, have been discovered. These fine and bulk chemicals and drugs have made up a market worth approximately $1.5 billion in 1983 (Emyanitoff and Weinert, 1984). Thus fermentation, or the production of useful biochemicals from cultures, has been an important and long-standing industry.

Revitalization of Interest in Fermentation

Recently, however, three major forces have caused a tremendous revitalization and diversification of the fermentation industry. The first force is the continuing development of the recombinant DNA technology started in the early 1970s. This has led to an entirely new market of microbially produced recombinant DNA products such as vaccines, polypeptide hormones, insulin, interferons, bulk enzymes, prescription enzymes, growth stimulating hormones, amino acids, and immunoregulators. The market for these "new" compounds is rapidly expanding, and is projected to be more than $300 million by 1985 and more than $1.3 billion by 1987 (Emyanitoff and Weinert, 1984). The challenge posed by the fermentation of recombinant microorganisms is to develop fermentation conditions that promote the stability of the cloned DNA molecules, control the expression of the cloned DNA, and obtain the maximal specific and volume productivities possible.

The second major force that is reshaping the fermentation industry is the development of animal and plant tissue culture systems. Both of these culture systems require fermentation conditions that are much different from those required for traditional fermentation of heterotrophic bacteria (Goldstein, 1983; Glacken et al., 1983). Plant tissue fermentation has importance in the continuing development of the basic physiology of crop plants, germ plasm preservation, genetic engineering, and plant secondary metabolites. Moreover, the regeneration of plants from tissue cultures will require new fermentation technology in order to deal with regenerated plant fidelity or trueness, the combined mutation, protoplast fusion and selection processes, and actual production of plant biochemicals in fermentors. Animal tissue culture fermentation has now become extremely important, mostly because of the development of hybridoma technology. Hybridomas are clones of fused tissue cells capable of the density-dependent production of monoclonal antibodies. These monoclonal antibodies are becoming economically significant in such diverse areas as enzyme or protein purification, diagnostics, therapeutic treatment of tumors, and treatment of some immunological diseases (Glacken et al., 1983). Mammalian cells are much more difficult to grow than bacterial cells, and certain cell lines require anchorage to surfaces for growth and production (Glacken et al., 1983). New fermentation systems are being developed to grow hybrid-omas using microcarriers or continuous-flow filtration systems (Glacken et al., 1983).

The final force driving the modernization of fermentation is the development of sophisticated computer-control systems (Oinas, 1982). The potential for these computer-based systems in terms of optimizing traditional fermentations, fermentation of organisms containing recombinant DNA, and for plant and animal tissue fermentations is enormous, as will be explained in this chapter. Computer control of fermentation has also opened the doors to new types of fermentation that are feedback-controlled, so that the reactions of the organisms themselves actually run the processes.

Production Optimization

Optimization of biological production can be separated into three areas: optimization of the producing organism itself, optimization of the media in which they are grown, and optimization of the fermentation process. Many of the "traditional" processes for producing fine chemicals such as penicillin or streptomycin, or bulk products such as glutamate and butanol, have undergone a series of culture and media optimizations so that the products could be obtained for the least cost. On the other hand, basic fermentation processes that had been first developed for acetone-butanol production in World War I and upgraded for the submerged culture of *Penicillium* in the 1940s have undergone only minor changes until the past ten years or so. The requirements for updated fermentation processes, for optimization of fermentation conditions, and for stringent control of the fermentation conditions have increased tremendously (Beaverstock and Trearchis, 1979). Much of the interest in fermentation processes today is placed on computer automation of fermentation processes, with special emphasis on control features (Oinas, 1982).

Development of Computer-Controlled Fermentation Processes

The chemical industry is largely responsible for the development of computer control designs, processes, and systems that are used in computer control of microbial and tissue cell fermentation today. The application of chemical process control to fermentation processes, however, was not a single evolution, largely because fermentation systems are poorly understood processes that are difficult to control, that have time-variant parameters, and because of the inherent need in fermentors of sterilizable, on-line, accurate sensors (Oinas, 1982). It is very often said that a fermentation system is only as good as its sensors. The use of computers to control and acquire data from fermentation processes has now become quite widespread in both industrial and academic settings. This is largely due to the decreased costs of computer control with greater control capabilities, the increased availability and reliability of microprocessors, and the strong development of useful sensory systems for fermentation (Oinas, 1982). The primary objective in the development of computer-controlled fermentation in industry has been to produce the fermentation product as economically as feasible (Oinas, 1982). Computer control has met this broad-based objective by providing quality control and time-saving features, by automatically calibrating sensory systems for quality control, by furnishing automatic documentation of all runs, by decreasing control loop costs, and in some systems, by using control to make the most efficient use of power input (Lundell, 1979; Oinas, 1982; Rolf and Lim, 1982). Computer control has also increased reproducibility in batches and saved energy and labor costs, which have been as important as improvements in product concentration and yield (Lundell, 1979). Most computer systems used thus far have been sophisticated enough to handle the usual tasks such as data acquisition, monitoring and documentation of parameters, continuous operation of set-point controls, and man-machine communications (Cooney, 1979). There are computer systems now, however, that are flexible enough to meet the needs of specialized fermentations, to handle a variety of hierarchical systems, and to apply very complex control and optimization strategies (Rolf and Lim, 1982).

HISTORICAL PERSPECTIVE

Computer-aided fermentation has made remarkable progress in the past twenty years since Yamashita and his colleagues in 1962 developed a direct digital control (DDC) system for glutamic acid fermentation in a pilot scale plant (Yamashita et al., 1969; Aiba, 1979). The history of computer-controlled fermentation can be traced back to the concurrent development in the 1940s of digital computers, submerged penicillin fermentations, and sensors such as pH probes and gas analyzers (Cooney, 1979). Controlled feed operations and direct digital control were first used in the fermentation industry in the early 1960s. By 1965 computer control had been developed for monosodium glutamate production, and in 1969 Dista Corporation reported using computer control of an entire penicillin production plant (Cooney, 1979).

Cooney (1979) showed a logarithmic growth in the 1970s for use of on-line computers in production systems, a trend that is continuing into the 1980s. Although long-term interest in computer-aided fermentations has been noted, the real advances in these processes have been made only in the past five to ten years (Cooney, 1979). In the seven years since Cooney's article, there has been proliferation of papers in the literature describing new computer-aided fermentation systems, and new uses for computer control of, and data acquisition from, fermentation processes. Relatively recent developments have opened doors to computer control and analyses of fermentation processes that were previously either unavailable, available only to relatively few, or available only at high costs. These developments include:

1. Inexpensive microchips (microprocessors) that can control processes using single board operations or that are the bases for microcomputers, and smart controllers, etc. (Hampel, 1979; Jefferies et al., 1979).

2. On-line quadrapole mass spectrometers for measurements of gases (Heinzle et al., 1983);

3. Enzyme electrodes (Barker and Somers, 1978) and volatile gas sensors (Mandenius and Mattiasson, 1983) for on-line measurement of metabolites, nutrients, or products;

4. Dialysis separation systems (Zabriski and Humphrey, 1978) and tangential flow membranes (Tutunjian, 1983) for on-line separation of culture biomass from soluble products;

5. Less costly and more versatile analog to digital (A/D) and D/A conversion systems (Titus et al., 1984);

6. Continuous cell density measurement devices (Ohashi et al., 1979);

7. Advanced electrical valving systems for gas-flow control (Mathers and Cork, 1984).

Another more intangible reason for recent advances in computer control of fermentations is the popularization of computers. This is especially true of microcomputers, which are now being used in schools, businesses, biological research labs, and homes. The boom in relatively inexpensive and easy-to-use microcomputers, caused by the development of the microchip industry, has hit its peak in the past four to five years. A simple demonstration of this phenomenon is that Drexel University in Philadelphia *requires* its students to obtain microcomputers for use in their studies, a requirement that would have been completely unheard of in past years.

WHY COMPUTER-CONTROL FERMENTATIONS?

This question is often asked during budgetary considerations, and must be properly answered for both the lab-bench researcher and the production fermentation engineer. The question is particularly important when it is considered that most current industrial fermentations are batch processes, which are not difficult to control using standard analog set-point controllers. The advantages offered by computer control of fermentation are inherently different for the lab-bench researcher and the production engineer. The primary difference is in philosophy: the researcher uses computer-controlled fermentations to answer questions about the metabolism of the organism and about the production of interesting enzymes or metabolites, whereas the production engineer uses the computer to maximize the efficiency of his fermentations, particularly minimization of their costs. Computer control of fermentations, however, offers the following general advantages to all users:

1. experimental variables can be manipulated any time of the day without the operator being present;

2. data can be logged at computer-determined variable frequencies to correspond with critical periods of the fermentation;

3. extensive data can be accumulated and stored on disks for later evaluation;

4. multivariable control strategies can be implemented to manipulate a process along an optimum trajectory;

5. experiments and runs can be reproduced much more accurately and consistently;

6. fermentations can be controlled through calculated variables in a real-time sense, i.e., automated control based on a derived model can be accomplished in absence of the operator.

7. the number of accessible data outputs from the fermentor can be increased because of the supervisory capabilities of computers;

8. statistical operations for end point analyses that are designed to save time are automatically done by the computer during the course of fermentations, which makes the process more cost-efficient.

9. the accuracy and reliability of data are increased due to the averaging of many points, the rejection of noisy and false signals, and the periodical internal recalibration that is included in most fermentation control programs.

THEORETICAL CONTROL AND MODELS TO PREDICT FERMENTATION PATTERNS

In developing a strategy for the computer control of fermentation, the primary questions are what parameters (or variables) should be measured, and how should they be measured? A strategy for overall control of microbial growth and production must be used as a base that can be altered for any particular organism with special needs. Several investigators have approached this question by developing growth models and predictions based on known growth phenomena (Nihtila and Virkkunen, 1977). For example, the models for growth and productivity and the parameters to be controlled for ethanol production by *Saccharomyces* (Nelligan and Calam, 1980) may be far different from those considered for antibiotic biosynthesis by *Streptomyces* spp. (Biryukov and Kantere, 1982; Votruba and Behal, 1984). Yet there are parameters that the two systems have in common, such as pH, pCO_2, biomass accumulation, and utilization of glucose or other sugars. On the other hand, the major metabolic control mechanisms of the cells in question and the regulatory "handles" that the researcher has on the particular organism should be considered. If the organism is sensitive to, and responds to, particular cultural changes, these should be incorporated into the control picture.

The generalized depiction for fermentation of cells shown in figure 1 is a conglomeration of models by Schmidt (1983), Glacken et al. (1983), and Bailey et al. (1983). Our model takes into consideration the various hierarchical levels in the fermentor, as follows:

Systems and Components of a Fermentor

Level I.   overall fermentor and its productivity capability;

Level II.   the major bulk phases in the fermentor;

Level III.   components of the bulk phases, including cell population;

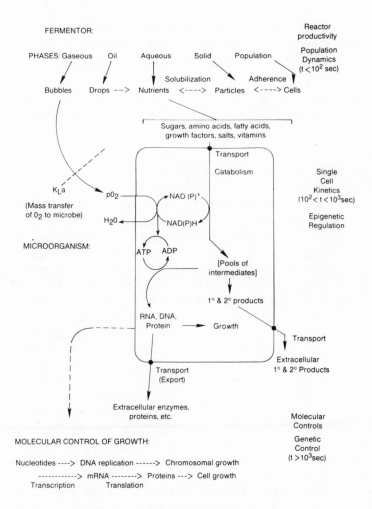

Fig. 1. Physiochemical, biochemical, and molecular components of the fermentor. The fermentor contains several interactive systems that must be considered when setting it up: the physical nature of the fermentor itself along with the type of fermentation process (continuous, batch, fed-batch, etc.); the bulk phase containing the medium components and the population of microorganism; the metabolic systems, their physiological regulation, and productivity of the organisms; and the molecular and genetic control of growth and stability. The turnover time for each level of control is given to the right: population dynamics can change in less than 100 sec, physiological or epigenetic controls can take place in 100 sec to ca. 17 min, and genetic controls take longer than 17 min. In actual culture systems, these times may vary considerably, depending on the growth rate of the microorganisms.

Level IV.   the entry of nutrients and oxygen into the individual cells;

Level V.   individual cell intermediary metabolism kinetics;

Level VI.   individual cell molecular controls on DNA, RNA, and protein synthesis;

Level VII.   kinetics of chromosomal or extrachromosomal genetic segregation.

Most classical fermentation engineers approach the above model system (fig. 1) from the standpoint of manipulating the major bulk phases so that the "mass transfer" of nutrients and oxygen into the cells is maximized (Spruytenburg et al., 1979). The second concept used by fermentation engineers uses the principle of material balances to indirectly measure the productivity (Mou and Cooney, 1983a, b; Stephanopoulos and San, 1982). Bailey and colleagues (1983), on the other hand, approach the model with the paradigm of: molecular level controls → single-cell kinetics → microbial population dynamics → reactor productivity. Their model is extremely important for fermentation of cells containing recombinant plasmids because the stability (lack of segregation and loss from the cells) of the extrachromosomal recombinant DNA is required for product formation, and therefore productivity (Bailey et al., 1983). The modern fermentation microbiologist must have the ability to understand and control fermentation processes using models on both ends of the spectrum because more fermentation organisms will contain recombinant DNA, or will be genetically complex. Domach and colleagues (1984) have recently developed a mathematical computer model to describe the glucose-limited growth of *Escherichia coli*. Their model considers all major aspects of cell metabolism, DNA synthesis and initiation controls, and most other factors considered in figure 1.

An important aspect of developing a general model such as in figure 1 is that it demonstrates the complexity of the fermentation process and the degree to which that process is made up of several subprocesses. If any of the subprocesses is left out of the control picture, the productivity suffers in the end. Just as control of fermentation must consider the subprocesses, the control system is only as good as its components. The following factors should be considered in selecting the components in a fermentation system: (1) sensors that measure variables of the fermentation process; (2) control algorithms and computer configurations; (3) fermentor configuration; and (4) type of fermentation.

SENSORS, DATA ACQUISITION, AND ON-LINE CALCULATIONS

Various research and review articles have described the requirements for a "fully (or properly) instrumented fermentor" (fig. 2; Harmes, 1972; Unden, 1979; Cooney, 1979; Hatch, 1982; Rolf and Lim, 1982). It is obvious that the ability to control a fermentation process is completely dependent on the degree to which parameters of that fermentation can be measured. Always included in assessments of properly instrumented fermentors are such parameters as measurement and control of pH, temperature, nutrient and gas feed rates, turbidity, pO_2, off-gases, and other such standard variables (Unden, 1979).

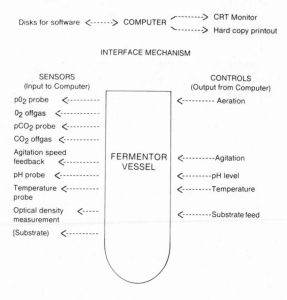

Fig. 2. A properly instrumented fementor, according to most fermentation engineers, should contain the output sensors as shown. Those sensors provide feedback to the computer for decisions on control. The fermentor should be controllable using nutrient flow rates, agitation, aeration, pH, temperature, and other additions.

Sensors

The computer control of fermentation is inherently limited by the feedback given to the computer by the sensors used. Thus, the sensors are, in many ways, the most critical components of the fully instrumented system. The requirements for a good sensor are:

1. It should be usable on-line (i.e., continuously or semi-continuously during the fermentation).
2. It should provide an electrical signal to the computer.
3. It should be stable under fermentation conditions, and not prone to electrical or mechanical drift.
4. It should be self-decontaminating or protectable from cell growth on the sensor;
5. If it measures biological or chemical parameters of the fermentation, it should have an *easily sterilizable* probe, electrode, light path, or analysis loop.

Table 1 provides a review of most measurable variables, and the sensors that measure each variable, that are available for modern instrumented fermentors. The measurable variables and their sensors shown are used to analyze the physical status of the fermentation system, the physiological status of the organism, and the status of production by the organisms. Some of these should be used for all fermentation processes, some for only special processes. In modern industry production facilities, the input of electrical power to the system and the analysis of power consumption by several facets of the fermentation are also analyzed. From these analyses, the most economical methods for maintaining proper fermentation conditions (heat and oxygen transfer among the most important) are calculated and translated by the control computers into large savings for the industry (Lundell, 1979; Oinas, 1982).

Gateway Sensors

Many of the variables shown in table 1 give information on the basic status of the organisms being grown, and give, with certain other experimentally determined values, a relatively clear picture of the fermentation. Often, however, certain variables cannot be directly measured because of interferences or lack of measurement mechanisms, and therefore must be measured indirectly and calculated. A good example of this is the mea-

TABLE 1

VARIABLES IN FERMENTATION AND THE USE OF SENSORS FOR MEASUREMENT AND/OR CONTROL OF FERMENTATION PROCESSES

Variable or Parameter	Sensor or Analysis	References
Physical parameters		
Temperature	Thermocouple — probe	(a)
Fermentation heat evolution .	Calorimeter/Analysis	(b)
Vessel head pressure.........	Pressure gauge or probe	
Agitation speed.............	Feedback tachometer	(a)
Power input (and consumption)	—	
Medium viscosity	—	
Air flow rate	Tylan mass flow controllers	(c)
Nutrient flow rate...........	Regulated by pumps	
Cells		
Concentration..............	Turbidimeters of various types	(a, d)
	Conductivity	(e)
Size, type, metabolic processes	Microfluorometers	(f)
Gas/liquid analyses		
Antifoam addition	Calculated by incrementor and known flow rate	
pH (including calculation of amount of acid or alkali added)	Electrodes; Additions of acid/alkali calculated using incrementor and known flow rate	
p[Select ion]	Select ion electrodes	
Redox potential (mV)	Redox electrode	(g)
pO_2......................	Galvanic or potentiometric electrode	
O_2 off-gas.................	Paramagnetic analyzer (or)	(h)
	On-line mass spectrometer	(i)
pCO_2	Electrode	
CO_2 off-gas	Infra-red analyzer (or)	(h)
	On-line mass spectrometer	(i)
Other off-gases (N_2, CH_4, etc.)	On-line Mass spectrometer	(i)
Cellular components (cytochromes, NADH, ATP, etc.)....................	On-line spectrophotometry	(j)
	On-line photometry	
	On-line fluorometry	(f)
Nutrient, metabolite, or product concentration:	Enzyme electrodes (or)	(k)
	Direct conductance probes	(e)
	Derivative spectrophotometry	(j)
	On-line gas chromatography	(l)
	Silicone tubing gas sensors	(m)

REFERENCES: (a) Titus et al., 1984; (b) Luong et al., 1983; (c) Mathers and Cork, 1984; (d) Ohashi et al., 1979; (e) Firstenberg-Eden and Zindulis, 1984); (f) Rolf and Lim, 1982; (g) Kjaergaard and Joergensen, 1979; (h) Mou and Cooney, 1983a; (i) Heinzle et al., 1983; (j) Woodrow and Spear, 1984; (k) Barker and Somers, 1978; (l) Comberbach and Bu'Lock, 1983; (m) Mandenius and Mattiasson, 1983.

surement of cell biomass in fermentation broths containing insolubles: the on-line biomass must be determined by means other than turbidimetric analyses. These indirect measurements can be carried out using "gateway" sensors (Hampel, 1979). These gateway sensors give information on the metabolism of the organisms that is directly relatable to biomass or some other immeasurable, but important, parameter. Good examples of these are the calculation of biomass of yeast by the stoichiometric production of ethanol Spruytenburg et al., 1979), the calculation of *Hansenula* biomass by utilization of methanol (Swartz and Cooney, 1979), or the production of glutamate by a corynebacterium using CO_2 evolution (Park et al., 1983). Biomass can also be related to heat of fermentation (Leiseifer and Schleser, 1983; Luong et al., 1983) or to the conductance in the fermentation vessel (Firstenberg-Eden and Zindulis, 1984). Likewise, under given conditions, the biomass of certain organisms may be calculated by its evolution of CO_2 and its utilization of O_2 (Wang et al., 1979). Real-time determinations of oxygen uptake and CO_2 evolution in a fermentor have been likened by Hampel (1979) to a differential respirometer, in which gas exchange rates can be computed.

Data obtained from gas balancing can be used to calculate two very important parameters, the volumetric gas transfer rate ($k_L a$; equation 1) and respiratory quotient (RQ; equation 2).

$$N_o = k_L a \, (C^* - C_L) \tag{1}$$

where N_o is the oxygen transfer rate (mmol O_2/liter-hour), $k_L a$ is the mass transfer coefficient (mmole O_2/atmosphere-liter-hour), C_L^* represents the concentration of oxygen in the gas phase in equilibrium with the saturation of oxygen in the liquid phase (atm), and C_L, which represents the dissolved oxygen in the fermentor (atm; another measurement obtained as above). The $k_L a$ is used to estimate the mass transfer of oxygen to the organisms in the fermentation system under given conditions. Oxygen mass transfer is considered by most fermentation engineers to be one of the most important factors that potentially limit growth, including growth of bacteria (Hampel, 1979), plant cells (Goldstein, 1983), and mammalian cells (Glacken et al., 1983). Respiratory quotient is a measure of the physiological status of the organism based on its ratio of CO_2 liberated to O_2 consumed. Several investigators have tried to link directly the RQ of an organism with various physiological states (Wang et al., 1979; Swartz and Cooney, 1979). The RQ of endogenous metabolism has been reported as about 0.8 to 0.7, and the RQ of oxidative metabolism is ca. 1.0 to 0.9 (Wang et al., 1979).

$$RQ = QCO_2/QO_2. \tag{2}$$

Finally, material balances can be constructed and calculated, using equation 3, from knowing the amount of O_2 used and CO_2 produced (Wang et al., 1979) and a glucose enzyme electrode probe to analyze the amount of substrate utilized (Chotani and Constantinides, 1982; Cleland and Enfors, 1983). Thus, using the "gateway sensors" showing the amount of O_2 and CO_2 gases exchanged, several very important characteristics of the culture can be calculated.

The principles of equation 3 explain the use of other gateway parameters to calculate biomass, substrate formation, or other physiological parameters:

$$CH_mO_l + aNH_3 + bO_2 \rightarrow ycCH_pO_nN_q + zCH_rO_sN_t + cH_2O + dCO_2 \tag{3}$$

[C-source] + [N-source] + [O_2] $\rightarrow$ Biomass + Product + H_2O + CO_2 + heat

These include heat evolved from fermentation (Luong and Volesky, 1982; Leiseifer and Schleser, 1983), utilization of substrate (e.g., glucose), and amount of acid produced during a fermentation (calculated by amount of alkali required to neutralize the culture system). The heat released during fermentation processes, a great economical problem of industrial processes at large scales (due to costs of cooling the fermentors), is a blessing for the bench fermentation researcher. The heat liberated from metabolism is directly correlated to the metabolism efficiency, and therefore may be used to calculate biomass and other fermentation parameters (Luong and Volesky, 1982; Luong et al., 1983; Leiseifer and Schleser, 1983).

Other sensors recently developed are electrodes measuring specialty compounds (Barker and Somers, 1978; Blum et al., 1983; Cenas et al., 1984), including antibiotics such as penicillin (Nilsson et al., 1978). Almost any enzyme that consumes or liberates electrons may be immobilized onto membranes and placed onto oxygen or redox electrodes (Barker and Somers, 1978; Blum et al., 1983; Cenas et al., 1984). These techniques may provide fermentation researchers with the ability to measure, on-line, a wide variety of metabolites and products in the fermentation broths. A major shortcoming of these enzyme electrodes at this time is the lack of methods to sterilize them for insertion into fermentors. Sterilizable electrodes and sensors that measure a variety of inorganic compounds, such as nitrate, ammonia, or sulfide have been developed recently and may be used in fermentation research as information contributors for feedback control.

On-line quadrapole mass spectrometers (QMS) are the best methods for measuring gases; up to eight different gases from eight different fermentors can be multiplexed into, and measured every few seconds on a Perkin-Elmer QMS. These gases included CO_2, O_2, CO, H_2, N_2, and CH_4 (Heinzle et al., 1983; Pungor et al., 1983). The QMS can also measure volatile compounds in the liquid phase of the fermentor, using a recently devised coupling system (Pungor et al., 1983). Moreover, methods now have been developed to continuously measure volatile compounds such as acetic acid, acetone, butanol, or ethanol from fermentations using on-line gas chromatography (Comberbach and Bu'Lock, 1983). With the recent development of micro-high performance liquid chromatography (μHPLC), there will soon likely be reports of on-line μHPLC measurements of highly specific, difficult-to-measure products or metabolites.

On-line flow fluorometry is another recent development in which an advancement made in cell metabolite measurement was coupled with fermentation technology (Rolf and Lim, 1982). The coupling of on-line flow fluorometry to fermentors has already been used in separating difficult classes of cells by their metabolic differences in mixed fermentations, and for determination of specific subcellular materials such as RNA, DNA, specific proteins, and NADH (Rolf and Lim, 1982). Woodrow and Spear (1984) have recently shown that on-line derivative spectrophotometry may be used to measure certain products in fermentation broths.

All of these sensors, including many more not covered here for reasons of brevity, enhance the use of computers to control fermentations. These sensors provide feedback information that is critical to the control processes and to the development of models to predict product formation. Once the models are developed and the fermentations using them are controlled, the sensors become the fulcrum on which the success of the fermentation control programs rely.

Finally, the development of tangential flow filters and hollow fiber filters allow the analyses of fermentation products from very crude fermentation broths (Tutunjian, 1983). The filters separate the metabolites or products from the fermentation broths and the cells are returned to the fermentors. On-line filtration systems such as these have recently been marketed by biotechnology firms for the continuous production of monoclonal antibodies from hybridomas. These on-line filtration/separation systems promise to be a fast-growing and exciting new use for modern fermentations. It is likely, for example, that the filtration will be coupled with certain sensors, such as the derivative spectrophotometry mentioned above, to analyze difficult-to-measure components in future fermentations.

PROCESS CONTROL

Once the basic fermentor is set up and a series of sensors are selected for monitoring various parameters of the fermentations, a strategy for control must be developed. This strategy is partially dependent on engineering expertise, equipment available to the laboratory, funding available, and requirements for control capability.

Low-level vs. Supervisory Control

There are two basic levels of control, low-level and supervisory control. Supervisory control is an off-line control in which the process has been optimized according to certain mathematical models, and the computer control is used to carry out the optimization strategy (Aiba, 1979; Rolf and Lim, 1982; Stephanopoulos and San, 1982). Supervisory control uses computations and calculations of certain culture parameters based on variables measured by sensors in the culture vessel. An example of supervisory control would be the automatic alteration of fermentation conditions based on the on-line computer computation of RQ from variables such as pO_2, ingas and offgas O_2, pCO_2, and ingas and offgas CO_2 (Wang et al., 1979). In this example, an optimized RQ value would be achieved by alterations of culture conditions in response to values received from the sensors and then calculated by the control computer. Supervisory control (or "modern control" as described by Rolf and Lim [1982]) of the fermentation would be directed by the computer in response to the model. Rolf and Lim (1982) divided "modern control" into three categories: (1) steady state optimization for continuous fermentations; (2) dynamic optimization for batch and fed-batch fermentations; and (3) estimation of fermentation parameters for feedback control. They (Rolf and Lim, 1982) and other investigators (Nihtila and Virkkunen, 1977; Mou and Cooney, 1983a, b; Domach et al., 1984; Gallegos and Gallegos, 1984) have well-documented important models for developing strategies for supervisory control from engineering points of view, so these will not be covered further in this chapter.

Low-level control, on the other hand, is exemplified by direct on-line control loops that feed back directly from a single sensor to the variable being controlled. Low-level control loops are the backbone of most computer-controlled fermentation operations and are of utmost importance to the development of computer control of fermentation.

Types of Low-level Control

The simplest type of control is the timing-coordinated on/off control of switches, relays, or valves. The computer simply controls the fermentation by activating bits at preprogrammed preset intervals or timetables that in turn control the switches. Included in this type of control would be the initiation of sterilization and cooling cycles, sampling, preprogrammed addition of nutrients (e.g., for fed-batch cultures), and harvesting (Hampel, 1979). These are operated using a binary output and a timing cycle set up by the computer algorithm.

Other types of computer control can be generally divided into two major categories, "hardware controller actuated control" (HCA) and "direct digital control" (DDC). The major difference is that for hardware controller actuated control, an analog controller acts as a sensor of the fermentation and itself provides control for the fermentation. The analog hardware controller sends the signals to the computer through an analog-to-digital (A/D) converter, and the computer generates signals that are converted to analog signals by a digital-to-analog (D/A) converter to provide set-point sequencing and timing for the analog hardware controller. Thus the computer actually acts as a supervisor for the hardware controller in this system. With direct digital control (DDC), on the other hand, the computer receives a signal from a sensor (through an A/D circuit) and then uses digital signals to control *directly* the disposition of the control devices. Thus in DDC, the computer *is* the controller. A major advantage of HCA is that the analog control devices can continue operation in case of computer failure, whereas DDC relies on the integrity of the computer. On the other hand, DDC allows greater flexibility, and the control is more precise than HCA because the control responses are modeled as direct algebraic functions (Rolf and Lim, 1982). Both types of control systems typically utilize two basic approaches to implement the control, set-point limit control and proportional-integral-differential control.

Approaches to Control Implementation

There are two major approaches in implementing control, setpoint limit control (SLC) and proportional-integral-differential (PID) control. A parameter, such as those given in table 1, may be controlled within setpoint limits. The simplest example of this is the control of pH by the pH meter-controller between two present values, such as 7.2 and 7.4. If the culture pH drops below 7.2, the controller turns on a pump to add sterile alkali to the culture to bring it back within the preset limits. If the culture then goes too alkaline, acid is added to bring it within the setpoint limits. The difficulty with this type of control is that it often causes overshoot of acid or alkali addition. If the setpoint limits are too close, it causes problems by keeping the two addition pumps alternating "on" nearly constantly. On the other hand, computer algorithms using this principle are simple to write and react in the same manner as simple controllers. The computer also adds the ability to calculate internally the amount of acid or alkali added to the fermentation, which provides the investigator with instant information about the amount of acid or alkaline products produced in the culture system, and thus about the metabolism of the organism.

PID control algorithms are much superior to the SLC control algorithms mentioned above. In PID control algorithms, there is one setpoint, such as pH 7.3, using the same example as above. If the pH of the culture goes above or below the setpoint, the computer algorithm turns on the proper pump to add acid or alkali. The difference here is that the amount of change made is proportional to the distance of the measured value of the setpoint. This provides greater stability in the control loops and reduces the "overshoot" often experienced with SLC types of control. Controller parameters using PID are difficult to set up, and they need to be retuned periodically to adjust for time-variant system changes. The PID control approach is the most common approach used for both HCA systems and DDC systems.

Hardware Controller Actuated Systems

The D/A-A/D converter is the bridge of the HCA system (Titus et al., 1984). The sensors described in table 1 generate analog signals, based on either current (0–40 mA etc.) or voltage (0–50 mV, 0–5 V, etc.). These signals cannot be read directly by the computer, which is directed by binary signals. For monitoring the fermentation, an A/D converter converts the analog signals from the sensors into signals readable by the computer. The computer can then control the fermentation by making decisions and/or calculations based on the input signals, and sending out a signal to alter the system. The microprocessors of the computer generate digital signals, so these signals must be converted to analog by the D/A converter before they can be understood by the hardware control elements (Servos, valving devices, pumps, etc.) hooked to the fermentors (Titus et al., 1984). These A/D-D/A converters are thus the heart of a HCA system. D/A-A/D converters are available with different sensitivities that are characterized by the number of bits (2^x bits). Hence, an 8-bit converter (2^8) gives 256 discrete steps for a given analog range, e.g., 256 steps covering a 0–5 volt signal (ca. 51 steps/volt). A 12-bit converter would give 4,096 discrete steps over the same range (ca. 819 steps/volt), and a 16-bit converter 65,436 steps. The steps represent the amount of fine tuning that can be done using each type of converter. The cost of the converter is directly related to its sensitivity. Eight-bit converters are rather inexpensive (ca. $50 to $500), but are not sensitive enough for most fermentation needs. The 12-bit converter is available at medium cost (ca. $500 to $5,000) and is sufficient for most fermentation A/D-D/A needs (Titus et al., 1984). Sixteen-bit converters are available at high cost, but are usually more than what is needed for fermentation processes.

Hardware controller actuated control systems have the disadvantage that they are limited to certain fermentation conditions. Other disadvantages are that the versatility of control approaches that can be implemented is limited, and that the A/D-D/A converters are needed. There is also some variability and limitation in response in the control mechanisms which also detract from use of hardware actuated control systems (Rolf and Lim, 1982).

Direct Digital Control

Direct digital control is an alternative form of control different from the HCA type of control. Here the computer reads signals (through an A/D converter) directly from sensors that monitor the fermentation, and then controls the fermentation using a control algorithm (often PID) by sending digital signals back to control devices that are read and acted upon. In this type of fermentation control, the computer is the controller.

Although DDC was first used in industry some 20 years ago, Lim and his associates have been among the leaders in the recent development and advancement of DDC control mechanisms (Mohler et al., 1979; Rolf et al., 1982; Rolf and Lim, 1982). This system will perhaps be the trend of the future because of the inherent advantages, as pointed out by Rolf and Lim (1982). These include:

1. tremendous versatility in control algorithms;

2. easy adjustment of controller parameters;

3. improved response to control parameters;

4. automatic documentation of control actions; and

5. the overall ease of measurements of manipulated variables.

The major disadvantage of DDC has been that it requires fidelity of the computer. Hence, the computer cannot be used for other functions while DDC control algorithms are running. The recent development of micro-computer-minicomputer hierarchical control systems, much by Lim and co-workers (Mohler et al., 1979; Rolf et al., 1982; Rolf and Lim, 1982), has alleviated most of the disadvantages of DDC. In these systems, the micro-computers carry out the low-level DDC control while the larger computers carry out the supervisory functions with greater efficiency.

PROGRAMMING

Computer-control of fermentation offers certain advantages over con-ventional simple analog controllers. The major advantages are:

1. the ability to monitor on-line progress of specified aspects of the fer-mentation;

2. the ability to adjudicate and change set-points during fermentations;

3. the ability to define functions and control strategies with software (high-level control); and

4. the ability to store great amounts of data during fermentation runs that can later be closely analyzed.

The computer language must have the capability to do all the above functions within short time periods and without completely dominating computer memory space. Most of the programming for computer control of fermentations has been done using FORTRAN or BASIC languages, both high-level languages. In the future, FORTH and Pascal are finding wider usage in computer control, as they are more efficient computer languages for control purposes than either BASIC or FORTRAN. Like-wise, process control languages may be used for certain systems such as those obtained from fermentor design companies or those not requiring alterations in control operations. The programming algorithms and lan-guage depend greatly on the type of control to be used, the desired results and the type of computer(s) used (see next section). There are, however, several important functions that all programming algorithms must entail. The software should contain the following parameters to be useful for modern computer-controlled fermentations:

1. a real-time priority based system (Rolf and Lim, 1982);

2. reliability under different fermentation conditions;

3. user-friendly structure, so that non-programmers can use it;

4. a system for displaying continuous updates of the fermentation progress and control variables;

5. a system for data collection into files (on disks) for further analysis;

6. a system for alteration of set-point variables or other control features during the course of a fermentation run; and

7. a fail-safe system for handling temporary power failure.

COMPUTER CONFIGURATIONS

Minicomputers equivalent to the Digital Electronics Corporation (DEC) PDP-11 series have been used for standard computer control of fermentations in most industrial and biotechnology research institution settings (Dobry and Yost, 1977; Hatch, 1982; Bowski et al., 1983). These systems have the power, the speed, the versatility, and the disk-operating systems that are optimal for controlling various processes, and have found usage in nearly every aspect of industrial process control (Dobry and Yost, 1977; Bowski et al., 1983). These minicomputers are relatively inexpensive for industry, as they represent only a small fraction of capital equipment costs (Hampel, 1979; Lundell, 1979). Yet their return is potentially enormous in terms of productivity, energy savings, efficiency of operations, and optimal control. The opposite problem exists for use of these minicomputers at the laboratory bench scale. Here the minicomputers represent major costs with a less justifiable return. The use of microcomputers, particularly in the research laboratory (Yano et al., 1981; Mori et al., 1983; Forrest and Tsao, 1983; Titus et al., 1984), has increased greatly. Microcomputers have been used in hierarchical systems in conjunction with minicomputers (Mohler et al., 1979; Rolf et al., 1982; Rolf and Lim, 1982), and more recently have been used increasingly in stand-alone configurations (Hampel, 1979; Titus et al., 1984). In the hierarchical systems, the microcomputers are often used as DDC arms with the minicomputers used as the supervisory bodies (Rolf and Lim, 1982). In stand-alone direct set-point control (DSC) type configurations, microcomputers are rapidly gaining popularity as biological researchers become more adept in programming, and the random access memory (RAM) of the microcomputers increases (Hampel, 1979). Although many research institutions and companies have, or have access to, mainframe computers, the priority time use required for computer control of fermentations is not normally available (Hampel, 1979; Rolf and Lim, 1982). The costs and availability of funds, memory requirements, disk-operating system requirements, advisory computing requirements, and the ability to program are all major factors in considering the size and type of computer for controlling fermentations (Hampel, 1979). Computers are items that are costing less each year while gaining in memory, disk-operating features, interfacing capabilities, and other important features. This combination should eventually allow even the most tightly budgeted fermentation laboratories to enjoy reasonable levels of computer control capability.

COMPUTER-CONTROLLED FERMENTATIONS

Many original papers and review articles have described the use of computers to control and monitor fermentations. In this section, a few examples of these will be reviewed in order to introduce various aspects and uses of computer-controlled fermentations. We shall attempt to focus on the more unusual or more speculative examples, rather than simply repeat the observations of so many others.

Microcomputer monitoring and control have recently been applied to several types of continuous fermentation processes. Mori and colleagues (1983b) utilized a microcomputer to control a variety of continuous cultures that they termed "nutristats." Their concept was to use feedback control from one or two of several possible variables in order to achieve non-chemostat type steady state. Their system employed the ethanol-utilizing yeast *Candida brassicae* and sensors that measured pO_2 and ethanol concentration in the vessel. They then grew the yeast in microcomputer-controlled vessels using either pO_2 or ethanol concentration, or both, to control feed rates and culture cell density (Mori et al., 1983b). We recently utilized a microcomputer with an A/D-D/A system to control turbidostat- and chemostat-type continuous cultures of *Vibrio natriagens* (Titus et al., 1984). The feedback in our system was provided by a flow-through optical density cell, and the culture was maintained in steady state at a given turbidity with pH and pO_2 also controlled. Gallegos and Gallegos (1984) have also developed a model for computer control of continuous culture systems, which should assist in the development in more complex process control optimization of continuous cultures.

Forberg and colleagues (1983) recently showed that immobilized, nongrowing cells of *Clostridium acetobutylicum*, fed with pulses of limiting concentration of glucose and other nutrients, produced acetone and butanol with very high specific productivity. Normal immobilized systems are based on the continuous addition of nutrients with at least one growth factor missing, so that the cells do not grow. This procedure, though, can lead to degeneration of the immobilized cultures and their productivity. The investigation by Forberg and colleagues (1983) used the pulsed addition of nutrients in order to maintain energy requirements without providing enough nutrients for growth. The periodicity of the pulses of nutrients in their system slowed with increased incubation time. They used a computer for monitoring their system, but did not take advantage of control features. If their system were used and computer control were adapted to it, the flow of nutrients could be controlled by a combination of the measurement of glucose in the outflow to maintain limitation, and the measurement of acetone and butanol products. These same concepts could be applied to the production of secondary metabolites by immobilized plant cells, a procedure that is gaining in popularity (Shuler et al., 1983).

We are currently using a microcomputer system to control the addition of nutrients to a fed-batch system using a preformed gradient from 0.1_x to 20_x nutrient concentration, as described by Srinivasan and colleagues (1977). This "gradient-feed" type of culture yields very high cell densities with a limited amount of culture volume. The concept is based on the fact that as the culture density increases, the nutrients added will be utilized at a faster rate, a rate that correlates with biomass. If the rate of nutrient addition resembles the growth rate of the organism, the culture is subjected to nearly limiting levels of nutrients. Under these conditions, culture densities of nearly $100 \, g \, l^{-1}$ may be achieved. The major limitation, of course, is pO_2. In our system, the dissolved oxygen is maintained at nonlimiting levels by a priority-based increase first in agitation, then aeration, and finally oxygenation. Similar types of computer-controlled fed-batch, repeated fed-batch, and modified fed-batch experiments have been carried out by Mori and colleagues (1983a) and Mou and Cooney (1983a, b).

Optimization of growth of mammalian cells, either anchored or suspended (Glacken et al., 1983), and of plant cells, either entrapped or suspended (Goldstein, 1983), will almost certainly be accomplished using computer controlled fermentations. Glacken and colleagues (1983) have already developed a computer paradigm for growth of mammalian tissue culture systems. Although there are several specific difficulties with scale-up of tissue cultures, many of these will be overcome by use of feedback from acute sensors and computer control of growth parameters. Some of the major problems include the riddance of toxic wastes to which tissue cultures are so sensitive, maintenance of oxygen mass transfer without the shearing problems caused by stirred tank reactors, and feed of nutrients.

A final area of importance for which computer control may have a great effect is the production of recombinant DNA products by bacteria containing controllable plasmid cloning vectors. Cloning vectors have been derived that at restrictive temperatures have low copy numbers (Uhlin et al., 1979). The growth of biomass can take place under those plasmid restrictive conditions. When the temperature of the culture system is raised to a permissive temperature, the plasmids start "run-away" replication, with concomitant production of enzymes or proteins that are encoded by the cloned DNA in the vectors. For computer-controlled fermentations, these runaway plasmid cultures could be controlled at restrictive temperatures until certain culture densities were reached, at which point the computer would automatically raise the temperature to permissive run-away plasmid replication conditions for the production phase.

These are but a few examples of the potential uses for computer control of fermentations in the near future. As sensor systems continue to be developed, and models begin to examine more in depth factors in culture growth and production, computer optimization and control of fermentations should come into its own.

ACKNOWLEDGMENTS

We sincerely thank Steven Schlasner for his advice and criticism of this manuscript.

LITERATURE CITED

Aiba, S. 1979. Review of process control and optimization in fermentation. Biotechnol. Bioeng. 9:269–81.
Bailey, J. E., M. Hjortso, S. B. Lee, and F. Srienc. 1983. Kinetics of product formation and plasmid segregation in recombinant microbial populations. *In* K. Venkatasubramanian, A. Constantinides, and W. R. Vieth (eds.), Biochemical engineering, 3:71–87. New York Academy of Science.
Barker, A. S., and P. J. Somers. 1978. Enzyme electrodes and enzyme based sensors. *In* A.

Wiseman (ed.), Topics in enzyme and fermentation biotechnology, 2:120–51. Halsted Press, Chichester, England.

Beaverstock, M. C., and G. P. Trearchis. 1979. Fermentation control systems: a time for change. Biotechnol. Bioeng. Symp. 9:241–56.

Biryukov, V. V., and V. M. Kantere. 1982. Optimization of secondary metabolite production using an on-line real-time computer. *In* V. Krumphanzyl, B. Sikyta, and Z. Vanek (eds.), pp. 687–701. Academic Press, London.

Blum, L., J. C. Bertrand, and Pr. R. Coulet. 1983. Amperometric sensor and immobilized enzyme electrode for the determination of enzymatic activities. Anal. Lett. 16(B7):525–40.

Bowski, L., C. R. Perley, and J. M. West. 1983. A minicomputer system for analyzing and reporting pilot plant fermentor data. Biotechnol. Bioeng. 25:1237–50.

Cenas, N., J. Rozgaite, and J. Kulys. 1984. Lactate, pyruvate, ethanol, and glucose-6-phosphate by enzyme electrode. Biotechnol. Bioeng. 26:551–53.

Chotani, G., and A. Constantinides. 1982. On-line glucose analyzer for fermentation applications. Biotechnol. Bioeng. 24:2743–45.

Comberbach, D. M., and J. D. Bu'Lock. 1983. Automatic on-line fermentation headspace gas analysis using a computer-controlled gas chromatograph. Biotechnol. Bioeng. 25:2503–18.

Cooney, C. L. 1979. Computer application in fermentation technology: a perspective. Biotechnol. Bioeng. Symp. 9:1–11.

Dobry, D. D., and J. L. Jost. 1977. Computer applications to fermentation operations. *In* D. Perlman (ed.), Annual reports on fermentation processes, 1:95–114. Academic Press, New York.

Domach, M. M., S. K. Leung, R. E. Cahn, G. G. Cocks, and M. L. Shuler. 1984. Computer model for glucose-limited growth of a single cell of *Escherichia coli* B/r-A. Biotechnol. Bioeng. 26:203–16.

Emyanitoff, R., and H. M. Weinert. 1984. Market-based analysis of the economics of bioprocess engineering. Genet. Engn. News 4:8.

Firstenberg-Eden, R., and J. Zindulis. 1984. Electrochemical changes in media due to microbial growth,. J. Microbiol. Meth. 2:103–15.

Forberg, C., S.-O. Enfors, and L. Haggstrom. 1983. Control of immobilized, non-growing cells for continuous production of metabolites. Eur. J. Microbiol. Biotechnol. 17:143–47.

Forrest, E. H., and G. T. Tsao. 1983. A research-oriented microcomputer-coupled fermentor with full feedback control capability. Am. Chem. Soc. Abstr. no. 48.

Gallegos, J. A., and J. A. Gallegos. 1984. Estimation and control techniques for continuous culture fermentation processes. Biotechnol. Bioeng. 26:442–51.

Glacken, M. W., R. J. Fleischaker, and A. J. Sinskey. 1983. Large-scale production of mammalian cells and their products: engineering principles and barriers to scale-up. *In* K. Venkatasubramanian, A. Constantinides, and W. R. Vieth (eds.), Biochemical engineering, 3:355–72. New York Academy of Science.

Golstein, W. E. 1983. Large-scale processing of plant cell culture. *In* K. Venkatasubramanian, A. Constantinides, and W. R. Vieth (eds.), Biochemical engineering, 3:394–408. New York Academy of Science.

Hampel, W. A. 1979. Application of microcomputers in the study of microbiol processes. Adv. Biochem. Eng. 13:1–33.

Harmes, C. S. III. 1972. Design criteria of a fully computerized fermentation system. Dev. Ind. Microbiol. 13:146–56.

Hatch, R. T. 1982. Computer applications for analysis and control of fermentation. Ann. Rep. Ferm. Proc. 5:291–311.

Heinzle, E., K. Furukawa, I. J. Dunn, and J. R. Bourne. 1983. Experimental methods for on-line mass spectometry in fermentation technology. Biotechnol. 1:181–88.

Jefferies, R. P. III., S. S. Klein, and J. Drakeford. 1979. Single-board microcomputer for fermentation control. Biotechnol. Bioeng. Symp. 9:2312–239.

Kjaergaard, L., and B. B. Joergensen. 1979. Redox potential as a state variable in fermentation systems. Biotechnol. Bioeng. Symp. 9:85–94.

Leiseifer, H. P., and G. H. Schleser. 1983. Calorimeter-fermentor combination for investigations on microbiol steady state cultures. Z. Naturforsch. 38c:259–67.

Lundell, R. 1979. Computer control of fermentation plants. Biotechnol. Bioeng. Symp. 9:381–83.

Luong, J. H., and B. Volesky. 1982. A new technique for continuous measurement of the heat of fermentation. Eur. J. Appl. Microbiol. Biotechnol. 16:28–34.

Luong, J. H. T., L. Yerushalmi, and B. Volesky. 1983. Estimating the maintenance energy and biomass concentration of *Saccharomyces cerevisiae* by continuous calorimetry. Enz. Microb. Technol. 5:291–96.

Mandenius, C. F., and B. Mattiasson. 1983. Improved membrane gas sensor systems for on-line analysis of ethanol and other volatile organic compounds in fermentation media. Eur. J. Appl. Microbiol. Biotechnol. 18:197–200.

Mathers, J., and D. G. Cork. 1984. Microcomputer interfacing to mass flow type gas controllers for biotechnology applications. Biotechnol. Lett. 6:87–90.

Mohler, R. D., P. J. Hennigan, H. C. Lim, G. T. Tsao, and W. A. Weigand. 1979. Development of a computerized fermentation system having complete feedback capabilities for use in a research environment. Biotechnol. Bioeng. Symp. 9:257–68.

Mori, H., T. Yamane, and T. Kobayashi. 1983a. Comparison of cell productivities among fed-batch, repeated fed-batch and continuous cultures at high cell densities. J. Ferment. Technol. 61:391–401.

Mori. H., T. Yamane, T. Kobayashi, and S. Shimizu. 1983b. New control strategies for continuous cultures based upon a microcomputer-aided nutristat. J. Ferment. Technol. 61:305–14.

Mou, D. G., and C. L. Cooney. 1983a. Growth monitoring and control through computer-aided on-line mass balancing in a fed-batch penicillin fermentation. Biotechnol. Bioeng. 25:225–55.

Mou, D. G., and C. L. Cooney. 1983b. Growth monitoring and control in complex medium: a case study employing fed-batch penicillin fermentation and computer-aided on-line mass balancing. Biotechnol. Bioeng. 25:257–69.

Nelligan, I., and C. T. Calam. 1980. Experience with the computer control of yeast growth. Biotechnol. Lett. 2:531–36.

Nihtila, M., and J. Virkkunen. 1977. Modelling, identification, and program development for a pilot scale fermentor. *In* Van Nauta Lomke (ed.), Computer applications to process control, pp. 423–29. IFAC and North-Holland Publishing Co.

Nilsson, H., K. Mosbach, S.-O. Enfors, and N. Molin. 1978. An enzyme electrode for measurement of penicillin in fermentation broth: a step toward the application of enzyme electrodes in fermentation control. Biotechnol. Bioeng. 20:527–39.

Ohashi, M., T. Watabe, T. Ishikawa, Y. Watanabe, K. Miaw, M. Shoda, Y. Ishikawa, T. Ando, T. Shibata, T. Kitsunai, N. Kamiyama, and Y. Oikawa. 1979. Sensors and instrumentation: steam-sterilizable dissolved oxygen sensor and cell mass sensor for on-line fermentation system control. Biotechnol. Bioeng. Symp. 9:103–16.

Oinas, R. 1982. Application of computers to the monitoring, control, and reporting of fermentation processes. Proc. Biochem. 17:10–11, 18, 36.

Park, S. H., K. T. Hong, J. H. Lee, and J. C. Bae. 1983. On-line estimation of cell growth for glutamic acid fermentation system. Eur. J. Appl. Microbial. Biotechnol. 17:168–72.

Pungor, E., Jr., E. J. Schaefer, C. L. Cooney, and J. C. Weaver. 1983. Direct monitoring of the liquid and gas phases during a fermentation in a computer-mass-spectrometer-fermentor system. Eur. J. Microbiol. Biotechnol. 18:135–40.

Rolf, M. J., P. J. Hennigan, R. D. Mohler, W. A. Weigand, and H. C. Lim. 1982. Development of a direct digital-controlled fermentor using a microminicomputer hierarchical system. Biotechnol. Bioeng. 24:1191–1210.

Rolf, M. J., and H. C. Lim. 1982. Computer control of fermentation processes. Enz. Microb. Technol. 4:370–80.

Schmidt, A. 1983. The technological possibilities for measuring of physiological population state features in fermentation practice Acta Biotechnol. 3:171–76.

Shuler, M. L., O. P. Sahai, and G. A. Hallsby. 1983. Entrapped plant cell tissue cultures. In K. Venkatasubramanian, A. Constantinides, and W. R. Vieth (eds.), Biochemical engineering, 3:373–82. New York Academy of Science.

Spruytenburg, R. I., J. Dunn, and J. R. Bourne. 1979. Computer control of glucose feed to a continuous aerobic culture of *Saccharomyces cerevisiae* using the respiratory quotient. Biotechnol. Bioeng. Symp. 9:359–68.

Srinivasan, V. R., M. B. Fleenor, and R. J. Summers. 1977. Gradient-feed method of growing high cell density cultures of *Cellulomonas* in a bench-scale fermentor. Biotechnol. Bioeng. 19:153–55.

Stephanopoulos, G., and K.-Y San. 1982. On-line estimation of the state of biochemical reactors. In: J. Wei and C. Georgakis (eds.), 7th Internat. Symp. Chem. React. Engin., ACS Symp. Ser. No. 196., Am. Cem. Soc., Boston.

Swartz, J. R., and C. L. Cooney. 1979. Indirect fermentation measurements as a basis for control. Biotechnol. Bioeng. Symp. 9:95–101.

Titus, J. A., G. W. Luli, M. L. Dekleva, and W. R. Strohl. 1984. Application of a microcomputer-based system to control and monitor bacterial growth. Appl. Environ. Microbiol. 47:239–44.

Tutunjian, R. S. 1983. Ultrafiltration processes in biotechnology. In K. Venkatasubramanian, A. Constantinides, and W. R. Vieth (eds.), Biochemical engineering, 3:238–53. New York Academy of Science.

Uhlin, B. E., S. Molin, P. Gustafsson, and K. Nordstrom. 1979. Plasmids with temperature-dependent copy number for amplification of cloned genes and their products. Gene 6:91–106.

Unden, A. 1979. Instrumentation for fermentation monitoring and control. Biotechnol. Bioeng. Symp. 9:61–63.

Votruba, J., and V. Behal. 1984. A mathmatical model of kinetics of the biosynthesis of tetracyclines. Appl. Microbiol. Biotechnol. 19:153–56.

Wang, H. Y., C. L. Cooney, and D. I. C. Wang. 1979. On-line gas analysis for material balances and control. Biotechnol. Bioeng. Symp. 9:13–23.

Woodrow, J., and K. Spear. 1984. The use of derivative spectrophotometry for the rapid measurement of substrate and product concentration in fermentation. Appl. Microbiol. Biotechnol. 19:177–180.

Yamashita, S., H. Hoshi, and T. Inagaki. 1969. Automatic control and optimization of fermentation processes: glutamic acid. *In* D. Perlman (ed.), Fermentation advances, pp. 441–63. Academic Press, New York.

Yano, T., T. Kobayashi, and S. Shimizu. 1981. Control system of dissolved oxygen concentration employing a microcomputer. J. Ferment. Technol. 59:295–301.

Zabriski, D. W. and A. E. Humphrey. 1978. Continuous dialysis for the on-line analysis of diffusible components in fermentation broth. Biotechnol. Bioeng. 20:1295–1301.

WALTER COLLI, NORMA W. ANDREWS,
ALEJANDRO M. KATZIN, VERA Y. KUWAJIMA,
AND BIANCA ZINGALES

Identification of the Plasma Membrane Antigen of *Trypanosoma cruzi* as a First Approach to Its Production by Using Recombinant DNA Technology

30

ABSTRACT

The effect of several monosaccharides on the interiorization capacity of *Trypanosoma cruzi* trypomastigotes (TM) (Y strain) into mammalian cells was studied. Only ClcNAc elicited a significant inhibitory effect, the inhibition reaching 70–80% of the control at concentrations around 50mM when LLC-MK$_2$ cells were employed as receptor cells. HeLa cells were even more susceptible to inhibition of penetration by GlcNAc. The inhibition does not seem to be metabolic in origin, being suggestive of a competition between GlcNAc and glycosylated structures at the cell surface level. Stimulation of the penetration upon preincubation with WGA further implies that GlcNAc residue on the host-cell and parasite surfaces may participate in the interaction mechanism.

Affinity chromatography of detergent-lysed TMs on WGA-sepharose yielded a glycoprotein of M$_r$ 85 kDa (Tc-85) that is resistant to neuraminidase treatment and is immunoprecipitated by serum of chagasic human patients. This glycoprotein is possibly identical with a 85 kDa TM-specific surface antigen uncovered by immunoprecipitation of surface-labeled TMs with anti-TM serum after removal of other antigens common to the epimastigote form.

By immunoprecipitation studies, we have demonstrated the existence of

TM-specific surface proteins with M_r about 95 kDa and in the range between the 80 kDa and 95 kDa antigens. The 80 kDa and 95 kDa antigens exist also on the epimastigote membrane surface. In addition, TMs, but not epimastigotes, possess an 82 kDa antigen that is precipitated by anti-epimastigote serum, possibly representing a protein sharing common epitopes with the 95 kDa and 80 kDa proteins. Epimastigotes also contain a specific 70–72 kDa antigen not present in TMs.

Most of the TM-specific antigens, including Tc-85, are modified when TMs are treated with 150 μg/ml trypsin. Concomitantly, cell invasion capacity is reduced by more than 90% when infection is performed immediately after enzymatic treatment. That the loss of infectivity is not due to a decrease in viability is clear from the fact that the interiorization capacity gradually recovers to normal levels within 3 h after trypsinization. Cycloheximide inhibits both processes; upon its removal, both protein synthesis and parasite interiorization proceed.

Serum from chagasic human patients inhibits interiorization of *T. cruzi* into mammalian cells. The same results were obtained with monovalent F (ab′) fragments purified from human chagasic serum IgG. Serum raised in rabbits against epimastigote plasma membranes was ineffective in inhibiting interiorization.

Taken as a whole, the results suggest that: (1) specific proteins on the surface of the TMs are involved in the interaction with the mammalian host-cell surface; (2) the most likely candidates for such proteins are those of apparent M_r above 95 kDa and those of apparent M_r 80 and 95 kDa; (3) the proteins of apparent M_r 95 kDa, 80 kDa, and 70–72 kDa are not involved in the interaction with the host-cell surface; (4) involvement of Tc-85, a GlcNAc-containing glycoprotein from the TM surface, in the host-parasite interaction is consistent with the inhibition of the interiorization process by GlcNAc.

It is proposed that inhibition of endocytosis of TMs by mammalian cells in the presence of GlcNAc may reflect the existence of a GlcNAc-specific receptor activity on the surface of either the parasite or the host cell. The existence of Tc-85 on the membrane of the TM points to the host-cell membrane as the locale for this GlcNAc receptor activity.

INTRODUCTION

Trypanosoma cruzi, the causative agent of Chagas's disease, is a parasite with at least three well-defined differentiation stages: the amastigote, which lives and reproduces inside vertebrate cells; the epimastigote, a dividing form that inhabits the midgut of a reduviid insect; and the trypomastigote (TM), a nondividing cell that, in nature, is the link between the vertebrate and the invertebrate phases of the protozoan biological cycle. These forms are distinct not only in terms of their host environments but also with respect to their morphologies and biological behavior (Brene, 1980; Colli, 1979).

Epimastigotes are unable to invade other cells, the exception being professional phagocytes where they are destroyed inside phagolysosomes (Behbehani, 1973). TMs, however, interiorize into a variety of cultured cell lines (Dvorak and Howe, 1976), including macrophages. Indeed, no cell line resistant to TM invasion *in vitro* has been described. Moreover, once interiorized, the TMs manage to escape attack by proteolytic enzymes of the endocytic vacuole and undergo differentiation into dividing amastigotes.

Since the primary event in TM interiorization by the host cell is necessarily contact between plasma membranes, it is reasonable to expect that stage-specific surface components may be involved. Identification of such components, in particular specific proteins involved in anchorage and subsequent interiorization of the TM into the host cell, would provide a basis for developing strategies for blocking these events, thus interrupting the biological cycle of the parasite. In the case of man, this might represent an effective means of preventing the onset of disease.

In the present work, we report evidence which implies that TM surface proteins indeed play a role in the phenomenon of parasite interiorization.

RESULTS

Effect of N-Acetyl-D-Glucosamine (GlcNAc) and
Wheat Germ Agglutinin (WGA) on
Parasite Interiorization into Mammalian Cells

The effect of several monosaccharides on the interiorization capacity of *T. cruzi* TMs into mammalian cells was studied, employing a method described previously (Andrews and Colli, 1982). The effect of increasing concentrations of N-acetyl-D-glucosamine, N-acetyl-D-galactosamine, D-glucose, and D-galactose on infection of LLC-MK$_2$ cells is compared in figure 1A. Only GlcNAc elicits a significant inhibitory effect, the inhibition being 80–90% relative to the control at concentrations around 50 mM. Several other sugars, including D-mannose, N-acetyl-D-mannosamine, x-methyl-D-mannoside, and buffered N-acetylneuraminic acid, had no sig-

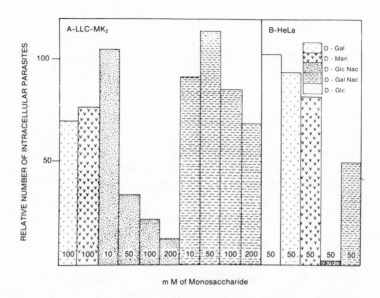

Fig. 1. Effect of monosaccharides on the interiorization of TMs in mammalian cells. Confluent monolayers of cells (8×10^4 cells/cm^2) were incubated with 7×10^6 TMs/ml (m.o.i. = 20) at 34°C in Waymouth's medium containing the indicated concentrations of monosaccharides (values are mM). (A) LLC-MK$_2$ cells; (B) HeLa cells.

nificant effect on infection of the cultured cells by the parasites when tested under the same conditions. HeLa cells were even susceptible to inhibition of penetration by GlcNAc (fig. 1B). With HeLa cells, some inhibition also was observed with GalNAc, an epimer of GlcNAc. By experiments performed with HeLa cells, we have shown that blocking of parasite penetration by GlcNAc is apparently time-independent, being observed at both short and long exposure periods to the parasites (fig. 2).

The GlcNAc had no detectable effect on cell-invasion capacity when host cells were preincubated with inhibitory concentrations of this sugar, followed by removal of the monosaccharide prior to incubation (fig. 3A). The same result (occasionally with a slight inhibition) was obtained upon preincubation of TMs with GlcNAc under the same conditions. This, together with the fact that endocytosis of polystyrene latex beads by LLC-MK$_2$ cells was not altered in the presence of 100 mM GlcNAc (not shown), suggests that the inhibitory action of GlcNAc is not metabolic in origin.

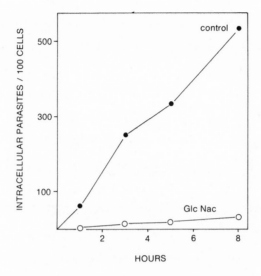

Fig. 2. Effect of GlcNAc on the interiorization of TMs into mammalian cells at various exposure times. Confluent monolayers of HeLa cells (8×10^4 cells/cm^2) were incubated with 10^7 TMs/ml (m.o.i. = 30) at 34°C in Waymouth's medium containing (0), or not containing (0), 100 mM N-acetyl-D-glucosamine, for the indicated periods of time. After the exposure period, the coverslips were washed 3 times with PBS, exposed to distilled water for 2 min, washed 3 times with PBS, fixed, and processed for optical microscopy. The values are the mean of triplicates.

Our results are thus suggestive of a competition of GlcNAc with glycosylated structures at the cell surface level.

In view of the above results and the high specificity of WGA for GlcNAc, we examined the effects of this lectin on infection *in vitro*. Other workers (Henriquez et al., 1981) have reported that invasion of Vero cells by *T. cruzi* TMs is inhibited upon treatment of the host cells with various lectins, including WGA, prior to infection; a fixed concentration of 10 µg lectin/ml and a preincubation period of 30 min were employed. Our results with the LLC-MK$_2$ cell line (a monkey kidney epithelial cell line, as are Vero cells) are somewhat different, at least with regard to WGA. Indeed, we find that preincubation of the cells for a 4 h period with 50 µg WGA/ml results in a stimulation of the interiorization process; no effect on interiorization was detected at lower lectin concentrations (fig. 3A). The stimulatory effect at

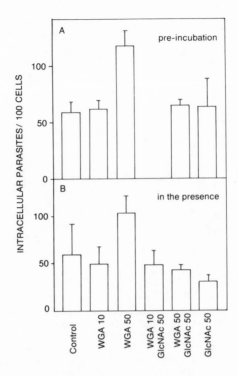

Fig. 3. Effect of WGA and GlcNAc on the interioration of TMs into LLC-MK$_2$ cells. Confluent monolayers of LLC-MK$_2$ cells (8×10^4 cells/cm^2) were incubated at $34°C$ with: (A) Waymouth's medium containing the indicated substances (values are mM for GlcNAc and ug/ml for WGA). After 4 h, cultures were washed 3 times with PBS and exposed to a suspension of 10^7 TMs/ml (m.o.i. = 30) in Waymouth's medium for an additional 4 h period. (B) Waymouth's medium containing the indicated substances and 10^7 TMs/ml (m.o.i. = 30) for 4 h. After exposure, processing was as in fig. 2. The values are the mean of triplicates, the bars representing S.D.

higher WGA concentrations is reversed by the addition of GlcNAc. The same effect is observed when the lectin is maintained in the medium throughout the infection period (fig. 3B). Stimulation of infection was observed even with short preincubation periods (30 min), implying that this effect is not a consequence of prolonged exposure of the cells to the lectin. The most probable explanation for this observation (cf. Piras et al., 1980) is that parasite interiorization increases during the period in which endocytosis of the lectin is active. Indeed, the longer infection period employed by us

(4 h) corresponds to the period in which endocytosis of bound lectin must be active. In this regard, it is known that for HeLa cells complete endocytosis of surface-bound WGA requires at least 2 h (Kramer and Canellakis, 1979).

As reported previously (Piras et al., 1980), treatment of the TMs with WGA before exposure to the host cells also increases infection rates.

The data are thus consistent with the presence of GlcNAc-containing structures on the cell surface of either TMs or host cells, or both, endocytosis in the presence of WGA being facilitated by bridging by this lectin between parasite and host-cell. A preliminary account of these results has been published (Andrews and Colli, 1981).

Effect of Normal and Immune Serum on Parasite Interiorization Into Mammaliam Cells

The experiments described in the previous section were all performed in serum-free medium (Waymouth's MAB medium (Gorham and Waymouth, 1965)). In addition to the finding that infection of HeLa cells, LLC-MK$_2$ cells, and 3T3 fibroblasts is successfully obtained in the absence of serum, we observed that normal serum from various sources induces a partial inhibition of adhesion and interiorization of *T. cruzi* TMs (Andrews and Colli, 1982).

Pre-incubation of TMs with sera from chagasic human patients (with maintenance during the infection period) induces an inhibition of mammalian cell infection varying from 30 to 60% as compared with normal human serum. Normal rabbit serum and serum obtained in rabbits against epimastigote plasma membrane vesicles have no such effect (Zingales et al., 1982).

The capacity of these human serum antibodies to block the invasion of mammalian cells in culture is further confirmed by comparison (fig. 4) of the effect of F (ab') monovalent fragments (isolated after pepsin digestion) from normal and chagasic serum. In fact, the infection rate of HeLa cells is significantly reduced upon increasing the concentration of F (ab') fragments from chagasic serum.

Some variability in the extent of inhibition has been observed with sera from different infected patients or animals. In part this might be correlated with the immunofluorescence and agglutination titers of the sera employed. However, it is possible that the apparent tendency of TMs to shed antibodies from their surfaces (Schumnis et al., 1978) is also responsible for

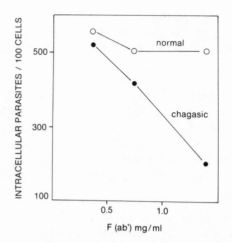

Fig. 4. Effect of F (ab') fragments from normal and chagasic human serum on the interiorization of *T. cruzi* into mammalian cells. Monovalent F (ab') fragments were obtained by reducing F (ab')$_2$ purified from pepsin-treated human IgG. Purified TMs (20 × 10^6 parasites/ ml) were preincubated at 4°C for 2 h with various concentrations of F (ab') fragments from normal (0) and chagasic (0) humans. HeLa cells, plated at 4 × 10^4 cells/cm^2 two days before, were then exposed to these parasites (34°C for 4 h) at m.o.i. = 50. Processing of the coverslips was as in fig. 2. The results are the mean of the triplicates.

the observed variability. At present we are actively seeking incubation conditions that minimize shedding in order to increase the effect of penetration blocking antibodies.

These results show that antibodies are raised under natural conditions against specific TM antigens presumably involved in the primary interaction with the host-cell membrane.

Specific Antigenic Pattern of Trypomastigote Surface Proteins

Immunoprecipitation studies were undertaken in order to analyze in detail the specific cell surface antigens of the TM form. Intact living parasites were radioiodinated, lysed with non-ionic detergent, and incubated with various antisera from immunized rabbits or chagasic human patients. The immunoprecipitates were absorbed with *S. aureus* and analyzed by SDS-polyacrylamide gel electrophoresis (Laemmli, 1970). The results are summarized schematically in figure 5.

Epimastigotes have three main surface antigens of apparent M$_r$ 95 kDa, 80 kDa and 70–72 kDa when immunoprecipitated with antisera raised

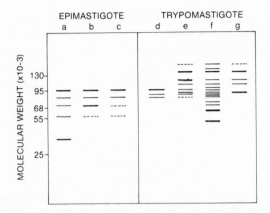

Fig. 5. Radioimmunoprecipitation of *T. cruzi* surface polypeptides by rabbit antisera. Analyses were made in SDS-polyacrylamide 7–14% linear gradient gels. Tracks (a) to (c) are derived from epimastigote extracts, and (d) to (g) from TM extracts; (a) and (f) represent total, nonprecipitated, cellular extracts; (b) and (d) are the polypeptides precipitated by anti-epimastigote sera; (c) and (e) are the polypeptides precipitated by anti-TM sera; (g) represents the polypeptides precipitated by anti-TM serum after previous immunoprecipitation of the extract with excess anti-epimastigote serum. Lower molecular weight bands have been omitted for clarity.

against a series of antigens of epimastigote origin (Zingales et al., 1982). With anti-TM serum, only the 95 kDa and 80 kDa antigens were recognized, suggesting that the 70–72 kDa polypeptide is an epimastigote-specific antigen (cf. Snary et al., 1981).

TMs immunoprecipitated with rabbit anti-TM or human chagasic sera have several cell surface antigens, the most conspicuous of which have apparent M_r of 130–40 kDa, 110 kDa, 95 kDa, and 82 kDa. In the region between 80 kDa and 95 kDa, several other sharp bands, corresponding to TM-specific antigens, can be distinguished in the autoradiographs.

The anti-epimastigote sera recognize the 95 kDa, 82 kDa, and 80 kDa antigens of the TM cell surface. The first and last of these thus appear to be common to both differentiation stages of *T. cruzi*. Although not specifically recognized in epimastigotes by the anti-epimastigote sera, the 82 kDa TM antigen may also be common in the sense of being related, by common epitopes, to the 95 kDa or 80 kDa antigens.

When radioiodinated TMs are immunoprecipitated with excess anti-epimastigote serum (in order to exhaust common antigens) and subsequently immunoprecipitated with anti-TM serum, a conspicuous antigen of apparent M_r 85 kDa can be visualized in the autoradiographs.

In summary, it can be concluded that several of the antigens on the TM surface are nonexistent on the epimastigote surface. These TM-specific antigens are localized in the region of apparent M_r above 95 kDa and between the antigens of M_r 80 kDa and 95 kDa, the most conspicuous being that of 85 kDa. Some of these proteins are glycoproteins since they are retained on Sepharose-Con A columns and can be subsequently eluted with a mixture of α-methyl-D-mannoside and α-methyl-D-glucoside (Katzin and Colli, 1982).

We have evidence (Zingales et al., 1982) that our 95 kDa and 80 kDa antigens correspond, respectively, to the 90 kDa (Snary and Hudson, 1979) and 75 kDa (Nogueira et al., 1981) antigens described by other workers. We attribute the differences in apparent molecular weight to the higher resolution of our slab gradient gels and to slight differences in the molecular weights assigned to protein markers.

A GlcNAc-containing Glycoprotein of M_r 85 kDa (Tc-85) Specific For The Trypomastigote Plasma Membrane

Both the inhibition of TM interiorization into mammalian cells by GlcNAc and the stimulation of penetration observed in the presence of WGA led us to search for GlcNAc-containing glycoproteins on the TM cell surface.

Detergent-solubilized proteins from surface radioiodinated TMs were chromatographed on WGA-sepharose columns. The retained material, which eluted with 0.1 M GlcNAc, contained a single band of apparent M_r 85 kDa (Tc-85) when analyzed by SDS-polyacrylamide gel electrophoresis. This band has not been found in the plasma membrane of epimastigotes or amastigotes (Katzin and Colli, 1982).

The Tc-85 does not lose the property of binding to WGA when TMs or the isolated glycoprotein are preincubated with 1.1 units/ml of neuraminidase for 60 min at 37°C. Trypsin treatment, however, appears to modify Tc-85 (Katzin and Colli, 1982).

Although isolated Tc-85 is immunoprecipitated by human chagasic serum, we as yet have insufficient evidence to confirm the identity between this glycoprotein and the antigen of apparent M_r 85 kDa uncovered by sequential immunoprecipitation with heterologous and homologous antisera (see "Specific Antigenic Pattern of Trypomastigate Surface Proteins" above). We plan to utilize bidimensional gel electrophoresis and peptide mapping after controlled proteolysis for intercomparison of these two macromolecules.

Effect of Trypsin-Treatment on Interiorization of Trypomastigotes into Mammalian Cells

Trypsin treatment of TMs (150 ug/ml) greatly simplifies the cell-surface iodination profile (fig. 6). Several proteins of apparent M_r above 80 kDa are modified, in particular Tc-85, since no protein is retained on WGA-sepharose columns following enzymatic treatment of the TMs. Nonetheless, several of the remaining polypeptide fragments are still immunoprecipitated with chagasic human serum or anti-TM rabbit serum.

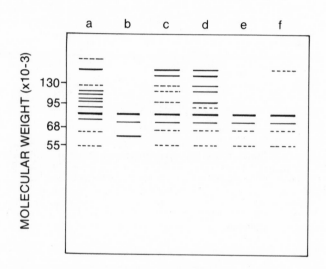

Fig. 6. Total pattern of radioiodinated TM surface proteins. Effect of trypsinization. Suspensions of 30 × 10⁶ TMs were washed twice with medium 199, iodinated by the iodogen technique, washed twice with the same medium and processed for electrophoresis after the following treatments: (a) BSA, 150 ug/ml, 20 min. at 37°C; (b) trypsin, 150 ug/ml, 20 min. at 37°C; (c) as in (b) and further incubated for 1 h in DME-2% calf serum; (d) as in (b) and further incubated for 3 h in DME-2% calf serum; (e) as in (c) except that DME contained 5 ug/ml cycloheximide; (f) as in (d), except that DME contained 5 ug/ml cycloheximide. Lower molecular weight bands have been omitted for clarity.

The cell invasion capacity of trypsinized TMs was reduced by more than 90% when infection was performed immediately after the enzymatic treatment. That this loss of infectivity is not due to a decrease in viability is clear from the fact that the interiorization capacity gradually recovers to normal levels within 3 hr after trypsinization (fig. 7).

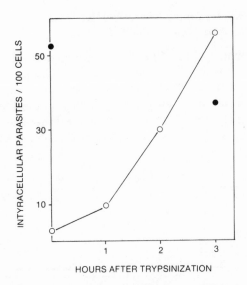

Fig. 7. Effect of trypsinization on cell-invasion capacity of TMs. TMs were washed twice with medium 199, treated with trypsin (0) or BSA (0), 150 ug/ml for 20 min. at 37°C. The reaction was stopped with soybean trypsin inhibitor. Parasites were centrifuged and resuspended in DME-2% calf serum at a cell density of 25×10^6 parasites/ml. At the time points indicated in the figure, 3 ml of each cell suspension were incubated for 1 h with sub-confluent layers of LLC-MK$_2$ cells (4×10^4 cells/cm^2). Processing of the coverslips was as in fig. 2. The results are the mean of triplicates.

Cell surface radioiodination of trypsin-treated TMs provides evidence for partial resynthesis of membrane proteins after 1 hr; three hr after trypsinization, the surface membrane pattern recovers its original complexity. When protein resynthesis is blocked with cycloheximide (5 μg/ml), no interiorization of TMs into host cells is observed. Upon removal of cycloheximide, however, both processes—protein synthesis and parasite interiorization—proceed.

Taken as a whole, these results suggest that specific trypsin-sensitive cell surface proteins of the TM stage are involved in the invasion mechanism of mammalian cells.

DISCUSSION AND PERSPECTIVES

Unlike macrophages, which indiscriminately interiorize both differentiation stages of *T. cruzi*, nonprofessional phagocytes are able to interiorize only TMs. This suggests that TMs might be endowed with mechanisms that either facilitate endocytosis or promote active penetration. Since the primary event in interiorization is intermembrane contact, it is reasonable to expect that cell surface components play a key role in the overall process.

All the evidence presented above is consistent with the existence of a set of surface proteins specific to the TM stage of parasite life cycle. Hence, it is possible that one or more of these surface proteins is involved in anchorage to, and/or invasion of, the mammalian cell. This hypothesis is reinforced by the fact that serum from human chagasic patients inhibits interiorization of TMs. Moreover, since serum raised against epimastigote plasma membrane proteins (Franco da Silveira et al., 1979) is inefficient in promoting such inhibition, one must conclude that the common antigens, i.e., the 95 kDa, 82 kDa, and 80 kDa proteins, and the epimastigote-specific 70–72 kDa protein do not participate in the interiorization mechanism. Good candidates for the surface proteins involved in interiorization are those of apparent M_r above 95 kDa and those migrating between 80 kDa and 95 kDa. One of these proteins, Tc-85, can be isolated by affinity chromatography with WGA-sepharose columns.

It has become widely accepted that exposed carbohydrate residues on glyconjugates serve as determinants of recognition *in vivo* and *in vitro* (Barondes, 1981). Our results with carbohydrate inhibition of TM interiorization can be explained by the existence of a GlcNAc-specific receptor activity on the membrane surface of either the host-cell or the parasite. The presence of Tc-85, a GlcNAc-containing glycoprotein, on the TM membrane surface is consistent with the hypothesis that endocytosis of *T. cruzi* involves an ubiquitously distributed GlcNAc receptor on the mammalian cell surface. Nonetheless, presently available evidence is insufficient to rule out the possibility of the existence of a GlcNAc receptor activity on the TM membrane surface.

Two principal strategies should permit verification of the involvement of the TM-specific membrane proteins in the parasite interiorization process:

1. Our results show that trypsinization of TMs results in loss of most of the surface antigens, with concomitant impairment of cell penetration capacity. Significantly, these trypsinized parasites do not lose viability and recover their full penetration capacity after being allowed to resynthesize the original complement of membrane proteins. Kinetic studies of this resynthesis of the surface antigens, with the objective of correlating the rate of reappearance of the various surface proteins with that for recovery of the interiorization capacity of the parasite, are in progress. Since some of these proteins are gly-

coproteins, selective inhibition experiments with tunicamycin might also prove to be quite useful.

2. Monoclonal antibodies against surface proteins are being raised in order to identify those that inhibit the parasite-host cell interaction. A particularly good candidate for the initial studies is Tc-85 since it contains GlcNAc, a carbohydrate that proved to be a potent inhibitor of interiorization. Moreover, Tc-85 has the additional advantage that it is obtained easily in almost pure form through affinity chromatography, permitting direct use in protection studies *in vitro* and *in vivo*. In this regard, preliminary experiments in our laboratory have shown that antibodies raised against Tc-85 do indeed exert some inhibition of TM interiorization into LLC-MK$_2$ cells.

The TM cell is a morphologically well-defined entity (Hoare, 1972; Pan, 1978). In experimental studies, TMs have classically been obtained from four main sources: tissue culture, the blood of infected animals, reduviid urine, and as a product of epimastigote differentiation in axenic liquid culture media (metacyclic TMs). The first two sources furnish homogenous populations of mature, fully infective parasites, but the latter two sources contain parasites in several stages of differentiation. Clearly, results obtained with parasites from these latter two sources would be more reliable if it were possible to separate them from the medium as a homogenous TM population having the additional key property shown by TMs obtained from the first two sources (and not yet considered in the classification of the TM stage): a fully infective capacity as of mammalian cells. In this regard, infectivity toward cells maintained in culture, which can be quantified easily, is superior to infectivity in laboratory animals as a screening tool.

In particular, differentiation of epimastigotes into metacyclic TMs in axenic liquid culture media is still erratic and uncontrolled. Consequently, it is difficult to perform systemic experiments designed to investigate the events that accompany the differentiative transition. Moreover, the differentiation may not be synchronous at the level of morphological and molecular modifications. Indeed, preliminary experiments in our laboratory indicate that the membrane pattern of metacyclic TMs frequently (though not always) resembles more closely that of the epimastigote predecessor, despite the fact that the metacyclic cell is morphologically indistinguishable from tissue culture or blood TMs. Similarly, some batches of metacyclic TMs obtained in our laboratory are fully infective to LLC-MK$_2$ cells when compared with tissue culture TMs, and others barely interiorize. An excellent discussion of the morphological appearance and ultrastructural modifications of transitional forms present during differentiation has been published (Pan, 1978).

Hence, generalization of conclusions drawn from experiments performed with metacyclic TMs is, in our opinion, premature (Nogueira et al., 1982). To our knowledge, no truly systematic studies have been carried out with the objective of comparing the membrane protein and antigen patterns of either metacyclic trypanosomes or reduviid urine trypanosomes with those obtained from tissue culture or blood of infected animals.

Nonetheless, identification of the factors that stimulate differentiation of epimastigotes in axenic culture media would be extremely valuable if mature metacyclic parasites containing a full set of surface membrane antigens and possessing full capacity for penetration of mammalian cells could be obtained. This would make available mass quantities of trypomastigotes for direct extraction and isolation of antigens.

Finally, although partial protection against experimental infection has been obtained with preparations that range from parasite extracts to purified molecules, we emphasize that there is as yet no definitive evidence that sterile immunity can be attained in *T. cruzi* infections. It is our hope that identification of the surface proteins involved in TM interiorization into mammalian cells will open the way for obtaining highly effective antigens for blocking of the mammalian phase of the biological cycle. Genetic engineering should then provide a powerful technology for antigen production.

ACKNOWLEDGMENTS

This investigation received financial support from the Fundacao de Amparo a Pesquisa do Estado de Sao Paulo (FAPESP) to B. Zingales and W. Colli and from the UNDP/World Bank/WHO Special Programme for Research and Training in Tropical Diseases, the Conselho Nacional de Desenvolvimento Cientifico e Tecnologico (Projeto CNPq/FINEP), The Financiadora de Estudos e Projetos (FINEP), and the Project PNUD/UNESCO RLA 78/024 to W. Colli. We are indebted to Nuncio F. Martin for invaluable assistance with antibody development and to Dr. Frank H. Quina for a critical reading of the manuscript.

Note added in proof: While this paper was in press, M. J. Crane and J. A. Dvorak ([1982] Mol. Biochem. Parsitol. 5:333–41) reported an inhibitory effect on GlcNAc on TM interiorization in bovine embryonic skin and muscle (BESM) cells.

REFERENCES

Andrews, N. W., and W. Colli. 1981. Interiorization of *T. cruzi* in cultured mammalian cells: inhibition by N-acetyl-D-glucosamine. Abstr. 104. Eighth Annual Meeting on Basic Research on Chagas' Disease, Cazambu, Brazil.

Andrews, N. W., and W. Colli. 1982. Adhesion and interiorization of *Trypanosoma cruzi* in mammalian cells. J. Protozool. 23:264–69.

Barondes, S. H. 1981. Lectins: their multiple endogenous cellular functions. Ann. Rev. Biochem. 50:207–31.

Behbehani, M. K. 1973. Developmental cycles of *Trypanosoma (Schuzotrypanum) cruzi* (Chagas, 1909) in mouse peritoneal macrophages *in vitro* Parasitology 66:343–53.

Brene, Z. 1980. Immunity to *Trypanosoma cruzi.* Adv. Parasitol. 18:247–91.

Colli, W. 1979. Chagas' disease. *In* D. F. H. Wallach (ed.), The membrane pathobiology of tropical diseases, pp. 131–53. (Tropical Diseases Research Series.) Schwabe & Co., Basel.

Dvorak, J. A., and C. L. Howe. 1976. The attraction of *Trypanosoma cruzi* to vertebrate cells *in vitro.* J. Protozool. 23:534–37.

Franco da Silveira, J., A. Abrahamsohn, and W. Colli. 1979. Plasma membrane vesicles isolated from epimastigote forms of *Trypanosoma cruzi.* Biochem. Biophys. Acta 550:222–32.

Gorham, L. W., and C. Waymouth. 1965. Differentiation *in vitro* of embryonic cartilage and bone in a chemically defined medium. Proc. Soc. Exp. Biol. Med. 119:287.

Henriquez, D., R. Piras, and M. M. Piras. 1981. The effect of surface membrane modifications of fibroblastic cells on the entry process of *Trypanosoma cruzi* trypomastigotes. Mol. Biochem. Parasitol. 2:359–66.

Katzin, A. M., and W. Colli. 1982. Lectin receptors in *Trypanosoma cruzi*: an N-acetyl-D-glycosamine-containing surface glycoprotein specific for the trypomastigote stage. Biochem. Biophys. Acta 727:403–11.

Kramer, R. H., and E. S. Canellakis. 1979. The surface glycoproteins of the HeLa cell: internalization of wheat germ agglutinin-receptors. Biochem. Biophys. Acta 551:328–48.

Laemmli, U. K. 1970. Cleavage of structural proteins during the assembly of the head bacteriophage T$_4$. Nature 227:680–85.

Nogueira, N., S. Chaplan, J. Tydings, J. Unkeless, and Z. Cohn. 1981. *Trypanosoma cruzi*: surface antigens of blood and culture forms. J. Exp. Med. 153:629–39.

Nogueira, N., J. Unkeless, and Z. Cohn. 1982. Specific glycoprotein antigens on the surface of insect and mammalian stages of *Trypanosoma cruzi.* Proc. Natl. Acad. Sci. USA 79:1259–63.

Pan, S. C-T. 1978. *Trypanosoma cruzi*: ultrastructural observations on morphogenesis *in vitro* and *in vivo.* Exp. Parasitol. 46:92–107.

Piras, R., D. Henriquez, and M. W. Piras. 1980. Studies on host-parasite interaction: role of fibro-blastic cell surface function and *Trypanosoma cruzi* forms in the infective process. *In* H. Van den Bossche (ed.), The host invader interplay, pp. 131–34. Elsevier/North Holland Biomedical Press.

Schumnis, G. A., A. Szarfman, R. Langenbach, and W. de Souza. 1978. Induction of capping in blood-stage trypomastigotes of *Trypanosoma cruzi* by human anti-*Trypanosoma cruzi* antibodies. Infec. Immun. 20:507–9.

Snary, D., M. A. J. Ferguson, M. T. Scott, and A. K. Allen. 1981. Cell surface antigens of *Trypanosoma cruzi*: use of monoclonal antibodies to identify and isolate an epimastigote specific glycoprotein. Mol. Biochem. Parasitol. 3:343–56.

Snary, D., and L. Hudson. 1979. *Trypanosoma cruzi* cell surface proteins: identification of one major glycoprotein. FEBS Lett. 100:166–70.

Zingales, B., N. W. Andrews, V. Y. Kuwajima, and W. Colli, 1982. Surface antigens of *Trypanosoma cruzi*: possible correlation with the interiorization process in mammalian cells. Mol. Biochem. Parasitol. 6:111–24.

Molecular Cloning and Partial Sequence Determination of Kinetoplast DNA Minicircles from *Trypanosoma cruzi*

31

Kinetoplast DNA (kDNA), the mitochondrial DNA that characterizes and defines the order Kinetoplastida, has a remarkable structure (Simpson, 1972; Borst and Hoeijmakers, 1979; Englund, 1980). It is composed of thousands of catenated circular DNA molecules forming a massive network that can be seen by light microscopy and represents 15–30% of total cellular DNA. There are two components in kDNA: *maxicircles*, which represent 5–10% of the mass of the network and have 20,000–40,000 base pairs (bp) depending on the species, and *minicircles*, which represent 90–95% of the network and have 800–2500 bp. There are about 50 maxicircles and 3,000–15,000 minicircles per network. Maxicircles are the equivalent of the conventional mitochondrial DNA of other eukaryotes; no function has yet been definitely ascribed to minicircles.

Besides reviewing some aspects of the structure, replication, and functioning of kDNA networks, we will present some recent results on the following subjects: (1) the use of restriction fragment analysis of kDNA in the identification and characterization of trypanosomatids; (2) molecular cloning and characterization of mini- and maxicircle DNA sequences.

We had previously shown that the digestion of kDNA with selected restriction endonucleases and the analysis of the resulting fragments by high resolution gel electrophoresis could be successfully used for the characterization of stocks, strains and clones of *Trypanosoma cruzi* (Mattei et al., 1977; Morel et al., 1980). We have now extended these observations to hundreds of samples of trypanosomatids (monogenetic insect trypanosomatids: Plessman et al., 1982; *T. cruzi* and *T. cruzi*-like protozoa: Goncalves, Chiari, Camargo, Dias, Krieger, Romanha, Mueller, Deane, Simpson, and Morel, unpublished; *Leischmania* and *Endotrypanum*: Lopes et al., 1980), and Lopes, Grimaldi, Marzochi, Schnur, and Roitman, Lima, and Morel, unpublished). The concept of *schizodemes*—populations having similar kDNA restriction fingerprints (Morel et al., 1980)—is proving to be a most valuable tool for the identification and characterization of kinetoplastid flagellates.

Recently, in collaboration with Maria Deane and co-workers in our Institute, this methodology was used in the study of experimental Chagas' disease. It could be shown that one strain of *T. cruzi* could infect mice previously infected with another strain having a different kDNA restriction fingerprint (Deane et al., 1985). By control experiments, we showed that this methodology could detect a composite restriction fingerprint even if one of the strains represented as little as 2–5% of the total mixture (Goncalves, Deane, and Morel, in preparation). The possibility that this result can be extrapolated to human Chagas' disease is now being considered in light of recent results of schizodeme analysis of *T. cruzi* stocks isolated from chagasic patients from the county of Bambui, MG, Brazil. Some patients carry *T. cruzi* "strains" that could be mixtures of at least two other strains circulating in the same region according to *Eco*rI restriction fingerprints (Goncalves, Chiari, Dias, Krieger, and Morel, unpublished). The analysis of cloned populations isolated from such putatively composite strains is being performed in order to investigate the possibility of natural mixed infection or reinfection of human beings with *T. cruzi*.

In another series of experiments, we have cloned maxicircle *Eco*rI fragments from the insect trypanosomatid *Herpetomonas samuelpessoai* (Gomes, 1982) and minicircle sequences from *T. cruzi* (van Heuverswyn et al., 1980; van Heuverswy et al., 1981). A 3,300 bp *Eco*rI fragment from *H. samuelpessoai* maxicircle DNA that contains the mitochondrial ribosomal RNA genes has been cloned into the *Eco*rI site of plasmid pBR325. In collaboration with L. Simpson and A. Simpson (Dept. Biology, UCLA, USA) (unpublished), we recently showed that this fragment hybridizes with *Leishmania tarentolae* mitochondrial ribosomal RNA genes and subcloned its *Sau*3AI subfragments into the single stranded DNA phages M13mp8 and mp9. DNA sequence analysis of these genes will be in order to investigate phylogenetic relationships among these parasites and to study the mitochondrial ribosome organization and functioning in trypanosomatids.

Minicirlces or minicircle fragments of *T. cruzi* have been cloned into the *Eco*rI site of pBR325, into the *Kpn*I site of M13*Gori*1 or into the *Eco*rI site of M13mp7 (van Heuverswyn et al., 1980; van Heuverswy et al., 1981; Morel, Ray, Hines, van Heuverswyn, Simpson, Calcagotto, Mueller, and Cardoso, unpublished). Preliminary sequence data will be reported for cloned full-length minicircles or minicircle fragments as well as for uncloned restriction fragments obtained from total kDNA by preparative gel electrophoresis (van Heuverswyn, Mueller, Calcagnotto, Cardoso and Morel, in preparation). These sequences are being analyzed by the Queen and Korn (1980) computer program. Our results confirm that *T. cruzi* minicircles are formed by a basic sequence repeated four times per minicircle, a unique situation among all the Kineoplastida species analyzed until today. This repetition has been corroborated recently by the hybridization under high stringency conditions of a 40 bp *Eco*RI fragment of known sequence from *T. cruzi* minicircles that has been cloned into M13mp7, against total kDNA/*Eco*RI Southern blots. This fragment hybridizes efficiently against all the typical *Eco*RI kDNA fragments of the *T. cruzi* CL strain (1/4, 2/4, 3/4, and full-length minicircles), clearly demonstrating the existence of a basic sequence repeat in *T. cruzi* minicircles. We hope that this work will throw some light on the structure and functioning of *T. cruzi* minicircles and also will help the interpretation of *T. cruzi* schizodeme analysis by defining the conserved versus the variable regions of kDNA minicircles in this parasite.

REFERENCES

Borst, P., and J. H. J. Hoeijmakers. 1979. Kinetoplast DNA. *Plasmid* 2:20–40.
Deane, M. P., M. A. Sousa, N. Pereira, A. M. Goncalves, H. Momen, and C. M. Morel. 1985. *Trypanosoma cruzi*: inoculation schedules and reisolation methods select individual strains from doubly infected mice, as demonstrated by schizodeme and zymodeme analyses. J. Protozool. (in press).
Englund, P. T. 1980. Kinetoplast DNA. *In* M. Levandowsky and S. H. Hunter (eds.), Biochemistry and physiology of protozoa, 4:333–383. 2d ed. Academic Press, New York.
Gomes, S. A. 1982. Clonagem molecular de sequencias genicas de maxicirculos do DNA cinetoplastico de *Herpetomonas samuelpessoai*. M.Sc. Thesis, Universidade Federal do Rio de Janeiro. 98 pp.
Lopes, U. G., H. Momen, G. Grimaldi, M. C. A. Marzochi, R. S. Pacheco, and C. M. Morel. 1984. Schizodeme and zymodeme characterization of *Leishmania* in the investigation of foci of visceral and cutaneous leishmaniasis. J. Parasitol. 70:89–98.
Mattei, D. M., S. Goldenberg, C. Morel, H. P. Azevedo, and I. Roitman. 1977. Biochemical strain of characterization of *Trypanosoma cruzi* by restriction endonuclease cleavage of Kinetoplast DNA minicircles. FEBS Letters 74:264–68.

Morel, C., E. Chiari, E. Plessman Camargo, D. Mattei, A. J. Romanha, and L. Simpson. 1980. Strains and clones of *Trypanosoma cruzi* can be characterized by restriction endonuclease fingerprinting of kinetoplast DNA minicircles. Proc. Nat. Acad. Sci. USA, 77: 6810–14.

Morel, C., and L. Simpson. 1980. Characterization of pathogenic trypanosomatidae by restriction endonuclease fingerprinting of kinetoplast DNA minicircles. Amer. J. Trop. Med. Hyg. 29 (5) Suppl:1070–74.

Plessmann Camargo, E., D. M. Mattei, C. L. Barbieri, and C. Morel. 1982. Electrophoretic analysis of endonuclease-generated fragments of k-DNA, of esterase isoenzymes, and of surface proteins as aids for species identifications of insect trypanosomatids. J. Protozool., 29:251–58.

Queen, C. L., and L. J. Korn. 1980. Computer analysis of nucleic acids and proteins. Methods in Enzymology 65:595–609.

Simpson, L. 1972. The kinetoplast of the homoflagellates. Intern. Rev. Cytol. 32:140–207.

Van Heuverswyn, H. V., D. S. Ray, J. C. Hines, R. Muller, A. M. Calcagnotto, R. Galler, M. V. Tedeschi, M. A. B. Cardoso, M. M. R. Magalhaes, and C. M. Morel. 1981. Molecular cloning of *Trypanosoma cruzi* minicircle DNA sequences into plasmid pBR322 and bacteriophage M13Gori I. Abstracts VIIth Ann. Meet. "Pesquisa Basica em Doenca de Chagas," Caxambu, MG, Brazil, 3–5 November, Abstr.. BQ40.

Van Heuverswyn, H., R. U. Mueller, A. M. Calcagnotto, and C. M. Morel. 1981. Nucleotide sequence analysis of kinetoplast DNA minicircle from *T. cruzi.* Arg. Biol. Tecnol. 24:25.

C. D. DENOYA, A. R. TREVISAN, AND
J. ZORZOPULOS

Detection and Characterization of Plasmids in Clinical Isolates of Pathogenic Bacteria

32

ABSTRACT

Several multiresistant bacterial clinical isolates from the Children's Hospital of Buenos Aires were examined for antibiotic resistance and adherence properties. Plasmid content of 25 *Klebsiella pneumoniae* (Kb.p.) strains were analyzed and classified. Large plasmids of 105, 85, 68, and 50 MDal. and a small plasmid of 2.2 MDal. were the most widespread. Antibiotic resistance markers and plasmid stability of several bacterial strains were also analyzed. Rapid loss of antibiotic resistance markers was observed for one *Escherichia coli* strain under non-selective growing conditions. Under similar conditions, two Kb.p. strains did not lose the antibiotic resistance markers. Furthermore, the adherence capacity and the corresponding plasmid content of the *Klebsiella* strains were also studied. Two of them from a patient with an infected shunt showed a higher capacity for adherence to ventricular catheters than a strain from a buttock abscess. The relationship between plasmid content and adherence capability was demonstrated to be rather complex.

INTRODUCTION

Several characteristics that determine pathogenicity of bacteria are plasmid specified (Cabello and Timmis, 1979; Clancy and Savage, 1981; Crosa, 1980; So et al., 1978; Zink et al., 1980). Plasmid-determined antibiotic resistance and plasmid-determined adherence properties are of particular importance for pathogenicity. First, we intended to establish the conditions for a proper analysis of extrachromosomal factors of different bacterial genera as well as the best conditions for bacterial sample collection. Second, we decided to examine the antibiotic resistance and adherence properties of several bacterial genera of medical importance in Argentinian hospitals. In this paper, we describe the distribution of plasmids in drug-resistant bacteria isolated from clinical patients, as well as the stability *in vitro* of plasmids. The capacity for adherence to catheters and the corresponding plasmid content of some clinical isolates also were examined.

MATERIALS AND METHODS

Bacterial strains and antibiotic susceptibility

Bacteria characterized in this study were collected from patients with nosocomial infections at the Children's Hospital of Buenos Aires. Bacterial isolates were identified according to standard diagnostic procedures. Antibiotic susceptibility of bacterial isolates was performed according to published recommendations (Isenberg et al., 1980). Bacterial strains were usually grown in Luria (L) broth media.

Isolation and electrophoresis of plasmid DNA

Plasmid DNA was isolated as described by Birboim and Doly (1979), with some modifications in order to obtain a quick procedure suitable for screening purposes (Denoya et al., in preparation). Agarose gel electrophoresis and visualization of DNA bands were performed as described by Meyers and colleagues (1976). Horizontal 0.8% agarose gels were used.

Antibiotic resistance and plasmid stability studies

Multiresistant bacterial strains were grown for at least five consecutive transfers in L-broth media containing two combined antibiotics. After that a portion of the culture was transferred to an antibiotic-free media and another portion to an antibiotic-containing media. Cells were grown at 37°C for 16 h, and a new transfer was performed. The transfers were continued for a total of eight times. Changes in the antibiotic-resistant organisms in each transfer were monitored by plating samples at appropriate dilutions both on L-agar plates and on L-agar plus two combined antibiotics plates. After incubation, the percentage of antibiotic-resistant organisms in each transfer was calculated. An analysis of the plasmid content on agarose gels was performed on each sample. The strains used in these experiments were: (a) *Escherichia coli* strains Ec.25. Resistant to several antibiotics including Ampicillin and Amikacin. The antibiotic containing liquid media for this strain contains 40 u gr/ml Ampicillin and 40 μg/ml Amikacin. For an unknown reason, this strain is more sensitive to the Amikacin in solid media than in liquid media. Therefore, antibiotic-containing solid media contains 40 μg/ml Ampicillin and 15 μg/ml Amikacin. (b) *Klebsiella pneumoniae* strains Kb 2 and Kb 8. These *Klebsiella* strains are resistant to several antibiotics (table 1), among them, Ampicillin and Gentamicin. Liquid and solid antibiotic-containing media contain 40 μg/ml Ampicillin and 20 μg/ml Gentamycin. These two strains have been cloned from a single colony containing all the plasmids detected in the analysis of the hospital isolated population.

DETERMINATION OF ADHERENCE CAPACITY

Bacterial strains were grown at 37°C overnight in natural cerebrospinal fluid (CSF). 0.1 ml of culture was injected into a ventricular catheter (Holter) and incubated 40 min at room temperature (22–24°C). Afterward, twenty washes of 10 ml each with sterile saline solution were made using a syringe pushed by hand at vigorous flow. 100 μl of each wash were plated on L-agar plates and the number of colonies recorded after overnight incubation at 37°C. In addition, the catheter was cut into two pieces; one was incubated in L-broth overnight at 37°C, and the other was processed for scanning electron microscopy (Guevara et al., 1981).

RESULTS AND DISCUSSION

Antibiotic Resistance Patterns and Plasmid Content

Epidemiological studies of pathogenic bacteria require a number of techniques for the identification and characterization of different strains (Isenberg et al., 1980). Recently, the use of agarose gel electrophoresis of plasmid DNA has been demonstrated as useful in performing a "molecular epidemiology" of nosocomial infections (Willshaw et al., 1979; Schaberg et al., 1981). Since *Klebsiella* strains with multiple resistance are frequently the cause of outbreaks of hospital infections (Datta et al., 1979) and the frequency of appearance of multiresistant strains of *Klebsiella* is high in Buenos Aires, we decided to examine the antibiotic-resistance pattern and plasmid content of representative strains recovered from patients acquiring nosocomial infections at the Children's Hospital during this year. All the strains studied have a very similar antibiotic resistance pattern (table 1). It

TABLE I

STRAIN	SOURCE	UNIT	Ap	Cd	Cx	Tb	Sm	Gm	Ak	Cm	105	85	68	50	2.2	Others
Kb 23	CSF	7	+	+	+	+	+	+	−	+				+		
Kb 24	CSF	7	+	+	+	+	+	+	−	+				+		
Kb 25	CSF	7	+	+	+	+	+	+	−	+				+		
Kb 21	URINE	8	+	+	−	+	+	+	−	+				+		
Kb 22	URINE	18	+	+	−	+	+	+	−	+				+		38.0
Kb 20	URINE	10	+	−	−	+	+	+	−	+				+		2.6
Kb 1	CSF	18	+	+	−	+	+	+	−	+				+		
Kb 8	ABCESS	8	+	−	+	+	+	+	−	+		(+)		+		
Kb 2	CSF	18	+	+	−	+	+	+	−	+		+		+		2.6
Kb 26	CSF	10	+	+	−	+	+	+	−	+	+	+		+		
Kb 12	BLOOD	2	+	+	−	+	+	+	−	+		+				1.2
Kb 3	URINE	4	+	+	−	+	+	+	−	+		+			+	1.2
Kb 4	URINE	4	+	+	−	+	+	+	−	+		+	+		+	6.0
Kb 9	URINE	7	+	+	+	+	+	+	−	+					+	6.0 3.7 3.4
Kb 16	URINE	ICU	+	+	−	+	+	+	I	+	+	(+)	+		+	
Kb 17	URINE	10	+	+	−	+	+	+	−	+	+	(+)	+	(+)	+	38.0
Kb 18	URINE	18	+	+	−	+	+	+	−	+	(+)		+			38.0
Kb 6	BLOOD	2	+	+	−	+	+	+	−	+	(+)		+		+	2.8 2.6 (1.8) 1.5 1.2
Kb 5	CSF	2	+	+	−	+	+	+	−	+	(+)		+			2.8 2.6 (1.8) 1.5
Kb 7	ABCESS	2	+	+	−	+	+	+	−	+			+			2.8 2.6 1.8 1.5
Kb 13	PLEURA	ICU	+	+	−	+	+	+	−	+	(95)					
Kb 14	URINE	4	+	+	−	+	+	+	−	+	(95)				+	
Kb 15	URINE	4	+	−	−	+	+	+	−	+					+	
Kb 19	CATHETER	1	+	−	−	+	+	+	−	+	+					2.8
Kb 10	KNEE	14	+	+	−	+	+	+	−	+	+					3.4

NOTE. Abbreviations for drugs: Ap = ampicillin; Cd = cephaloridine; Cx = cefotaxime; Tb = tobramycin; Sm = sisomycin; Gm = gentamicin; Ak = amikacin; Cm = chloramphenicol. Other abbreviations: ICU = intensive care unit; MDal = Megadalton. Presence of the resistance marker and plasmid is denoted by +, its absence by −. Intermedium level of resistance is denoted by I. When a plasmid is present at a low concentration, it is denoted by (+).

is important to point out that Ampicillin and Gentamicin are the most frequently used antibiotics in this hospital. Three of the strains (Kb 23/24/25 from one patient, and Kb 26 and 9 from two others) are resistant to one of the new third generation cephalosphorines, Cefotaxime. In addition, one of the strains (Kb 16) possesses intermediate levels of resistance to the aminoglycoside Amikacin (M.I.C.:16). Recently we have detected two strains resistant to this antibiotic.

Figures 1, 2, and 3 contain the plasmid profiles of 25 strains, and Table 1 contains the estimated molecular weights of the plasmid present in each strain.

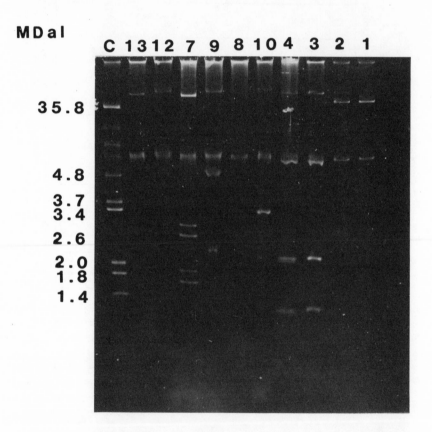

Fig. 1. Agarose gel electrophoresis of plasmid DNA extracted from clinical *Klebsiella pneumoniae* strains.

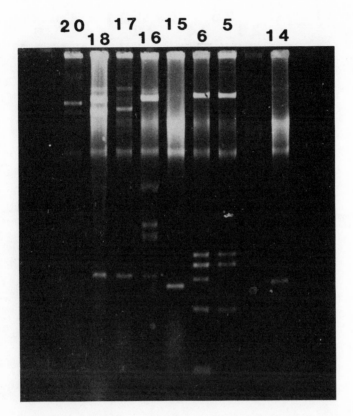

Fig. 2. Agarose gel electrophoresis of plasmid DNA extracted from clinical *Klebsiella pneumoniae* strains.

Several conclusions can be drawn from the results shown in table 1: (1) The plasmid banding is a very sensitive technique to fingerprint clinical isolates. It can be observed that the technique has great reproducibility because isolates taken at different days from the same patient (e.g., Kb 23-Kb 24-Kb 25, Kb 3-4, Kb 14-15) have the same pattern profile. Particularly interesting is the demonstration that a patient may possess at the same time a population of closely related bacteria, differing only in one or two plasmids (table 1, Kb 6 from blood and Kb 5 from CSF). It could be relevant that each strain was obtained from different sources. With isolates obtained from the same patient one week later, we demonstrated the predominance of only one of the variants. (2) Molecular epidemiology surveys of plasmids carried on in hospitals from the United States and Europe have

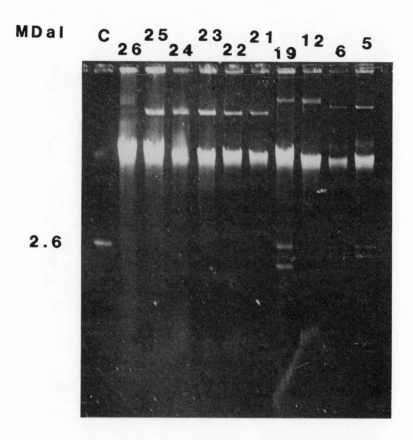

Fig. 3. Agarose gel electrophoresis of plasmid DNA extracted from clinical *Klebsiella pneumoniae* strains.

shown the existence of a group or a small number of groups of isolates with different plasmid profiles (Willshaw et al., 1979; Schaberg et al., 1981; Datta et al., 1979). In contrast, by our study in Argentina, we showed a very complex situation as far as bacterial plasmid content is concerned. It seems to be almost impossible to classify unique characteristic strains. They showed a complex network of relationships, even within the same nosocomial unit or source. However, some information seems to emerge from this chaotic picture. Some plasmids are more widespread than others. Within the large plasmid group, the 105 MDal (31%), 85 MDal (42%), 68 MDal (31%), and 50 MDal (42%) plasmids are the most widespread. Strains having the 50 MDal plasmid constitute a group that has no small plasmids. Exceptions to this rule are Kb 1 and Kb 2, which have a 2.6 MDal plasmid and seem to be a sort of connection between this group and the 85 MDal and 68 MDal plasmid-containing bacteria group. In addition, the 2.2 MDal is the most frequent plasmid within the small ones (42%). The diversity of plasmid profiles may reflect the lack of a good program for the administration and control of antibiotics in Argentina. However, further studies are necessary in order to document the distribution and evolution of R-factors in this part of the world.

Antibiotic Resistance and Plasmid Stability

During antibiotic therapy, bacteria bearing R-factors have a distinct advantage with respect to non-plasmid-containing bacteria (Hartley and Richmond, 1975; Petrocheilou et al., 1976). However, in the absence of antibiotic challenge, it has been suggested that the presence of a plasmid will be of disadvantage as a result of the plasmid maintenance cost (Anderson, 1974; Lacey, 1975). This simple hypothesis predicts that in the absence of selective pressure a population of antibiotic-sensitive bacteria will replace the R-factor-bearing bacterial population. This seems to be the case for some of the antibiotic-resistant bacterial strains that we have studied.

For example, figure 4 is of the progressive loss of antibiotic resistance of a multiresistant *E. coli* through successive transfers in antibiotic-free media. This *E. coli* strains possesses two detectable plasmids of 75 and 50 MDal (fig. 5, lane a).

After six transfers in antibiotic-free media, the 75 MDal plasmid is strongly diminished (fig. 5 lane b). Conversely, no significant loss has been detected in the 50 MDal plasmid, which is probably not involved in the antibiotic-resistance character. Other authors have presented similar results for other plasmid-bacteria systems (Helling et al., 1981; Wouters et al., 1980; Godwin and Slater, 1979). However, this is not a general rule, and sometimes the absence of antibiotic pressure has not resulted in a significant loss of the antibiotic resistance (Melling et al., 1977; Wouters et al., 1978). Indeed, this is the case for the multiresistant *Klebsiella* strain Kb 2 (fig. 5 lane d). This bacterial strain contains three detectable plasmids of 85, 50, and 2.6 MDal. None of these plasmids suffers any significant variation during the successive transfers in antibiotic-free media (fig. 5 lane e). The plasmid stability in this case may result from plasmid-determined selective advantages different from antibiotic resistance such as bacteriocin production (Adams et al., 1979) or other metabolic capabilities (Gunsales et al., 1975). The multiresistant *Klebsiella* strain Kb 8 deserves special mention

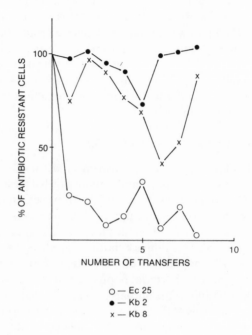

Fig. 4. Percentage of antibiotic-resistant cells in L-broth batch cultures after different transfers starting from L-broth plus antibiotic grown cells.

because it contains a detectable plasmid of 50 MDal (fig. 5 lane g). The concentration of this plasmid decreases below the detection limits of our analysis in agarose gels after only one transfer in antibiotic-free media (fig. 5 lane n). However, it seems that this plasmid remains in the bacterial cells in a very low copy number since the loss of antibiotic through the successive transfers in antibiotic-free media was never under 40% (fig. 4). Furthermore, when these apparently plasmid-free bacteria were transferred to a media containing antibiotic, the plasmid was detected again in amounts comparable to the amount of plasmid present in bacteria that have been grown continuously in antibiotic-containing media (fig. 5 lane j). Therefore, the copy number of this plasmid seems to be determined by the presence of antibiotic in the growing media. In any case, it seems to be clear that no simple explanation exists to predict the stability of the plasmid DNA

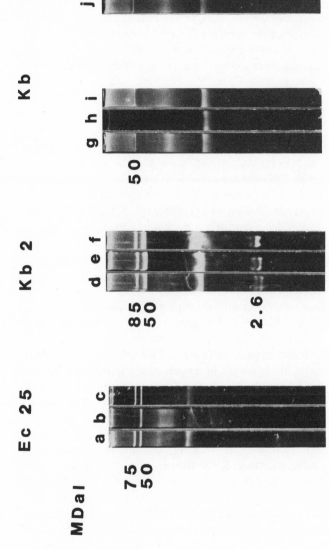

Fig. 5. Agarose gel electrophoresis of plasmid DNA of samples taken from different transfers in L-broth or L-broth plus antibiotic media. *a.* Ec. 25: starting cells; *b.* Ec. 25: transfer no. 6 in L-broth; *c.* Ec. 25 transfer no. 6 in L-broth plus antibiotic; *d.* Kb2: starting cells; *e.* Kb2: transfer no. 6 in L-broth; *f.* Kb2: transfer no. 6 in L-broth plus antibiotic; *g.* Kb8: starting cells; *h.* Kb8: transfer no. 1 in L-broth; *i.* Kb8: transfer no. 1 in L-broth plus antibiotic; *j.* Kb8: cells retransferred from L-broth to L-broth plus antibiotic.

bacterial content during environmental changes. Finally, there is a practical conclusion that our results on plasmid stability suggest. If a study of plasmid-determined properties of bacteria taken from natural environments is required, the selective pressure for the characteristic of interest should be maintained all along the isolation procedures and laboratory growing conditions. If this precaution is kept, the plasmid stability seems to be guaranteed.

Adherence to Catheters and Plasmid Content

It is well known that some bacteria have the ability to stick to inert surfaces (Costerton et al., 1978). Multiply-resistant *Klebsiella* strains are peculiarly associated with catheterized urological patients (Casewell et al., 1977) as well as with hydrocephalus patients with implanted ventriculoperitoneal (V-P) shunts for the control of cerebrospinal fluid (CSF) pressure (Guevara, et al., 1981). As part of our program, we started a study on adhesion of different nosocomial *Klebsiella* strains to catheters currently used in V-P shunts.

Two strains, one isolated from an infected CSF of a patient with an implanted V-P shunt (Kb 1) and the other obtained from a buttock abscess of an unrelated patient, were tested for the capacity to adhere to a ventricular catheter after incubation *in vitro* and successive washes with saline solution. In preliminary experiments, it was evident that Kb 1 possesses a higher capacity for adherence to catheters than Kb 8 strain. It was decided to isolate unique colonies to assure homogeneous populations. Several colonies of each strain were picked up, and each was controlled for its plasmid content. Surprisingly, one of the clones of Kb 1 lacked the 68 MDal plasmid (colony no. 1) (fig. 6, lane B). An experiment was performed in order to compare adhesivity of colony no. 1 and colony no. 5 of Kb 1 and colony no. 5 of Kb 8. Bacteria derived from both colonies of Kb 1 again had higher capacity for adherence than Kb 8. However, bacteria derived from colony no. 1 attached better than bacteria derived from colony no. 5. It seems then that the presence of the 68 MDal plasmid mediated some kind of interference with the attachment properties of Kb 1. It can be concluded that different strains of Kb.p. may have a diversity of phenotypes as far as adherence to surfaces is concerned. A study is in progress to understand the relationship that seems to exist between the presence or absence of different plasmids detected in the strains shown in figure 7 and their adherence capacities.

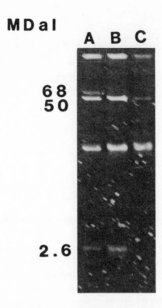

Fig. 6. Plasmid content of colony no. 1 (B) and colony no. 5 (A) of Kb1 and colony no. 5 (C) of Kb8.

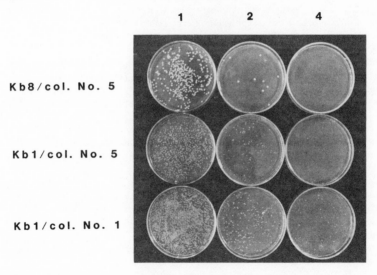

Fig. 7. Adherence capacity of bacteria derived from unique clones of *Klebsiella* strains Kb1 and Kb8.

ACKNOWLEDGMENTS

We thank Drs. A. J. Guevara and G. Zuccaro for providing us with ventricular catheters and cerebrospinal fluid. We are grateful to Dr. C. A. Guardiano (Lab. Britania) for his generous gift of several antibiotics used throughout this study, as well as the expert secretarial assistance of M. Denegri.

REFERENCES

Adams, J., T. Kinney, S. Thompson, L. Rubin, and R. B. Helling. 1979. Frequency-dependent selection for plasmid-containing cells of *Escherichia coli*. Genetics 91:627–37.

Anderson, J. D. 1974. The effect of R-factor carriage on the survival of *Escherichia coli* in the human intestine. J. Med. Microbiol. 7:85–90.

Birboim, H. C., and J. Doly. 1979. A rapid alkaline extraction procedure for screening recombinant plasmid DNA. Nucl. Acid Res. 7:1513–23.

Cabello, F., and K. N. Timmis. 1979. Plasmids of medical importance. *In* K. N. Timmis and A. Puhler (eds.), Plasmids of medical, environmental, and commercial importance, pp. 55–69. Elsevier-North Holland, Amsterdam.

Casewell, M. W., M. T. Dalton, M. Webster, and I. Phillips. 1977. Gentamicin resistant *Klebsiella aerogenes* in an urologic ward. Lancet 2:444–46.

Clancy, J., and D. C. Savage. 1981. Another colicin V phenotype: *in vitro* adhesion of *Escherichia coli* to mouse intestinal epithelium. Infect. Immun. 32:343–52.

Costerton, J. W., G. G. Geese, and K. J. Cheng. 1978. How bacteria stick. Scient. Amer. 288:86–95.

Crosa, J. H. 1980. A plasmid associated with virulence in the marine fish pathogen *Vibrio anguillarum* specifies an iron-sequestering system. Nature (London) 284:566–68.

Datta, N., V. M. Hughes, M. E. Nugent, and H. Richards. 1979. Plasmids and transposons and their stability and mutability in bacteria isolated during an outbreak of hospital infection. Plasmid 2:182–96.

Godwin, D., and J. H. Slater. 1979. The influence of the growth environments on the stability of a drug resistance plasmid in *Escherichia coli* K 12. J. Gen. Microbiol. 111:201–10.

Guevara, J. A., J. L. La Torre, C. D. Denoya, and G. Zuccaro. 1981. Microscopic studies in shunts for hydrocephalus. Child's Brain 8:284–93.

Gunsalus, I. C., M. Hermann, W. A. Toscano, D. Katz, and G. K. Garg. 1975. Plasmids and metabolic diversity. *In* Schlessinger (ed.), Microbiology 1974, pp. 207–12. American Society for Microbiology, Washington, D.C.

Hartley, C. L., and M. H. Richmond, 1975. Antibiotic resistance and survival of *E. coli* in the alimentary tract. Brit. Med. J. 4:71–74.

Helling, R. B., T. Kinney, and J. Adams. 1981. The maintenance of plasmid-containing organisms in populations of *Escherichia coli* Gen. Microbiol. 123:129–41.

Isenberg, H. D., J. A. Washington II, A. Balows, and A. C. Sonnenwirth. 1980. Collection, handling, and processing of specimens. *In* E. H. Lennett, A. Balows, W. J. Hausler, Jr., and J. P. Truant (eds.), Manual of clinical microbiology, pp. 52–82. American Society for Microbiology, Washington, D.C.

Lacey, R. W. 1975. Antibiotic resistance plasmids of *Staphylococcus aureus* and their clinical importance. Bacteriol. Rev. 39:1–32.

Melling, J., D. C. Ellwood, and A. Robinson. 1977. Survival of R-factor carrying *Escherichia coli* in mixed cultures in the chemostat. FEMS Microbiol. Lett. 2:87–89.

Meyers, J. A., D. Sanchez, L. P. Elwell, and S. Falkow. 1976. Single agarose gel electrophoretic method for the identification and characterization of plasmid deoxyribonucleic acid. J. Bacteriol. 127:1529–37.

Petrocheilou, V., J. Grinsted, and M. H. Richmond. 1976. R-plasmid transfer *in vivo* in the absence of antibiotic selection pressure. Antimicrob. Agents Chemother. 10:753–61.

Schaberg, D. R., L. S. Tompkins, and S. Falkow. 1981. Use of agarose gel electrophoresis of plasmid deoxyribonucleic acid to fingerprint gram-negative bacilli. J. Clin. Microbiol. 13(6):1105–8.

So, M., W. S. Dallas, and S. Falkow. 1978. Characterization of an *Escherichia coli* plasmid encoding for synthesis of heat-labile toxin: molecular cloning of the toxin determinant. Infect. Immun. 21:405–11.

Willshaw, G. A., H. R. Smith, and E. S. Anderson. 1979. Application of agarose gel electrophoresis to the characterization of plasmid DNA in drug-resistant enterobacteria. J. Gen. Microbiol. 114:15–25.

Wouters, J. T. M., F. L. Driehuis, P. J. Polaczek, H. A. Van Oppenraay, and J. G. van Andel. 1980. Persistance of the pBR322 plasmid in *Escherichia coli* K12 grown in chemostat cultures. Antonie van Leeuwenhoek 46:353–62.

Wouters, J. T. M., C. Rops, and J. G. van Andel. 1978. R-plasmid persistance in *Escherichia coli* under various environmental conditions. Proc. Soc. Gen. Microbiol. 5:61.

Zink, D. L., J. C. Feeley, J. G. Wells, C. Vanderzant, J. C. Vickery, W. D. Roof, and G. O'Donovan. 1980. Plasmid-mediated tissue invasiveness in *Yersinia enterocolitica*. Nature (London) 283:224–26.

R. R. BRENTANI AND L. L. VILLA

The Cloning of Human Insulin

33

ABSTRACT

A bacterial clone containing DNA complementary (cDNA) to human proinsulin was constructed. The 410 base pairs of DNA corresponding to human proinsulin, inserted in the Hind III site of pBR 322, have been manipulated in order to be subcloned in a plasmid containing part of the β galactosidase operon sequences and transformed in *Escherichia coli*. The isolation and purification of insulin produced by the recombinant DNA technology is discussed in detail.

DISCUSSION

Insulin, the hormone that regulates glucose homeostasis in vertebrates, is composed of two polypeptide chains (A and B) joined by disulfide linkages. Both chains are derived from one precursor, proinsulin, having a connecting peptide (C) between the A and B chains, which is excised before the secretion of insulin from the pancreatic B cells (Tager and Steiner, 1974). Moreover, the immediate translation product of the insulin gene is preproinsulin, a single polypeptide that contains about 20 additional residues at the N-terminal end (Chan et al., 1976). This prepeptide facilitates the transit of the insulin precursor into the endoplasmic reticulum and is cleaved away during this process (Blobel and Doberston, 1975).

Although the physiological effects of insulin are known in many aspects, for several years no specific data have been reported about the insulin gene(s). The availability of recombinant DNA technology, however, has resulted in the description of both rat (Ullrich et al., 1977; Villa-Komaroff et al., 1978; Cordell et al., 1979) and human (Bell et al., 1979, 1980) proinsulin-coding DNA sequences. The structural description of these genes allows study of insulin gene expression and its regulation both in normal and pathological states, such as diabetes.

Diabetes mellitus is a complex disease that may affect as many as 5% of the human population (Notkins, 1978) and whose etiology is very poorly understood. Although administration of the hormone ameliorates diabetic symptoms in many cases, such disease and its complications can be death-dealing. Insulin obtained from the pancreas glands of cow and pig is currently used in the treatment of many diabetics. However, since animal insulin is slightly different in structure from the human hormone, sometimes immunological problems can arise. Therefore, the human product may be preferable in treatment. In view of these facts, several efforts have been made to develop an alternative system for the production of insulin in large scale.

One of the procedures described deals with the production and purification of separated insulin chains A and B (Crea et al., 1978). Two recombinant clones were constructed containing chemically synthesized DNA fragments corresponding to polypeptides A and B. After recovery and purification, such products are oxidized in order to get the disulfide bridges reestablished between chains. However, because several orientations are allowed, most of the final product is not biologically active insulin, requiring further purifications.

Alternatively, recombinant clones have been described in which a hybrid protein of β lactamase and proinsulin are synthesized (Villa-Komaroff et al., 1978). From such product, human proinsulin can be purified, but at present high levels of expression still have not been attained.

In Brazil about 13 kg of bovine insulin per month are produced by Bio-Farm (Montes Claros, MG). Nevertheless, the bovine pancreas are imported from Argentina, causing some supply difficulties and making the product more expensive. Moreover, the porcine hormone is not produced in Brazil.

From all the data described above, we decided to develop new recombinants that would, in the near future, supply our diabetic population with the hormone. Such efforts resulted in the construction of a recombinant clone containing DNA complementary to human proinsulin (Villa et al., 1980, 1981), as described below.

Total poly A(+) messenger RNA was isolated from a human insulinoma by a thiocyanate guanidine procedure and purified by oligo d(T) column chromatography. cDNA and double-stranded cDNA were synthesized, using the mRNA as a template, with the aid of reverse transcriptase. After treatment with S_1 nuclease, the ds cDNA was ligated to Hind III linkers in the presence of T4 DNA ligase. Sticky ends were generated by Hind III digestion, and the material was annealed to pBR 322 linearized with the same endonuclease, followed by treatment with alkaline phosphatase. The recombinant plasmid was used to transform *E. coli* HB 101. By hybridization with a DNA fragment corresponding to rat insulin (Bolivar, F., unpublished results), we have identified one clone containing a 410 base pair insert, designated pI$_2$. Restriction mapping of this insertion with Eco RII and Hae III revealed that our recombinant contains all the proinsulin region by 5 or 10 base pairs.

To get expression of the proinsulin sequence cloned in *E. coli*, we had to construct a new recombinant plasmid having the cDNA insertion adjacent to part of the bacterial *lac* operon. Initially the 410 base pair insertion of pI$_2$ was cleaved with Eco RII and treated with DNA polymerase (Klenow fragment) to get blunt-ended extremities. After purification by gel electrophoresis and electroelution, such fragment was ligated to a chemically synthesized DNA oligonucleotide coding the first ten amino acids of the proinsulin polypeptide and having, at the NH$_2$ terminal, the codon for methionine and an Eco RI restriction site. The ligation was followed by digestion with Eco RI and Hind III; the resulting fragment was inserted in pBR 322 linearized with the same endonucleases and cloned in *E. coli* HB 101. After positive characterization, a DNA sequence carrying the *lac* promoter, operator, and β galactosidase structural gene was inserted into the Eco RI site of the proinsulin recombinant plasmid and again transformation was carried out. The clones obtained are expected to produce a fused protein of β galactosidase and proinsulin, separated by a methionine residue that can be cleaved by treatment with cyanogen bromide *in vitro*. Finally, the purified proinsulin can have the peptide C excised, and oxidation of chains can be accomplished through conventional methodologies.

It is our hope that these results will be helpful in studying the improvement of human insulin expression in bacteria. Moreover, it opens the possibility of programming the expression of various eukaryotic genes in alternative hosts.

REFERENCES

Bell, G. I., et al. 1979. Nucleotide sequence of a cDNA clone encoding human proinsulin. Nature 282:525–27.
Bell, G. I., et al. 1980. Sequence of the human insulin gene. Nature 284:26–32.
Blobel, G., and B. Dobberstein. 1975. Transfer of proteins across membranes. I. Presence of

proteolytically processed and unprocessed nascent immunoglobulin light chains on membrane-bound ribosomes of murine myeloma. J. Cell Biol. 67:835–51.

Chan, S. J., et al. 1976. Cell-free synthesis of rat preproinsulins: characterization and partial amino acid sequence determination. Proc. Natl. Acad Sci. USA 73:1964–68.

Cordell, B., et al. 1979. Isolation and characterization of a cloned rat insulin gene. Cell 18:533–44.

Crea, R., et al. 1978. Chemical synthesis of genes for human insulin. Proc. Natl. Acad. Sci. USA 75:5765–69.

Notkins, A. L. 1979. The causes of diabetes, Sci. Ameri. 241:56–57.

Tager, H. S., and D. F. Steiner. 1974. Peptide hormones. Ann. Rev. Biochem. 43:509–38.

Ullrich, A., et al., 1977. Rat insulin genes; construction of plasmids containing the coding sequences. Science 196:1313–19.

Villa-Komaroff, L., et al. 1978. A bacterial clone synthesizing proinsulin. Proc. Natl. Acad. Sci. USA 75:3727–31.

Villa, L. L., et al. 1980. Construction of bacterial recombinants containing human insulin cDNA. Eur. J. Cell Biol. 22:18.

Villa, L. L., et al. 1981. Insulina humana: obtencao de uma bacteria contendo a informacao para a sua sintese. Arg. Bras. Endocrinol. Metabol. 25:61–64.

OMAR O. BARRIGA AND JULIUS P. KREIER

The Use of New Biotechniques in the Development of Vaccines

34

INTRODUCTION

Parasitic diseases exact a tremendous toll on human health (Peters, 1978) and animal resources (Griffiths, 1978). Less apparent than many viral and bacterial diseases, they have been historically regarded with less alarm. Only the decline in the prevalence of acute viroses and bacterioses in the last decades has permitted an unbiased evaluation of the more chronic damage caused by parasitoses and encouraged the initiation of actions to prevent them. The most widely known of parasitic infections is possibly malaria. This disease affected 300 million people and killed 3 million per year only four decades ago (Russell, 1943). The advent of the residual insecticides during World War II, the first synthesis of antimalarial drugs in 1944 (Woodard and Doering, 1944), and the involvement of the World Health Organization in international operations of control since 1948 produced a radical and quick reduction in prevalence of malaria. By 1978, 37 countries had eradicated malaria, and many others had made important advances toward its control (Wernsdorfer, 1980).

Malaria eradication or control requires large sums of money and efficient organization. Vast regions of the world that lack these basic elements, particularly in Africa and Asia, have lagged in malaria control. Recent estimates put the number of people still affected by malaria at 150 million, and the number of deaths in Africa alone at 1 million (Noguer et al., 1978). Furthermore, the traditional methods of control of malaria have been hampered of late by the development of strains of mosquitos resistant to insecticides (Wernsdorfer, 1980) and strains of *Plasmodium* resistant to the conventional antimalarial drugs (Wernsdorfer and Kowznetsov, 1980). All these considerations have encouraged the search for alternative, and maybe simpler, means of fighting malaria. Immunization has been proposed as a possible alternative method of malaria control (Diggs, 1980; 1984).

The first attempts at vaccination against malaria were made in birds (Sergent and Sergent, 1910; Russell et al., 1941) and only much later were studies in rodents initiated (Corredetti et al., 1966; Nussenzweig et al., 1967; 1967a). Assays in monkeys (Sadun et al., 1969) and in man (Clyde et al., 1973) soon followed.

The success of the ecological methods of control of malaria before the 1970s, the demonstration of *Plasmodium*-induced immunodepression (McGregor and Barr, 1962), and the realization that the parasite changes its antigenic make-up in the course of the infection (Brown, 1977) discouraged work toward the development of a vaccine against malaria. The subsequent failure of the traditional methods of control in Africa and Asia, the development of powerful new biotechnologies, and particularly the strong support of the World Health Organization since 1976 have caused a large surge and a fundamental change in the direction of malaria research in the last decade. The development of an efficacious immunizing procedure is probably the first priority now. (Smith and Erickson, 1980; Warren, 1980) The results of this new direction are just becoming evident, and the production of an effective vaccine against malaria is hoped for in the near future (Diggs, 1984).

The immunology of malaria and the efforts to develop a vaccine have been abundantly reviewed (e.g., Cohen and Mitchell, 1978; Desowitz and Miller, 1980; Kreier, 1980; Jayawardena, 1981; Deans and Cohen, 1983). We will restrict our review here to the basic concepts of some of the new biotechniques that may make the production of vaccines of practical use against parasitic infections, particularly malaria, a feasible enterprise.

Taken to its bare bones, the rational production of a vaccine involves three major steps: identification of the antigens responsible for protection; their purification and characterization; and their production in useful quantities. In parasitology, these tasks have been largely facilitated by techniques of recent development. A major advance in the formulation of anti-parasite vaccines has been the recent development of the enzyme-linked immunoelectrotransfer blot technique that allows the separation of a complex mixture of proteins into its individual components and the identification of those that are relevant for immunization. Affinity chromatography, particularly when combined with monoclonal antibodies, may be the most powerful tool currently available for the isolation of individual antigens. The production of monoclonal antibodies by hybridoma technology has revolutionized the conventional methods of isolating and evaluating antigens. Finally, recombinant DNA technology and new methods of protein synthesis suggest that the antigens could be manufactured in the laboratory in the virtual absence of the corresponding parasite.

THE ENZYME-LINKED IMMUNOELECTROTRANSFER BLOT TECHNIQUE (EITB)

The identification of the antigens of a parasite is classically done by immunoelectrophoresis. First the parasite is solubilized (usually by physical methods) and the corresponding extract is injected into a rabbit. Once antibodies are produced, the proteins of the extract are separated by an electric field in a porous matrix, and then allowed to diffuse toward the antiserum and precipitate with the respective antibodies. This method has contributed much to our present understanding of parasite antigens but has two major drawbacks: the resolution of the individual proteins of the extract is limited, and the sensitivity of their reaction with the antibodies is low. The latter characteristic almost prevents the use of serum from infected hosts since many of its antibodies are too dilute to react by precipitation. The use of serum from animals immunized with an extract, on the other hand, often reveals parasite and host components that do not behave as antigens during an actual infection. This is particularly true when the immunized species is not the normal host of the parasite.

Techniques that are able to separate the proteins of a mixture with a very high degree of resolution were developed in the late 1960s (e.g., sodium dodecyl sulfate polyacrylamide gel electrophoresis, or SDS-PAGE; [Shapiro et al., 1967]). Weber and Osborn (1975) have done a comprehensive review of this technique. Very sensitive serologic tests that do not demand the attachment of extraneous groups to the reagents (as in the radioimmune assays) were devised shortly afterward (e.g., enzyme-linked immunosorbent assay, or ELISA [Engval and Pearlman, 1971; Van Weemen and Schuurs, 1971]). A recent review of the methodology of ELISA has been published by the World Health Organization (1976). It was almost a decade later that Renart et al. (1979), utilizing the pioneering work of Southern (1975) and Towbin et al. (1979), succeeded in putting together the two methods in the modern version of the EITB technique. In this procedure, the proteins of a complex parasite extract are separated by electrophoresis in a slab containing a polyacrylamide gel gradient, the resolved proteins are transferred from the slab onto a nitrocellulose sheet by a second electrophoresis perpendicular to the plane of the slab, and an ELISA with serum from infected hosts is performed to detect the antigenic proteins adsorbed on the sheet. Tsang et al. (1983) have described the use of this technique for the study of *Schistosoma* antigens. The EITB not only resolves proteins with great discrimination and pairs them with the corresponding antibodies at any time of infection that the researcher selects but also may simultaneously give information on the nature of the antigen. Running the parasite extract in parallel with molecular weight standards will allow an estimation of the molecular weight of selected antigens. Specific staining may also suggest the chemical nature of the antigen. The use of new metallic stains (Merill et al., 1981; Eastman Kodak, 1983) will permit detection of as little as 0.01 nanogram of protein.

AFFINITY CHROMATOGRAPHY

Affinity chromatography is the process by which a molecular species is separated from other molecules by adsorption with a complementary binding substance (ligand). Traditionally, affinity chromatography is called "positive" when the desired material is the one that becomes bound to the ligand, and "negative" when the contaminating materials are the ones that are absorbed by the ligand. The positive technique is used more commonly as a purification method. For this method to be of practical use, the binding must be reversible so that the desired material can be recovered after adsorption on the ligand, and specific so that it comes out uncontaminated. The presence of the ligand on an insoluble support (matrix) greatly facilitates recovery. Because of its specificity, affinity chromatography is much more efficient in purification of molecular species than more traditional methods that rely on non-specific physico-chemical properties of the molecules (such a salting-out, molecular sieving, ion exchange or isoelectric focusing): 1,150-fold concentration and 95% yield have been reported with this method (Staehelin et al., 1981).

Affinity chromatography was used for the first time at the beginning of the century for the adsorption of amylase onto insoluble starch. It was not until the late 1960s, however, that techniques were developed for the binding of amino groups to a matrix (Axen et al., 1967), and the method became popular only in the second part of the 1970s (Wilchek & Hexter, 1976).

The matrix most commonly used in immunology is a skeleton of agarose or polyacrylamide that is chemically activated to react with primary amino groups, since these groups are commonly available in proteins, which are the main concern of the immunologist. Any protein ligand will then attach to the activated matrix and will be ready to bind complementary molecules. The attachment of the ligand to the matrix is a rather simple process, and ready-made materials are available commercially for the investigator to prepare his own reagents (from Bio-Rad, Richmond, California, or Pharmacia, Uppsala, Sweden, for example). Although the reaction between the immobilized ligand and the desired molecule can be done in a breaker (batch method), it is more common to place the adsorbant in a column and pass the mixture containing the desired material through it. After rinsing out the non-adsorbed molecules, the bound substance can be eluted by methods mild enough to preserve its biological activity (Danielson et al., 1982; Bureau & Daussant, 1983).

The use of antigens or antibodies as ligands permits the researcher to take advantage of the exquisite specificity of the antigen-antibody reaction to separate the complementary molecules in a highly purified form. Even cells carrying these molecules may be collected (Basch et al., 1983). The recent advent of monoclonal antibodies has allowed the production of affinity chromatography columns with these reagents as ligands and the consequent isolation of a virtually pure antigenic molecular species (Roth & Casell, 1983). Protein A from *Staphylococcus* has a combining site specific for the Fc fragment of immunoglobulins, particularly IgG (Goudswaard et al., 1978). The use of this protein in affinity chromatography has become popular for separation of immunoglobulins. These may be collected from fluids either alone or as part of antigen-antibody complexes (Langone, 1983). Now that sepharose-protein A conjugates are widely available commercially, their use is greatly facilitated. An example of the ingenious uses to which affinity chromatography may be put is that of Bout et al. (1976), who used the technique to isolate the substances that are targets for antischistosome drugs. Later these substances were used successfully for the production of protective antibodies.

HYBRIDOMA TECHNOLOGY

One of the major limitations in the study of antibodies and antigens has been an inability to produce homogeneous antibodies that react exclusively with an individual antigenic determinant. Parasites release many antigenic molecules during infection, and each one of these usually has several antigenic determinants. Furthermore, even a single antigenic determinant gives rise to a number of distinct antibodies with different degrees of fit between determinant and antibody combining site. The general result is that our purest antiserum contains a collection of heterologous antibodies that react with a variety of different antigenic molecules. This causes two major problems. One is that morphological and functional studies of antibody molecules are actually studies of heterologous populations, and the other is that it is impossible to distinguish between determinants that elicit protective immunity and those that are irrelevant or even induce immunosuppression. The solution of this problem stems from the development of the hybridoma technique that permits isolation of a single antibody-producing cell and replication of it *in vitro* (Kohler and Milstein, 1975). Because of the genetic uniqueness of the monoclonal population so obtained, the antibodies produced are homogeneous.

Antibody-producing cells are terminal cells that survive *in vitro* only for a few days and do not replicate. Cells of certain lymphoid tumors (myelomas), however, replicate *in vitro* without limit. Many of these tumors produce antibodies that, because of the monoclonal origin of most tumors, are coded by a single genome and are, therefore, homogeneous. The so-called myeloma proteins have been very helpful in elucidating the structure of immunoglobulins. They have not assisted in the study of immunoglobulin function or of their respective antigens, however, because their specificity is unknown in most cases and cannot be directed at our will. When Kohler and Milstein (1975) fused mouse antibody-producing cells with mouse myeloma cells that were not antibody secreters, the resulting hybrid (hybridoma) had the tumor cells' ability to grow *in vitro* vigorously, and the lymphocyte's capacity to produce specific antibodies. Galfre and Milstein (1981) have recently described the technique in detail as it applies to mouse or rat hybrids. In essence, the spleen cells of an animal immunized by injection of an antigen or by infection are cultured and fused by treatment with polyethylene glycol with myeloma cells (from a commercial source, for example) of the same species.

The myeloma cells selected for hybridoma formation are mutants lacking certain enzymes, which prevents them from growing in a particular medium called HAT. When the fused cell population is cultured in this medium, the unfused myeloma cells and the lymphocytes will die, and only the hybrids that share the genome of the tumor and the normal cells will proliferate vigorously. After 2–3 weeks, the presence of antibody is determined, usually in the supernatant medium of the hybrid culture. Many methods are available for this assay but the ELISA (Douillard and Hoffman, 1983) is the one preferred by many workers.

Commonly, a large number of hybrids (and, therefore, hybrids producing antibodies of many antigenic specificities) is formed in the first culture. The next step, then, is to select those hybrids that produce the antibodies required by the investigator. A first step in isolation of hybridomas of single specificity is done by subculturing suspensions prepared by making twofold dilutions of the original culture. The subcultures of the most diluted suspensions (those containing fewer cells) that still produce the desired antibodies are then seeded in semi-solid medium. Each single cell in this medium is then assayed by the plaque-forming cell test (Bankert, 1983), or transferred to a fluid culture. In the latter case, the supernatant fluids from these cultures are tested by ELISA for antibody to the antigen of interest. If the antigen important to the worker is known but unpurified at this time, the corresponding antibody in the cultures can be identified by EITB. In cases in which the investigator is looking for properties of the antibody other than antigen binding (protection, for example), each clone is replicated and their respective antibody is tested individually (by passive immunization, for example). Large amounts of antibody can be produced by *in vitro* culture or by injecting the hybrids to produce antibody-forming tumors in the animal species from which the myeloma was obtained. At this time, aliquots of the cultures can be preserved in liquid nitrogen for later replication.

Intraspecies human hybridomas have been produced (Croce et al., 1980; Olsson and Kaplan, 1980), but the ethical restrictions on human experimentation have limited greatly the value of the results obtained. The production of human-mouse chimeras (reviewed by Olsson and Kaplan, 1983) is likely to contribute importantly in the near future.

In parasitology, monoclonal antibodies have been used to identify which components of plasmodia (Nussenzweig, 1982; Nardin et al., 1982; Holder and Freeman, 1981), trypanosomes (Pearson et al., 1979) and schistosomes (Vervaerde et al., 1979, Deelder, 1982) induce immunity.

Holder and Freeman, for example, produced an array of monoclonal antibodies to blood stage of *Plasmodium yoelii* that were screened for their ability to protect mice from infection. Two that protected mice were then studied further. In addition to various chemical evaluations, including molecular weight determination, the eluted antigens were shown in an immunization trial to induce a protective immune response. In subsequent studies by immunoelectronmicroscopy, the antibodies to one of these antigens were shown to be specific for the merozoite surface coat, and the antibodies to the other for the contents of the rhoptories (OKA et al., 1984).

The studies by Nussenzweig and colleagues showed that monoclonal antibodies against sporozoites neutralized infectivity of *P. berghei* sporozoites and protected mice by passive transfer. Radioimmunoprecipitation studies with these antibodies revealed reactivity with parasite material of 44,000 molecular weight, and immunoelectronmicroscopy revealed that the antigen is surface coat material (Fine et al., 1984). The demonstration by monoclonal antibody techniques that a specific sporozoite antigen induces immunity to infection and preparation of the antigen in pure form with the aid of an affinity column to which the monoclonal antibody was bound has permitted sequencing of the antigen and synthesis of segments of it. A synthetic dodecapeptide containing the epitope to which the protective monoclonal antibody was directed induced immunity to sporozoite challenge. (Gysin et al., 1983).

RECOMBINANT DNA

Affinity chromatography, particularly in association with specific monoclonal antibodies, affords the opportunity of obtaining antigens in a quantity and degree of purity undreamed of only a few years ago. The method is quite adequate for the research laboratory but does not permit the easy production of antigens in the huge amounts demanded by industry. Recombinant DNA technology, often referred to as "genetic engineering," may allow us to do just that.

Life depends ultimately on proteins. These are the compounds that catalyze all organic reactions, including those that will result in the formation of new structures. The amino-acid composition of each protein, which is the determinant of its specific activity, is coded by the sequence of triplets of purine (adenine and guanine, A and G), and pyrimidine (cytocine and thymine, C and T) bases in the chromosomal DNA molecules. In essence, the DNA codes are transcribed into messenger RNA molecules (mRNA, where thymidine is replaced by uracyl, U), and these are translated into the actual proteins on the ribosomes. The details of these processes have been explained in a very readable book by Watson and colleagues (1983).

The base sequences that code for each amino-acid began to be established in the early 1960s (Nirenberg and Matthaei, 1961) and were all known by the late 1960s. This afforded the possibility of deducing the sequence of a DNA fragment that coded for a protein (gene) if the amino-acid sequence of the protein was known, and vice versa. Sequencing of a complete protein is still a formidable task, but practical methods for the sequencing of DNA were developed in the last half of the 1970s (Sanger and Coulson, 1975; Maxam and Gilbert, 1977; Sanger et al., 1977). The DNA molecules of eukaryotic (nucleated) organisms are enormous, with sequences, sometimes overlapping, that represent many genes and with fragments that act as regulators of gene expression. In addition, they are tightly complexed with proteins.

At the current stage of technology, it is difficult to isolate the specific portion of DNA that codes for a particular protein from these large chromosomes. On the other hand, chemical synthesis of a gene is an inefficient method for the production of genes of large proteins and requires detailed previous knowledge of the amino-acid composition of the protein. So far, it has been used only for the preparation of human insulin (Goeddel et al., 1979), a small peptide whose 51 amino-acid sequence was known and which has a complex process of formation and secretion (involving pre-proinsulin and proinsulin) that makes the reproduction of the normal synthesis inefficient.

Messenger RNA molecules, on the contrary, are smaller, code for a single protein (in the eukaryotes), and bear long tails made of adenine residues. Based on this last property, mRNA can be isolated by affinity chromatography with oligo-d thymidine, which is the base complementary to adenine. The m-RNA is obtained from extracts of parasite stages that produce large amounts of the desired protein. The Nussenzweigs, for example, used messenger RNA from sporozoites in their studies to identify the antigen important in host defense (Yoshida et al., 1981). Up to 40% of the mRNA of sporozoites in the mosquito was specific for surface coat protein, greatly facilitating the production of appropriate DNA for genetic manipulation.

Once the mRNA has been isolated, it can serve as a template for the enzyme reverse transcriptase to synthesize the corresponding strand of DNA. Subsequent degradation of the RNA portion followed by treatment with DNA polymerase will result in the production of a double-stranded DNA molecule, complementary to the mRNA (cDNA). A loop formed at one end of this molecule is removed with S1 nuclease.

Once this gene is created, it is necesssary to incorporate it into the chromosomes of an organism easy to culture. Traditionally the gene is incorporated into the circular chromosomes (plasmids, pBR322, for example) of *Escherichia coli*. The usual way of linking the gene to the plasmid is to create "tails" of complementary bases at the ends of the gene and at the ends of an open plasmid: adenine on the gene and thymidine on the plasmid, for example. A common method of incorporating these tails is to add the corresponding segments to the DNA with the enzyme terminal transferase. Another method uses restriction enzymes. By this later method, a short segment of DNA is added at the end of the gene, and this segment is subsequently cut with an appropriate restriction enzyme. Many of these enzymes cleave double-stranded DNA in a staggered form leaving a short tail of bases at the end of one strand. Since the enzymes are very specific for the site of cleavage, the tail left on one side of the cut will be complementary to the tail left on the opposite side. In this way, one end of the gene will combine with the opposite end of the cleaved plasmid. Incidentally, the restriction enzyme also serves to open the circular plasmid. The tails of the gene and the plasmid are then joined permanently with DNA ligase so that the artificial gene becomes a part of the new hybrid micro-chromosome. This plasmid is then used to infect *E. coli* cells and becomes a permanent part of the bacterium's genome. Replication of the bacterium now will result in the production of many replicates of the gene. If the gene includes appropriate promoter sequences, then the gene may program synthesis of the desired product.

Because it is important to distinguish the bacterial cells that carry the hybrid plasmid from the cells that do not, the extra gene is often inserted in plasmids that also code for resistance to some antibiotic and the plasmid is introduced into nonresistant bacteria. In this way, the bacteria carrying the hybrid plasmid can be selected by their antibiotic resistance.

In practice, it is difficult to isolate a single species of mRNA, so different hybrid plasmids often contain diverse inserted genes. In other cases, the researcher prefers to cut into fragments all the DNA of the donor parasite and incorporate all the fragments into plasmids to produce a "parasite gene library." In either case, the bacterium producing the desired protein must be identified and cloned individually.

Although plasmids were the original vectors to introduce DNA fragments into host cells, bacteriophages and viruses of eukaryotes have also been used. Vaccinia virus has been used to insert the genes for synthesis of the surface coat of the sporozoite stage of *Plasmodium knowlesi* into cells *in vitro*, and *in vivo* in rabbits (Smith et al., 1984) and indicates the potential of this technology for immunization against malaria and other diseases. There are several major advantages to this system of immunization. The immunized individual, for instance, generates the desired antigens himself under the direction of the recombinant genome of the infecting vaccinia virus. The recombinant virus can carry a variety of introduced genes and thus a single vaccination can give immunity to a variety of developmental stages of a parasite or to a variety of organisms. Perhaps most important, the cost of production of the recombinant vaccinia virus once constructed is low in comparison with the cost of production of the antigens from the organism or even from genetically engineered bacteria.

SUMMARY AND CONCLUSIONS

In this brief review, we have described some of the recent technologies that will surely facilitate the development and production of vaccines against parasites in the near future. These procedures have been the result of advances in protein chemistry, cell biology, immunology, and genetics, and usually combine assets from several of these areas. For the new technologies to be fruitful, however, they must be applied with a clear understanding of the biological system under study. The selection of the antigen to purify or to replicate is still a question of basic parasitology. Should we use sporozoites, merozoites, or gametes for an anti-malaria vaccine or schistosomula, adult worms, males or females, or eggs for an anti-schistosoma preparation?

The evaluation of the antigen or its antibodies subsequent to purification is also a parasitological problem. Would *in vitro* assays be significant? Are animal models truly representative of the human infection? Is protection persistant? Would the immunogen cause autoimmunity?

The new technologies are powerful tools of inherent beauty so it is easy to fall in love with them. Still they do not replace completely the older and more conventional methodology but rather supplement it, and certainly do not dispense one from knowing the biology of the parasite. Future advances will be made only by those that grasp this and successfully incorporate the new technologies within our body of established biological knowledge.

REFERENCES

Axen, R., J. Porath, and S. Ernbach. 1967. Chemical coupling of peptides and proteins to polysaccharides by means of cyanogen halides. Nature (London).

Bankert, R. B. 1983. Rapid screening and replica plating of hybridomas for the production and characterization of monoclonal antibodies. Methods in Enzymology 92:182–95.

Basch, R. S., J. W. Berman, and E. Lakow. 1983. Cell separation using positive immunoselective techniques. Jour. Immunol. 56:269–80.

Bout, D., H. Dupas, and M. Capron. 1976. Drugs as ligands of immunogenic molecules in parasites: An approach to the isolation of target antigens. *In* H. van der Bossche (ed.), Biochemistry of parasites and host-parasite relationships, pp. 325–33. Elsevier, Amsterdam.

Brown, K. N. 1977. Antigenic variation in malaria. Adv. Exper. Biol. and Med. 93:5–25.

Bureau, D., and J. Daussant. 1983. Desorption following immuno-affinity chromatography: generalization of a gentle procedure for desorbing antigen. Jour. of Immun. Meth. 57:205–13.

Clyde, D. F., H. Most, V. McCarthy, and J. P. Vanderberg. 1973. Immunization of man against sporozoite induced falciparium malaria. Amer. Journ. of Med. Sci. 266:169–72.

Cohen, S., and G. H. Mitchell. 1978. Prospects for immunization against malaria. Curr. Top. Microbiol. and Immun. 80:97–137.

Corradete, A., F. Verolini, and A. Bucci. 1966. *Plasmodium berghei* da parti di ratti albinii frequentemente immunization con *Plasmodium berghei* irradiato. Parassitologa 8:133–45.

Croce, C. M., A. Linnenbach, W. Hall, Z. Steplewski, and H. Koprowski. 1980. Production of human hybridoma secreting antibodies to measles virus. Nature (London) 288:488–89.

Danielson, E. M., H. Sjostrom, and O. Noren. 1982. Hypotonic elution, a new desorption principle in immuno-absorbent chromatography. Journ. Immun. Meth. 52:232–33.

Deans, J. A., and S. Cohen. 1983. Immunology of malaria. Ann. Rev. Microbiol. 37:25–49.

Deelder, A. M. 1982. The application of monoclonal antibodies to studies of schistosmiasis. *In* Properties of monoclonal antibodies produced by hybridoma technology and their application to the study of diseases, pp. 99–109. UNDP/World Bank/WHO Special Program for Research and Training in Tropical Diseases, Geneva, Switzerland.

Desowitz, R. S., and R. H. Miller. 1980. A perspective on malaria vaccines. Bull. World Health Organ. 58:897–908.

Diggs, C. L. 1980. Prospects for development of vaccines against *Plasmodium falciparum* infection. *In* J. P. Kreir (ed.), Malaria, 3:299–315. Academic Press, New York.

Diggs, C. L. 1984. Research towards vaccination against malaria; an update. *In* M. Ristic, P. Ambrosie-Thomas and J. P. Kreier (eds.), Malaria and Babesosis, pp. 95–102. Martinus Nijhoff Dordrecht, The Netherlands.

Douillard, J. Y., and T. Hoffman. 1983. Enzyme-linked immuno-absorbent assay for screening monoclonal antibody production using enzyme-labeled second antibody. Methods in Enzymology. 92:168–74.

Eastman Kodak. 1983. Kodavie, electrophoresis visualization. Veterinary Technical Bull., Rochester, N.Y.

Engval, E., and P. Perlmann. 1971. Enzyme-linked immunosorbent assay (ELISA): quantitative assay of immunoglobulin G. Immunochem. 8:871–74.

Gysin, J., V. Nussenzweig, U. Schlesinger, and R. S. Nussenzweig. 1983. Synthetic peptide of circumsporite antigen induces antibodies to malaria parasites. Proceedings 2d Int. Conf. Malaria and Babesiosis, Annecy, France. P. 122.

Fine, E., M. Arkawa, A. H. Cochrane, and R. S. Nussenzweig. 1984. Immunoelectrophoretic observations on *Plasmodium knowlesi* sporozoites: localization of protective antigen and its precursors. Amer. Journ. Trop. Med. and Hygiene 33:220–26.

Galfre, G., and C. Milstein. 1981. Preparation of monoclonal antibodies: strategies and procedures. Methods in Enzymology 73:3–46.

Goeddel, D. V., D. G. Kleid, F. Boluard. 1979. Expression in *Escherichia coli* of chemically synthesized genes for human insulin. Proc. Nat. Acad. Sci. USA 76:106–10.

Goudsward, J., J. A. van der Donk, A. Noodzij. 1978. Protein A reactivity of various mammalian immunoglobulins. Scandinavian Journ. Immun. 8:2128–32.

Griffiths, R. B. 1978. Veterinary Aspects. pp. 41–46. *In* A. E. R. Taylor (ed.), The revelance of parasitology to Human Welfare Today, pp. 41–46. Blackwell, Oxford.

Holder, A. A., and R. R. Freeman. 1981. Immunization against blood stage rodent malaria using purified parasite antigens. Nature (London) 294:361–64.

Jayawardena, A. N. 1981. Immune responses in malaria. pp. 85–136. *In* J. M. Mansfield (ed.), Parasite diseases, 1:85–136. Marcel Dekker, New York.

464   *Biotechnology of Plants and Microorganisms*

Kohler, G., and C. Milstein. 1975. Continuous cultures of fused cells secreting antibodies of predefined specificity. Nature (London) 256:495–97.

Kreier, J. P. ed. 1980. Malaria, Vol. 3, Immunology and Immunization. Academic Press, New York. 346 pp.

Langone, J. J. 1983. Applications of immobilized protein A in immunochemical techniques. Journ. Immun. Meth. 55:277–96.

Maxam, A. M., and W. Gilbert. 1977. A new method of sequencing DNA. Proc. Nat. Acad. Sci. USA 74:560–64.

McGregor, I. A., and M. Barr. 1962. Antibody response to tetanus toxoid inoculation in malarious and nonmalarious Gambian children. Transact. Royal Soc. Trop. Med. and Hygiene 56:364–67.

Merrill, C. R., D. Goldman, S. A. Sedman, and M. H. Ebert. 1981. Ultrasensitive stain for proteins in polyacrylamide gels shows regional variation in cerebrospinal fluid proteins. Science 211:1437–38.

Nardin, E. H., V. Nussenzweig, R. S. Nussenzweig, W. E. Collins, K. T. Harinasute, P. Tapchaisri, and Y. Chomcharm. 1982. Circumsporite proteins of human malaria parasites *Plasmodium falciparum* and *Pasmodium vivax*. Journ. Exper. Med. 156:20–30.

Nirenberg, M. W., and J. H. Matthaei. 1961. The dependence of cell-free protein synthesis in *E. coli* upon naturally occurring or synthetic polyribonucleotides. Proc. Nat. Acad. Sci. USA 47:1588–1602.

Noguer, A., W. Wernsdorfer, R. Koutzentzov, and J. Henipel. 1978. The malaria situation in 1976. World Health Organ. Chron. 32:9–17.

Nussenzweig, R. S. 1982. Vaccination against malaria: the use of monoclonal antibodies for the characterization of protective antigens. Pontificiae Academiae Scientiarum Scripta Varia. 47:33–38.

Nussenzweig, R. S. 1967. Increased nonspecific resistance to malaria produced by administration of *Corynebacterium parvum*. Exper. Parasit. 21:224–31.

Nussenzweig, R. S., J. Vanderberg, H. Most, and C. Orten. 1967a. Protective immunity produced by injection of x-irradiated sporozoites of *Plasmodium berghei*. Nature (London) 216:160–62.

Oka, M., M. Aikawa, R. R. Freeman, and A. A. Holder. 1984. Ultrastructural localization of protective antigens of *Plasmodium yoelii* merozoites by use of monoclonal antibodies and ultrathin cryomicroscopy. Amer. Journ. Trop. Med. and Hygiene 33:342–46.

Olssen, C., and H. S. Kaplan. 1980. Human-human hybridomas producing monoclonal antibodies of predefined antigenic specificity. Proc. Nat. Acad. Sci. USA 77:5429–31.

Olssen, L., and H. S. Kaplan. 1983. Human-human monoclonal antibody-producing hybridomas: technical aspects. Methods in Enzym. 92:3–16.

Pearson, T. W., M. Pinder, G. E. Roelants, S. K. Kar, L. B. Lundin, K. S. Mayor-Whitey, and R. S. Hewett. 1979. Methods for derivation and analysis of anti-parasite monoclonal antibodies. *In* Hybridoma technology with special reference to parasite diseases, pp. 105–29. UNDP/World Bank/WHO Special Program for Research and Training in Tropical Diseases. Geneva, Switzerland.

Peters, W. 1978. Medical aspects: comment and discussion 2. pgs 25–40. *In* A. E. R. Taylor (ed.), The Relevance of Parasitology to Human Welfare Today, A. E. R. Taylor, ed. Blackwell, Oxford.

Renart, J., J. Reiser, and G. R. Stark. 1979. Transfer of proteins from gels to diazobenzyloxymethyl-paper and detection with antisera: a method for studying antibody specificity and antigenic structure. Proc. Nat. Acad. Sci. USA 76:3116–20.

Roth, R. A., and D. J. Casell. 1983. Insulin receptor: evidence that it is a protein kinase. Sci. 219:299–301.

Russel, P. F., H. W. Mulligan, and B. N. Mohan. 1941. Specific agglutinagenic properties of inactivated sporozites of *Plasmodium gallinaceum*. Journ. Malaria Inst. India 4:15–24.

Russel, P. F. 1943. Malaria and its influence on world health. Bull. New York Acad. Med. 19:599–630.

Sadun, E. H., B. T. Welde, and R. L. Hickman. 1969. Resistance produced in owl monkeys (*Aotus trivirgatus*) by inoculation with irradiated *Plasmodium falciparum*. Military Med. 134:1165–75.

Sanger, F., and A. R. Coulson. 1975. A rapid method for determining sequences in DNA with DNA polymerase. Molec. Biol. 94:444–48.

Sanger, F., S. Nicklen, and A. R. Coulson. 1977. DNA sequencing with chain terminating inhibitors. Proc. Nat. Acad. Sci. USA 74:5463–67.

Sergent, E., and E. Sergent. 1910. Sur l'immunité dans le paludism des oiseaux. Conservation *in vitro* des sporozoites de *Plasmodium relictum*. Immunite relative obtenue par inoculation de ces sporozoites. Compte Rendu. Acad. Sci. (Paris) 154:407–9.

Shapiro, A. L., E. Vinuela, and J. V. Maizel, Jr. 1967. Molecular weight estimation of polypeptide chains by electrophoresis in SDS polyacrylamide gels. Biochem. Biophys. Res. Commun. 28:815–20.

Smith, E. A., and J. M. Erickson. 1980. The agency for international development program for malaria research and development. *In* J. P. Kreier (ed.), Malaria, 3:331–37. Academic Press, New York.

Smith, G. L., G. N. Godson, V. Nussenzweig, R. S. Nussenzweig, J. Barnwell, and B. Moss. 1984. *Plasmodium knowlesi* sporozoite antigen: expression by infections recombinant vaccinia virus. Science 224:397–99.

Southern, E. M. 1975. Detection of specific sequences among DNA fragments separated by gel electrophoresis. J. Molec. Biol. 98:503–17.

Staehelin, S., D. S. Hobbs, and H. F. Kung. 1981. Purification and characterization of recombinant human leukocyte interferon (IFLNA) with monoclonal antibodies. J. Biol. Chem. 256:9750–54.

Towbin, H., T. Staehlin, and J. Gordon. 1979. Electrophoretic transfer of proteins from polyacrylamide gels to nitrocellulose sheets: procedure and some applications. Proc. Nat. Acad. Sci. USA 76:4350–54.

Tsang, V., J. M. Peralta, and A. R. Simons. 1983. Ensyme-linked immunotransfer plate techniques (EITB) for studying the specificities of antigens and antibodies by gel electrophoresis. Methods of Enzymology 92:378–91.

Van Weemen, B. K., and A. H. W. M. Schuurs. 1971. Immunoassay using antigen-enzyme conjugates. FEBS Letters 15:232–36.

Varwaerde, C., J. M. Grzychd, H. Bazin, M. Capron, and A. Capron. 1979. Production of monoclonal antibodies against *Schistosoma mansoni*, preliminary studies on their biological activities. *In* Hybridoma technology with special reference to parasitic diseases, pp. 139–43. UNDP/World Bank/WHO Special Program for Research and Training in Tropical Diseases, Geneva, Switzerland.

Warren, K. S., 1980. The great neglected diseases of mankind. *In* J. P. Kreier (ed.), Malaria, 3:335–38. Academic Press, New York.

Watson, J. D., J. Tooze, and D. T. Kurtz. 1983. Recombinant DNA, Short Course. W. H. Freeman, New York. 260pp.

Weber, K., and M. Osborn. 1975. Proteins and sodium dodecul sulfate: molecular weight determinations on polyacrylamide gels and related procedures. *In* H. Neurath, R. L. Hill, and G. L. Baeder (eds.), The proteins, pp. 179–222. Academic Press, New York.

Wernsdorfr, W. H. 1980. The importance of malaria in the world. In J. P. Kreier (ed.), Malaria, 1:1–93. Academic Press, New York.

Wernsdorfer, W. H., and R. L. Kouzenetsov. 1980. Drug resistant malaria—occurrence, control and surveillance. Bull. World Health Org. 58:341–52.

Wikchek, M., and C. S. Hexter. 1976. The purification of biologically active compounds by affinity chromatography. Meth. Biochem. Anal. 23:347–85.

Woodward, R. B., and W. F. Doering. 1944. The total synthesis of quinine. J. Amer. Chem. Soc. 66:849.

World Health Organization. 1976. The enzyme linked immunosorbent assay (ELISA). Bull. World. Health Org. 54:129–39.

Yoshida, N., P. Potocnjak, V. Nussenzweig, and R. S. Nussenzweig. 1981. Biosynthesis of Pb 44, the protective antigen of sporozoites of *Plasmodium berghei*. J. Exp. Med. 154: 1225–36.

Contributors

M. L. R. Aguiar-Perecin, ESALQ, Universidade de Sao Paulo, 13.400 Piracicaba, Brazil

N. J. Alexander, USDA, Agric. Microb. Fermentation Laboratory, 1815 North University Street, Peoria, Illinois 61606

Paulo F. C. de Araujo, ESALQ, Universidade de Sao Paulo, 13.400 Piracicaba, Brazil

J. L. Azevedo, Universidade de Brasilia, Instituto de Biologia Animal, 70.910 Brasilia, Brazil

Y. P. S. Bajaj, Punjab Agricultural University, Tissue Culture Laboratory, Ludhiana 141004, India

C. Ball, Panlab Genetics, Inc., 110 110th Avenue, N. E., Bellevue, Washington 98004

O. Barriga, Ohio State University, Department of Microbiology, 484 West Twelfth Avenue, Columbus, Ohio 43210

D. Boulter, University of Durham, Durham DH1 3HP, England

R. R. Brentani, Universidade de Sao Paulo, 13.400 Piracicaba, Brazil

A. Clausi, General Foods Corporation, White Plains, New York 10625

O. J. Crocomo, University of Sao Paulo, Department of Chemistry, ESALQ, 13.400 Piracicaba, Brazil

E. Chartone-Souza, Universidade Federal de Minas Gerais, Caixa Postal 1621/2— PAMPULHA, Belo Horizonte MG, Brazil

E. C. Cocking, University of Nottingham, Department of Botany, Nottingham NG7 2RD, England

W. Colli, Universidade de Sao Paulo, Instituto de Quimica, Caixa Postal 20780, 01000 Sao Paulo, Brazil

C. D. Denoya, Instituto SIDUS, Larrea 926, 117 Capital Federal, Argentina

C. E. Flick, DNA Plant Technology Corporation, 2611 Branch Pike, Cinnaminson, New Jersey 08077

D. M. Glover, Imperial College of Science and Technology, Department of Biochemistry, London SW7 2AZ, England

H. Heslot, Institut National Agronomique, 16, rue Claude Bernard, 75231 Paris, France

J. Janick, Purdue University, Department of Horticulture, West Lafayette, Indiana 47907

R. M. Kottman, Vice-President Emeritus, Ohio State University, College of Agriculture, 2120 Fyffe Road, Columbus, Ohio 43210

F. S. Lara, Universidade de Sao Paulo, Instituto de Quimica, Cidade Universitaria "Armando de Salles Oliveira," 05508 Sao Paulo, Brazil

W. H.-T. Loh, DNA Plant Technology Corporation, 2611 Branch Pike, Cinnaminson, N.J. 08077

D. Martins da Silva, Universidade de Sao Paulo, Instituto de Quimica, ESALQ, 13.400 Piracicaba, Brazil

C. M. Morel, Fundacao Oswaldo Cruz, Brazil

F. G. Nobrega, Universidade de Sao Paulo, Departmento de Bioquimica, Cidade Universitaria "Armando de Salles Oliveira," 05508 Sao Paulo, Brazil

I. Roitman, Universidade de Brasilia, Departmento de Biologia Celular, 70.910 Brasilia, Brazil

A. P. Ruschel, CENA, Universidade de Sao Paulo, 13.400 Picacicaba, Brazil

W. R. Sharp, DNA Plant Technology Corporation, 2611 Branch Pike, Cinnaminson, New Jersey 08077

W. R. Strohl, Ohio State University, Department of Microbiology, 484 West Twelfth Avenue, Columbus, Ohio 43210

F. C. A. Tavares, Universidade de Sao Paulo, ESALQ, 13.400 Piracicaba, Brazil

R. J. Whitaker, DNA Plant Technology Corporation, 2611 Branch Pike, Cinnaminson, N.J. 08077

C. E. Winter, Instituto Butantan, Servico de Genetica, CP 65, 01000 Sao Paulo, Brazil

Index